Annals of Scientific Society for Assembly, Handling and Industrial Robotics 2024

Martin-Christoph Wanner ·
Thorsten Schüppstuhl · Kirsten Tracht
Editors

Annals of Scientific Society for Assembly, Handling and Industrial Robotics 2024

Editors
Martin-Christoph Wanner
Fraunhofer IGP
Rostock, Mecklenburg-Vorpommern, Germany

Thorsten Schüppstuhl
Hamburg University of Technology
Hamburg, Germany

Kirsten Tracht
University of Bremen
Bremen, Germany

ISBN 978-3-031-91462-1 ISBN 978-3-031-91463-8 (eBook)
https://doi.org/10.1007/978-3-031-91463-8

This Springer imprint is published by the registered company Springer Nature Switzerland AG
The registered company address is: Gewerbestrasse 11, 6330 Cham, Switzerland

Contents

Human–Robot Interaction

Development of a Simplified Augmented Reality-Assisted Programming for Flexible Robot Systems

Fabian Adler, Anne Blum, and Rainer Müller

Abstract

Flexibility and reconfigurability are potential enablers for addressing current challenges for production. Some of these challenges include for example skills shortages, global competition policies, small quantities and the level of company digitalization. Many suppliers have recognized the potential of more flexible and reconfigurability production systems, such as flexible robot systems, to address these challenges. However, small and medium-sized companies still lack acceptance and integration of such systems, often due to a lack of expertise in their implementation and operation. The objective of this work is to simplify the commissioning and programming of flexible robot systems through the development of augmented reality (AR)-assisted programming. The Microsoft HoloLens 2 is used for this purpose, with the aim of avoiding operator restriction and utilising it as an extension of the conventional process. The assistance system is designed to enhance intuitive operation and present information

Certain partial results have been produced within the project "Mittelstand-Digital Zentrum Saarbrücken". The "Mittelstand-Digital Netzwerk" offers comprehensive support for digitalization with the "Mittelstand-Digital Zentren" and the "Initiative IT-Sicherheit in der Wirtschaft". Small and medium-sized enterprises benefit from concrete practical examples and customised, provider-neutral offers for qualification and IT security. The Federal Ministry for Economic Affairs and Climate Protection enables free use and provides financial subsidies of the Mittelstand-Digital services. Further information can be found at www.mittelstand-digital.de.

F. Adler (✉) · A. Blum
ZeMA gGmbH, Forschungsbereich Montagesysteme, Saarbrücken, Germany
e-mail: fabian.adler@zema.de

R. Müller
Lehrstuhl für Montagesysteme, Universität des Saarlandes, Saarbrücken, Germany

M.-C. Wanner et al. (eds.), *Annals of Scientific Society for Assembly, Handling and Industrial Robotics 2024*, https://doi.org/10.1007/978-3-031-91463-8_1

in a more user-friendly manner. To achieve this, calibration and referencing strategies are being developed for robots, AR glasses, and the production system. The study investigates the suitability of AR glasses as an input device for robot programming. An accuracy study will evaluate the impact of various sources of error. Unity was used as the programming software for the HoloLens 2, with ROS as the underlying operating system. As a result, the developed programming system is not a complete alternative to conventional programming options, but rather a supplement and support for conventional programming at present.

Keywords

Flexible robot • Augmented reality • Intuitive programming

1 Introduction

In traditional assembly line production fixed robot cells with short cycle times can be used due to the large quantities with few variants. However, the trend today is moving towards product customisation, which increases product diversity and therefore the number of variants [1]. Traditional automation systems are unable to fulfil this requirement, necessitating the need for more flexible systems [2]. While large companies are facing this issue more and more, small and medium-sized enterprises (SMEs) have long been confronted with low quantities and a high number of variants, which significantly hinders automation [3]. For the robot system to be economically viable, it must be flexible in terms of location and able to quickly adapt to changing operational environments and tasks [4]. Additionally, commissioning and operation time must be drastically simplified. Flexible robot systems, including human–robot collaboration, are seen as a potential solution to this problem [5]. Despite their advantages and technical availability, companies often lack the expertise or technology to use them effectively. To maximize the potential of adaptability, location flexibility, and resulting usage time, it is crucial that flexible robot systems are made as accessible as possible for companies [6, 7].

The objective of this work is therefore to simplify the commissioning and programming of such flexible robot systems. To achieve this, an assisted robot programming system is to be developed using augmented reality (AR). A head-mounted AR device, Microsoft Hololens 2, is used for this purpose. Figure 1 illustrates a three-step workflow of the realised assistance system. In order to enable such a system, it is necessary to implement calibration and referencing strategies for the robot, AR glasses, and production system. The key aspect of robot programming is to verify the suitability of the AR glasses as a programming input device. This requires conducting an accuracy study by analysing individual influencing variables.

After a brief outline of the state of the art and the technical basics, the scenario and the components used are presented. Subsequently, the developed assistance system is

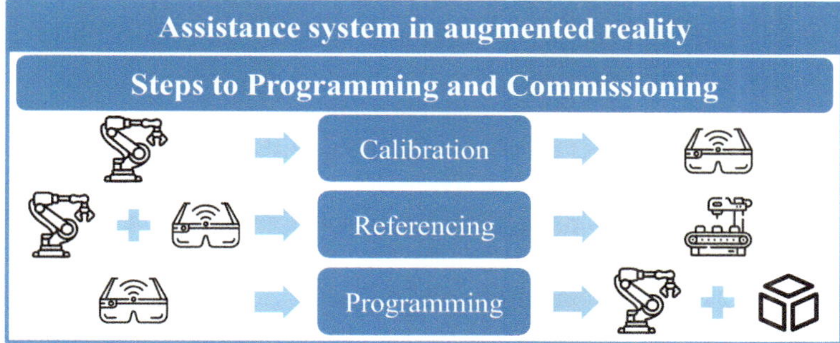

Fig. 1 Overview of the system to be developed and necessary steps

presented, including the necessary steps such as calibration, referencing, and programming for building an AR-assisted robot programming method. Finally, the accuracy and overall accuracy of the system is determined for each step, as illustrated in Fig. 1.

2 Related Work

Adaptable production systems afford greater flexibility in the scheduling of resources in response to changing conditions. There are two types of production systems with this characteristic: flexible and reconfigurable. Flexible production systems ensure that operating resources work within defined limits. In contrast, reconfigurable production systems can be modified at any time to adapt to changing requirements. In this case a flexible production system is used. [2, 8, 9]

To implement such flexible production systems, physical security devices are often removed to increase accessibility and mobility, but programming remains a challenge, particularly for non-experts [6, 10]. Augmented reality has the potential to facilitate the intuitive and natural interaction of non-experts with complex systems, such as industrial robots, by projecting digital information into the real world [11, 12].

2.1 Robot Programming with Augmented Reality

When implementing robot programming in AR, the use of waypoints like for conventional robot programming remains unchanged. The AR application expands and visualizes available information for humans and their interaction with robots. Among others, Blankemeyer et al. [13] investigated how the HoloLens supports humans during a pick and place task using optical markers. The HoloLens utilises optical markers to identify the corresponding component, generate a virtual image, and permit human manipulation. Similarly,

Rudorfer et al. [12] developed a pick and place application in AR. By utilising a calibrated camera positioned above the workspace, placed objects in the robot workspace could be recognized and displayed as a virtual object in AR. This facilitated a more expedient and intuitive programming approach, but accuracy suffered due to calibration issues between the HoloLens and the robot as well as object recognition and hand tracking of the HoloLens. Ostanin et al. [14] utilised a HoloLens to display the robot being used in augmented reality. Waypoints are set using defined gestures. The potential trajectory was displayed in AR between the waypoints. The operator could then decide whether the robot moves first as a simulation in the AR or directly as the physical robot. The implementation is analogous to the procedure in this work, but lacks a more holistic approach to using a flexible robot system efficiently [12–14].

The publications on this topic repeatedly highlight the aspects of accuracies and coordinate transformations of the AR-based systems.

2.2 Accuracy Influences Due to Augmented Reality

As previously observed by Rudorfer et al. [12], the use of AR devices for programming robots introduces inaccuracies into the system. Soares et al. [15] examined the hand tracking of a HoloLens 2 in more detail. Various scenarios were explored, such as hand detection without movement or with a certain hand movement speed, head movement, different hand sizes or use of the left and right hand. An average accuracy of approximately 20 mm was determined, with the influencing factors identified as having a significant impact on the accuracy. The repeatability was approximately 6–7 mm [15].

Another factor to consider is the drift of placed virtual objects in augmented reality. This can occur due to rapid user movement or an insufficient number of features during the mapping process of the real environment by the HoloLens 2. As a result, the virtual objects in the room may shift from their original position. Scargill et al. [16] investigated the inaccuracies that may arise due to drifts when the HoloLens 2 is moved during the AR application. The study defined a series of scenarios, including pausing the application, continuously focusing on an object while moving it, or looking away and then back again. The results showed that movements during the application generate a larger drift, with a mean of 16–17 mm. Conversely, only changes in gaze or pausing the application result in smaller deviations, with a mean of 3–5 mm [16].

The sources of error described in the HoloLens 2, which result from hand tracking of approximately 20 mm and the drifting of virtual objects during movement of around 16–17 mm, are detrimental to the typical accuracy requirements of less than one millimetre in industrial robots.

3 AR-Assisted Programming Scenario

Flexible robot systems are highly versatile and can be deployed across various production systems to perform a range of tasks. To simplify the programming of augmented reality-assisted flexible robot systems, several steps must be taken. Figure 2 shows the scenario under consideration and the necessary steps. The mobile robot system should be deployable at stations A-C (blue lines), must be capable of referencing itself at various stations (yellow lines) and programming the corresponding process at each station (green line). An augmented reality assistance system will be developed to support these steps. Calibration between the robot and a HoloLens 2 must be performed (red line). It is noteworthy that the application should be designed to be independent of the robot. This constraint will have a significant impact on the calibration process, as the calibration must also interpret the robot's dimensions.

Figure 3 illustrates an example of a flexible robot system, an assembly station, and additional peripheral devices. In order to differentiate from existing approaches, such as those described in Sect. 2, as few additional parts as possible but as necessary were used for the technical realisation. The challenge was to achieve a balance between a reduced number of parts and enhanced accuracy.

The robot system used is a UR10e robot from Universal Robots, while as augmented reality device the HoloLens 2 from Microsoft is used. To ensure safe operation, the robot has been equipped with additional safety measures, including a safety skin and two diagonally arranged safety floor scanners. Although not discussed in detail in this work, these measures are crucial for safe and efficient operation. The robot was calibrated with the HoloLens 2 using a QR code attached to the gripper (see Fig. 3b). A mechanical referencing system was utilized for referencing. This system is attached to the robot flange,

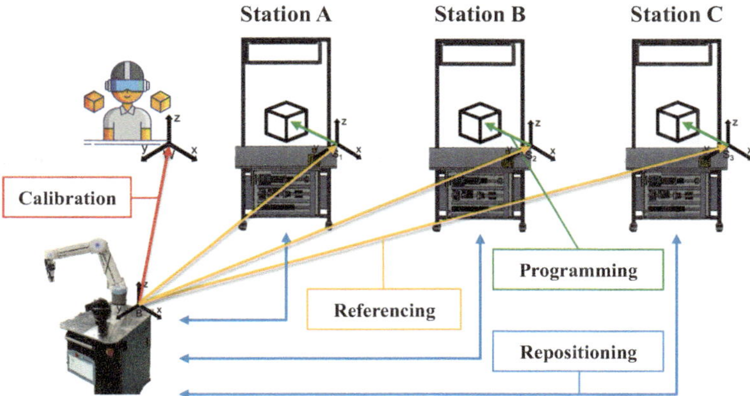

Fig. 2 Illustration of the use of the augmented reality-assisted programming

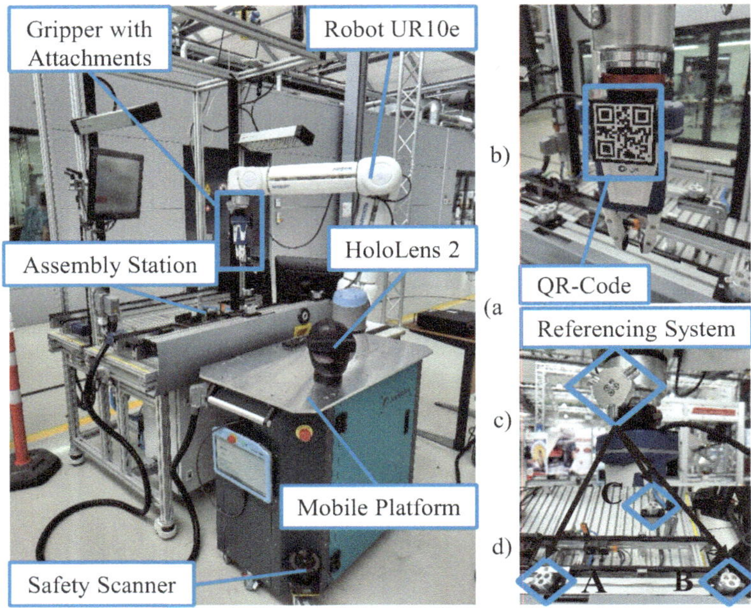

Fig. 3 Demonstrator setup of the flexible robot system

in a similar manner to the QR code. The referencing markers are located at the assembly station (see Fig. 3c and d).

The Robot Operating System (ROS) was used for communication between the robot and the HoloLens 2. ROS is renowned for its adaptability to diverse robot systems and integration with other research initiatives. Moreover, ROS offers an interface to Unity, which utilized for developing the AR assistance system. Unity operates in a distinctive manner compared to other robot systems, as it uses Cartesian coordinate systems. The x-axis represents the horizontal, the y-axis represents the vertical, and the z-axis represents the depth [17].

4 Results for Calibration, Referencing and Programming

To implement the scenario described in Sect. 3, functions for calibration, referencing, and programming were developed and implemented for the AR assistance system (see Fig. 4). The following sections provide a detailed description of the approach and results. The accuracy of alternative programming methods is crucial for the meaningful implementation of this technology as an alternative to existing programming possibilities. As previously stated in Sects. 2 and 3, errors can occur during calibration and referencing,

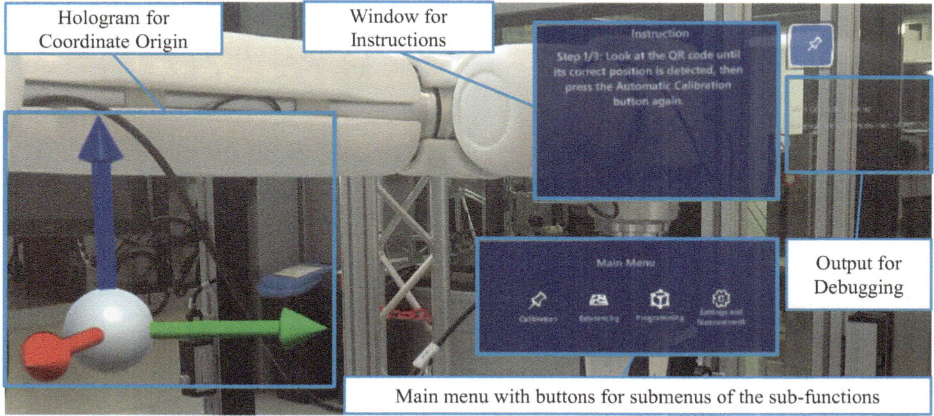

Fig. 4 Overview of the AR assistance system for calibration, referencing and programming

as well as with the technology used, which can affect accuracy. Investigated error sources include user inputs, such as hand tracking as well as drifts, QR code detection, and the mechanical referencing system.

4.1 Calibration Between AR and Robot

The calibration process is designed to accurately determine the robot's location within the augmented reality environment. The primary objective is to establish the base coordinate system of the robot using the AR assistance system and connect it to the AR world. This allows for the creation of waypoints for the robot by placing holograms relative to the determined coordinate origin.

The calibration method used is the three-point method, as illustrated in Fig. 5. The robot is moved to three predefined positions, with the first point serving as the starting point for defining the coordinate system. The robot then moves along the x-axis of the base coordinate system, followed by a shift along the y-axis from the previous position. The positions of the respective robots are captured within the AR environment by scanning the QR code on the gripper (see Fig. 3b). Vectors are defined between the determined starting point and the positions along the x- and y-axes to form the corresponding axes of the coordinate system. The z-axis is determined by the cross product of these vectors. Finally, the calibrated coordinate system is aligned by considering the known displacements of the predefined points within the robot's base.

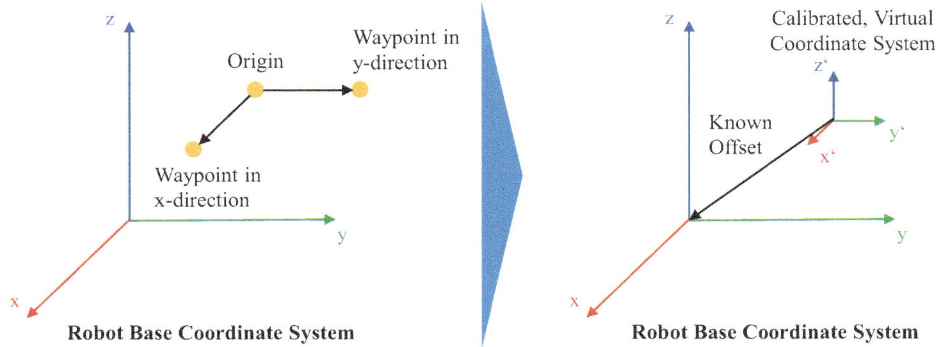

Fig. 5 Three-point calibration of the robot coordinate system to the AR environment

4.2 Referencing Between Robot and Assembly System

Rueckert et al. [18] used a visual system for referencing between a robot and an assembly station, resulting in an average accuracy of 5,9 mm. However, to achieve the highest precision in robot programming, a mechanical referencing system is employed (see Fig. 3c and d). Similarly to robot calibration, a three-point calibration is also used to determine the reference coordinate system of the assembly system [18].

To perform the referencing, the position of the referencing system on the robot is stored as a known point. To capture reference points A-C (see Fig. 3d), the robot is manually guided to the reference markers of the referencing system on the robot and clamped into place. Subsequently, the coordinates of the reference markers relative to the robot's base coordinate system are transmitted to the AR application. Consequently, the position and orientation of the calibrated reference coordinate system at the station are always uniquely defined, whereas the position and orientation of the robot's base coordinate system depend on its positioning.

4.3 Accuracy Influences for Programming in Augmented Reality

In order to gain a deeper insight into the underlying causes of errors, a simplified error chain of influences was established during the calibration, referencing, and programming steps (see Fig. 6). It is important to note that the inherent inaccuracies of the robot are not considered in this case, as they are also present in conventional programming and would have only a minimal impact on the result. The relevant error sources for the sub-functions of the assistance system include the mathematical algorithm and QR code reading during calibration, the mechanical referencing system during referencing, and drift as well

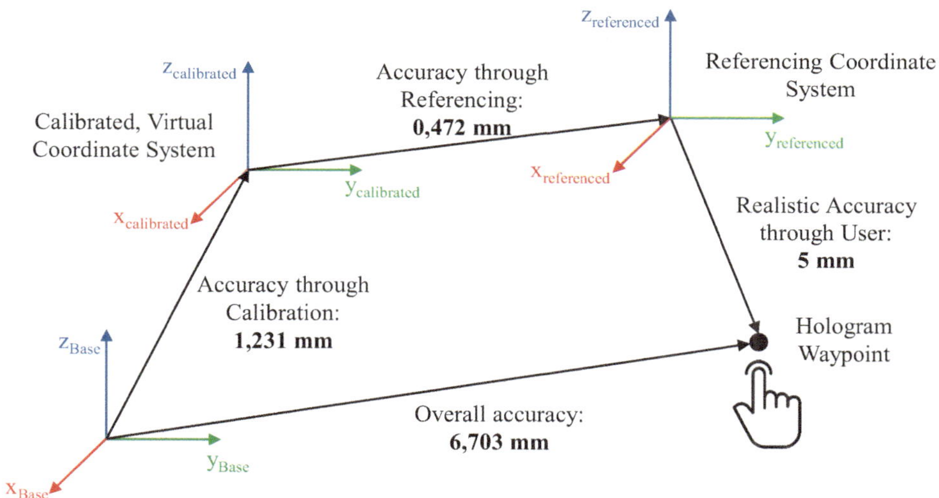

Fig. 6 Composition of the programming accuracy

as user input during waypoint placement during programming. To determine the influences, tests were consistently conducted with a sample size of n = 50. The procedure for collecting the measurement data is described separately for calibration, referencing, and programming.

The calibration's accuracy is verified by scanning the QR code used during calibration. After calibration, the robot is moved to a random position in space and the point is read using both the QR code and the robot controller. The distance between the two coordinates determines the mean calibration accuracy, which is 1,231 ± 0,9 mm. As stated by Microsoft [19], the act of reading QR codes an estimated ± 2,5 mm of inaccuracy into the system. Consequently, an investigation was conducted to ascertain the extent to which the QR code affects the precision of the calibration. It was determined that the source of the QR code reading error has an average accuracy of 1,309 ± 0,233 mm, which is largely responsible for the inaccuracy of the calibration [19].

To determine the accuracy of the referencing, a known waypoint was established in the AR application relative to the referencing coordinate system. Subsequent to the referencing process, the coordinates of the known waypoint and the referencing coordinate system were determined in AR relative to the calibrated base coordinate system. The position of the known point relative to the referencing coordinate system remains constant. However, the coordinates of both the known point and the referencing coordinate system change relative to the calibrated base coordinate system of the robot. By calculating the difference between the known waypoint's coordinates in the base coordinate system and those obtained through the referencing coordinate system, the mean accuracy of the referencing process is determined to be 0,472 ± 0,309 mm.

As shown in Fig. 6, the accuracies resulting from calibration and referencing can be quantified by calculating the mean accuracy, which is 1,703 mm when these values are added together. However, the accuracy during programming is highly dependent on the user. Although a theoretical maximum mean accuracy of 37 mm has been reported for drift and hand tracking with the HoloLens 2, this is not a blanket statement [15, 16]. Initial research with AR-trained individuals has shown that accuracies of 5 mm can be realistic.

5 Conclusion

The paper outlines a simplified augmented reality-assisted programming approach for flexible robot systems, with a particular focus on methods for calibration, referencing, and programming in a production environment. Intuitive usability and increased information provision were central to the assistance system. Furthermore, an important aspect of using such a programming alternative is the accuracy of such a system, which varies from 7 mm (from practical experience) to 40 mm (from literature) depending on the user. After initial attempts to assess the overall accuracy, a value of 6,7 mm was determined. However, further validation is required. The developed programming system is not yet a complete alternative to conventional programming methods. Rather, it can be seen as an extension and support to conventional programming.

In general, further research is required to assess the extent of inaccuracies introduced by user input. The use of assistive tools in AR-based robot programming may help to minimise the impact of human involvement on accuracy. In addition to accuracy assessments, a study should be conducted to determine the user experiences and opinions of such a programming system. Initial tests have shown that experienced robot programmers tend to rely more on conventional programming methods, whereas newcomers perceive added value in utilising AR-based robot programming.

References

1. Bauernhansl, T.: Die vierte industrielle revolution. Der Weg in ein wertschaffendes Produktionsparadigma. In: Vogel-Heuser, B., Bauernhansl, T., Hompel, M. (eds.) Handbuch Industrie 4.0, vol. 4, 2nd edn., pp. 1–31. Springer Vieweg, Berlin, Heidelberg (2017)
2. Beauville dit Eyneaud, A., Klement Nathalie, Roucoules, L., Gibaru, O., Durville Laurent: framework for the design and evaluation of a reconfigurable production systems based on movable robot integration. Int. J. Adv. Manuf. Technol. **118**, 2373–2403 (2022)

3. Krot, K., Kutia, V.: Intuitive methods of industrial robot programming in advanced manufacturing systems. In: Burduk, A., Chlebus, E., Nowakowski, T., Tubis, A. (eds.) Intelligent Systems in Production Engineering and Maintenance, pp. 205–214. Springer Cham, Basel (2018)
4. Hedelind, M., Hellström, E., Jackson, M.: Robotics for SME's—investigating a mobile, flexible, and reconfigurable robot solution. In: Proceedings of 39th International Symposium on Robotics (ISR). 39th International Symposium on Robotics (ISR), Seoul, Südkorea, pp. 56–61. (2008)
5. Maier, H.: Grundlagen der Robotik. VDE Verlag, Berlin (2016)
6. Luisa Hornung: Ergebnisse der Umfrage Mensch-Roboter-Kollaboration https://www.tourings.eu (2021)
7. Hüppi, R., Nielsen, E., Grüninger, R., Brom, C.: Effizienter Robotereinsatz schon bei kleinen und mittleren Serien
8. Demeester, F., Dresselhaus, M., Essel, I., Jatzkowski, P., Nau, M., Pause, B., Plapper, P.W., Schmitt, R.H., Schönberg, A., Voss, H.: Referenzsysteme für wandlungsfähige Produktion. In: Tagungs zum Aachener Werkzeugmaschinen-Kolloquium 2011. In: Aachener Werkzeugmaschinen-Kolloquium 2011, Aachen, Deutschland, pp. 449–477. Shaker (2011)
9. Müller, R., Hörauf, L., Kuhn, D., Karkowski, M., Holländer, M.: Wandlungsfähige Montagesysteme für die nachhaltige Produktion von morgen. wt Werkstatttechnik online **110**, 579–584 (2020)
10. Bauer, W., Bender, M., Braun, M., Rally, P., Scholtz, O.: Leichtbauroboter in der manuellen Montage - einfach einfach anfangen. Erste Erfahrungen von Anwenderunternehmen. Fraunhofer IAO, Stuttgart (2016)
11. Gruenefeld, U., Prädel, L., Illing, J., Stratmann, T., Drolshagen, S., Pfingsthorn, M.: Mind the ARm. In: Preim, B., Nürnberger, A., Hansen, C. (eds.) Proceedings of the Conference on Mensch und Computer. MuC'20: Mensch und Computer 2020, Magdeburg Germany, 06 09 2020 09 09 2020, pp. 259–266. ACM, New York, NY, USA (2020). https://doi.org/10.1145/3404983.3405509
12. Rudorfer, M., Guhl, J., Hoffmann, P., Krüger, J.: Holo Pick'n'Place. In: Proceedings of 2018 IEEE 23rd International Conference on Emerging Technologies and Factory Automation (ETFA). 2018 IEEE 23rd International Conference on Emerging Technologies and Factory Automation (ETFA), Turin, Italien, pp. 1219–1222. IEEE. (2018)
13. Blankemeyer, S., Wiemann, R., Posniak, L., Pregizer, C., Raatz, A.: Intuitive robot programming using augmented reality. In: Proceedings of 7th CIRP Conference on Assembly Technologies and Systems. 7th CIRP Conference on Assembly Technologies and Systems, Tianjin, China, pp. 155–160. Elsevier. (2018)
14. Ostanin, M., Klimchik, A.: Interactive robot programming using mixed reality. In: Proceedings of 12th IFAC Symposium on Robot Control SYROCO 2018. 12th IFAC Symposium on Robot Control SYROCO 2018, Budapest, Ungarn, pp. 50–55. Elsevier. (2018)
15. Soares, I., Sousa, R.B., Petry Marcelo, Moreira, A.P.: Accuracy and repeatability tests on HoloLens 2 and HTC Vive. Multimodal Technol. Inter. **5** (2021)
16. Scargill, T., Premsankar, G., Chen, J., Gorlatova, M.: Here to stay: a quantitative comparison of virtual object stability in markerless mobile AR. In: Proceedings of 2022 2nd International Workshop on Cyber-Physical-Human System Design and Implementation (CPHS). 2022 2nd International Workshop on Cyber-Physical-Human System Design and Implementation (CPHS), Mailand, Italien, pp. 24–29. IEEE. (2022)
17. Microsoft: Unity—Entwickeln Ihres ersten Spiels mit Unity und C#. https://learn.microsoft.com/de-de/archive/msdn-magazine/2014/august/unity-developing-your-first-game-with-unity-and-csharp. Last accessed 8 April 2024
18. Rückert, P., Adam, J., Papenberg, B., Paulus, H., Tracht, K.: Calibration of a modular assembly system for personalized and adaptive human robot collaboration. In: Proceedings of 7th

CIRP Conference on Assembly Technologies and Systems. 7th CIRP Conference on Assembly Technologies and Systems, Tianjin, China, pp. 199–204. Elsevier. (2018)
19. Microsoft.: Übersicht zur QR-Codenachverfolgung. https://learn.microsoft.com/de-de/windows/mixed-reality/develop/advanced-concepts/qr-code-tracking-overview. Last accessed 8 April 2024

An Enhanced AI-Based Approach for Contextualization and Prediction of Long-Term Human Activities in Industrial Robot Applications

Sebastian Krusche, Jayanto Halim, Shuxiao Hou, Mohamad Bdiwi, and Steffen Ihlenfeldt

Abstract

Predicting long-term human activities and movement patterns in industrial environments with content semantics using RGB color images and 3D point clouds can enhance the efficiency and effectiveness of human–robot collaboration applications. This work aims to develop methods for predicting complex, long-term human activities and estimating movements within industrial processes. The main contributions of this work are: 1. Designing a multi-layered structure with various methods for predicting long-term human intentions and walking paths in an industrial context, 2. Conducting empirical experiments with 60 subjects to collect data on human actions and activities in an industrial environment, and 3. Training a comprehensive end-to-end multitask learning system for long-term prediction of human activities and movement patterns. These approaches were integrated into the proposed framework and evaluated across six scenarios in an industrial use case.

Keywords

Activity prediction • Future prediction • Human motion prediction • Motion forecasting • Multi-task learning

S. Krusche (✉) · J. Halim · S. Hou · M. Bdiwi · S. Ihlenfeldt
Fraunhofer Institute for Machine Tools and Forming Technology, Chemnitz, Germany
e-mail: sebastian.krusche@iwu.fraunhofer.de
URL: https://www.iwu.fraunhofer.de/

M.-C. Wanner et al. (eds.), *Annals of Scientific Society for Assembly, Handling and Industrial Robotics 2024*, https://doi.org/10.1007/978-3-031-91463-8_2

1 Introduction

Flexible production lines that can adapt to changes in the process chain and product diversity are essential to meet the future challenges of industrial production. Agile machine systems must be quickly converted to new production processes based on order demands, posing a significant challenge for safety and control technology. One solution for high efficiency and adaptability is to design a production line with fenceless workstations, enabling human–robot collaboration [1] and meeting high safety and control technology demands. The control system must safely detect humans in the robot's workspace and predict their future intentions. To meet this requirement, methods are needed to predict human activities and movements reliably over extended periods, exceeding 1 s [2, 3]. Long-term activity prediction enables humans to work alongside robots in the same workspace during specific product steps, avoiding collisions or hazards. The robot's path can be adapted or replanned based on these predictions. Data-driven approaches, such as those described in [4], integrate sensor results to enable the robot to adapt its trajectory in response to environmental changes. As a result, there would be no loss of machine utilization, as the robot could continue moving at high speed while maintaining the appropriate safety distance [5]. In human activity prediction, short-term predictions are defined as those lasting less than 0.5 s, while long-term predictions extend beyond 1 s second [2, 3]. While a highly reliable short-term prediction can ensure human safety by stopping the robot, causing a deliberate process interruption, long-term predictions allow the control system to plan the robot's movements without interrupting the process.

This paper proposes an enhanced AI-based method for contextualizing and predicting long-term human activities in industrial robotics. Our approach aims to predict complex sequential actions based on partially observed images and motion sequences by analyzing the relationships and interactions between humans and dynamic agents such as robots, or AGVs. We employ up to four 3D sensors from different perspectives to ensure comprehensive coverage of the action sequences within the surveillance space. Our contributions can be summarized as follows: 1. design of a multi-layer structure with various methods for predicting long-term human intentions and walking paths in an industrial context, 2. conducting empirical experiments with 60 subjects to collect data on human actions and activities in an industrial environment, 3. Training a comprehensive end-to-end multi-task learning system for long-term prediction of human activities and movement patterns. All these procedures are integrated into the proposed framework and evaluated across six scenarios of an industrial use case involving robots.

2 Related Work

Most studies on human action prediction focus on detecting temporal or spatial changes in 2D color images using neural networks [6–8], often involving early-stage detection or classification of actions. Paper [8] introduces a method for predicting human interactions using deep temporal information from videos. Flow coding represents low-level motion information, while a deep convolutional neural network extracts detailed temporal features. To improve the anticipation of hard-to-predict actions at an early stage, approach [9] incorporates an additional memory module that records sequences and provides predictions.

In contrast to traditional two-stream architectures, the model in paper [10] trains both input and forget gates jointly across modalities, rather than treating the two streams as separate entities without mutual information. Paper [11] presents an action anticipation model that predicts future actions by synthesizing visual and temporal information, moving away from traditional methods that separately predict video features before anticipation. These methods require a continuous video stream to record spatial and temporal changes, enabling the prediction of future actions.

Predicting human movement is challenging due to the need for 3D information, especially in industrial robotics contexts. Another common approach involves predicting future human activity and movement using skeleton-based methods. Work [12] describes a joint prediction method that incorporates both motion-based features from previous activities and object features present in the scene. Paper [13] addresses challenges in accurately predicting future human motion due to incomplete spatiotemporal information. To overcome this, the authors propose an activity-driven attention-multilayer perceptron (MLP) association method and introduce a novel activity-driven optimization supervision approach to enhance prediction accuracy.

Paper [14] presents an adaptive sampling-based strategy for dynamically recognizing and predicting human activities, using a new adaptive cost function to train a Gated Recurrent Unit (GRU). Paper [15] introduces the Dropout Autoencoder LSTM (DAELSTM), a novel method for learning predictive spatiotemporal motion models from data. This model synthesizes natural-looking motion sequences over long time horizons by combining a recurrent neural network for temporal aspects with an autoencoder to recover the spatial structure of the human skeleton.

The approaches [12–17] assume human activity is strictly skeleton-based, which can be challenging to ensure in industrial contexts. In some cases, recognition is used to predict future actions directly related to the current action. Hierarchical division of activity into two levels allows for predicting future actions based on combinations of simpler actions [18]. Another approach for long-term prediction involves analyzing group sports activities to forecast future positions and plays [19].

Besides the disadvantages of each approach mentioned above, recognizing and tracking multiple people in videos remains a significant challenge, especially in crowded and cluttered scenes. None of the existing approaches utilize a multimodal 3D sensor system for long-term human activity prediction. A multi-sensor system allows for viewing the activity scene from multiple perspectives, enabling the creation of more accurate hypotheses through data fusion. Additionally, these approaches predominantly focus on non-industrial settings where they predict human actions without explicitly accounting for interactions with the environment or other dynamic objects.

3 Approach

Manufacturing environments are complex, especially when humans and robots collaborate in a confined workspace. The sensor system must infer future actions and activities from the movements of dynamic agents such as humans, robots, and AGVs. Our approach involves training an existing end-to-end multitasking learning system using rich visual features of human interactions and activities of multiple agents from various sensor perspectives. Inspired by previous work [20], we aim to adapt the pedestrian path prediction approach to industrial production. Our goal is to apply this approach to the long-term prediction of activities and actions in human–robot cooperation scenarios within an industrial context.

The overall architecture of our approach is illustrated in Fig. 1. We employ a multimodal sensor system for action sequence detection, observing the scene from multiple perspectives in real-time. Using color images and point clouds, we perform valid segmentation and tracking of humans and dynamic objects with a multi-layer structure of various DNN classifications [21, 22]. The tracks of individual dynamic agents are fed into the end-to-end multitask learning system to predict human activity or walking paths [20]. By partially observing the interactions between dynamic agents and the environment, future human behavior regarding activity and position is predicted.

3.1 Data Annotation

Training context-based AI classifiers requires a large number of contextual datasets featuring industrial actions and activities. Our work employs an automatic "Human Annotation" tool to label human actions in multimodal datasets of color images and 3D point clouds efficiently. The core of the tool features a multi-layered structure of various DNN classifiers, ensuring reliable recognition of human movements and actions in industrial environments. It also processes synchronized data from a multi-sensor system observing the scene from multiple perspectives. Fusing individual sensor data increases confidence

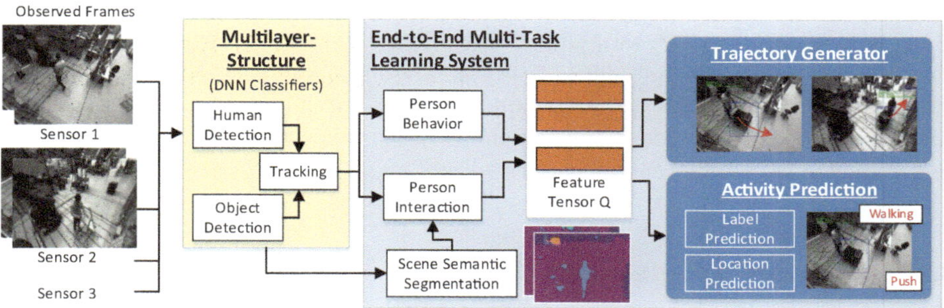

Fig. 1 Overall architecture of AI-based approach for contextualization and prediction of long-term human activities (Illustration adapted Fig. 2 from [20])

in the hypotheses and ensures that human actions are tracked over a larger, more complex working area. The "Human Annotation" tool [23] can localize, track, and classify people and dynamic objects in the sensor data. Reliable classification of humans is based on image data from camera sensors that capture the scene from multiple angles. Various AI-based classifiers are used for human pose estimation [21, 22], with their hypotheses tested using a plausibility filter. Subsequently, in the 3D workspace, people or objects are segmented into 3D bounding boxes based on point clouds from four sensors. In addition to 2D/3D bounding boxes, the annotation tool results include labeling human activities [24] and key points of the human skeleton, as shown in Fig. 2.

The dataset consists of 110 videos featuring six different activity sequences, each performed by ten subjects and observed synchronously from two perspectives at 10 FPS. The resolution of the color images and point clouds is 518×385 pixels. The dataset was

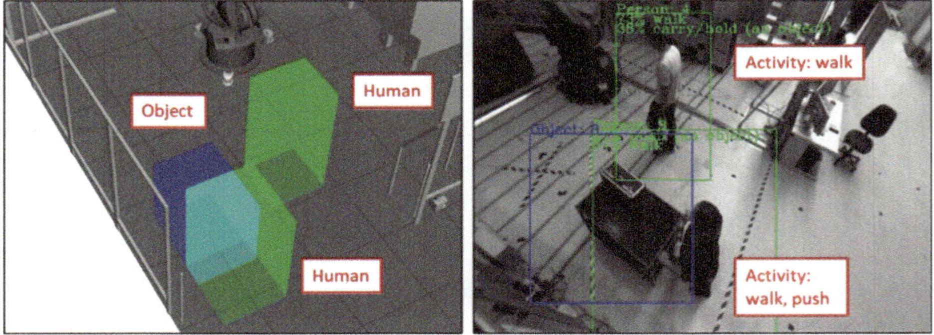

Fig. 2 Automatic human annotation tool results—(Left image) 3D workspace with 3D bounding boxes for human (green) and object (blue); (Right image) 2D sensor view with 2D bounding boxes for human (green) including action marker and object (blue)

divided into 70% for training, 15% for validation, and 15% for testing. Following [20], the videos are initially reduced to 2.5 FPS to allow for 3.2 s (8 frames) of observation and 4.8 s (12 frames) of trajectory prediction. During training, we aim to optimize the parameters to ensure real-time, long-term prediction capability in an industrial control engineering context. The goal is to maintain the video frame rate and operate observation and prediction within the sensor system's 10 FPS update cycle.

3.2 Network Architecture

The overall architecture of the end-to-end multitask learning system is shown on the left in Fig. 1. Unlike most existing works [24–26] that represent a person as a point in space, our end-to-end multitask learning system uses two modules to encode extensive visual information about each person's behavior and their interaction with the environment: The **Person Behavior Module** encodes detailed visual information for people within a scene by considering both appearance and body motion, using techniques such as object detection [27] for appearance and skeletal-based pose detection [28] for motion. This module uses pre-trained models and LSTM encoders to generate feature representations that capture changes in appearance and body movements. The **Person Interaction Module** includes person-scene and person-object modeling. For person-object modeling, it determines the spatial relationships between the person and other objects at each point. For person-scene modeling, it encodes the semantic features of the surrounding scene using a pre-trained scene segmentation model [29].

Based on this encoded visual information, the **Trajectory Generator** summarizes the visual features and predicts the future trajectory using an LSTM decoder with focused attention [30]. **Activity prediction** uses extensive visual semantics to forecast the person's future activity. The scene is also divided into a multi-scale discretized grid, called the Manhattan grid, enabling robust prediction of activity location through classification and regression. As described in the Data Preparation section, our work aims to train the existing end-to-end multitask learning system with context-based datasets to evaluate its suitability for predicting future human activity and movement patterns, and, if necessary, adapt it to the existing control technology requirements of the production environment.

4 Experiments

Training our end-to-end multitask learning system requires a large amount of data featuring various actions and activities in an industrial context. To gather this data, numerous experiments were conducted where people performed various actions in an uncontrolled industrial environment. Table 1 summarizes the test scenarios, including the types of actions and the number of test subjects. A multimodal 3D sensor system captured the

Table 1 Test scenarios with corresponding action types and subject number

No.	Title	Action types	Subjects
1	Person walks into robot cell	Standing (static), walking	1
2	Person walks with item	Standing (static), walking, setting up ladder	1
3	Person pushes a transport cart	Standing (static), walking, pushing transport cart	1
4	2 persons walk into robot cell	Standing (static), walking	2
5	2 persons hand over an item	Standing (static), walking, handing over item	2
6	2 persons with a transport cart	Standing (static), walking, Pushing transport cart	2

scenes from multiple perspectives during the experiments. The color images and point clouds from the 3D sensors were recorded synchronously and merged into a dataset. The test scenarios range from very simple to highly complex, involving various actions related to the environmental structure.

The automated annotation framework enables the generation of numerous datasets for training our end-to-end multitask learning system. Capturing the scene from multiple sensor perspectives increases the number of valid sequences. The annotation of actions and activities revealed that simple actions like walking and standing are the most common, while complex activities such as setting up a ladder or pushing a transport cart are less frequent. The prediction results from the annotation framework, shown in Fig. 3, demonstrate that valid verification of extensive datasets can be achieved with minimal resources using a multi-layered structural model composed of various DNN classifiers.

The results of the automatic annotation of human activities are summarized in the diagram in Fig. 4 for evaluation. Based on the six scenarios, numerous activity scenes were identified for each activity class. General activities like walking and holding an object are more frequently represented than specialized activities such as pushing or pulling an object.

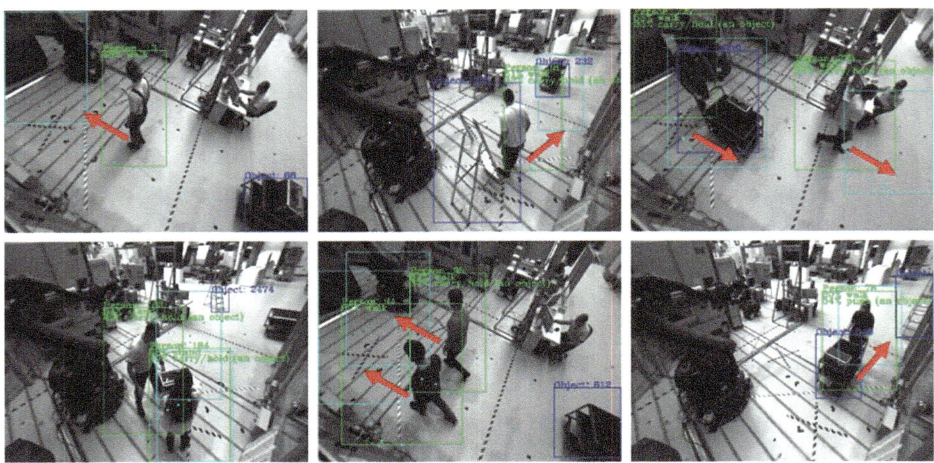

Fig. 3 Activity overview—first row: (Left) single person walking, (Center) setting up a ladder, (Right) single person pushing transport cart; second row: (Left) two person walking, (Center) handing over an item, (Right) two person pushing transport cart; object (blue bounding box) and Human (green bounding box); predicted human walk path (turkey dotted line) human walk path direction (red arrow)

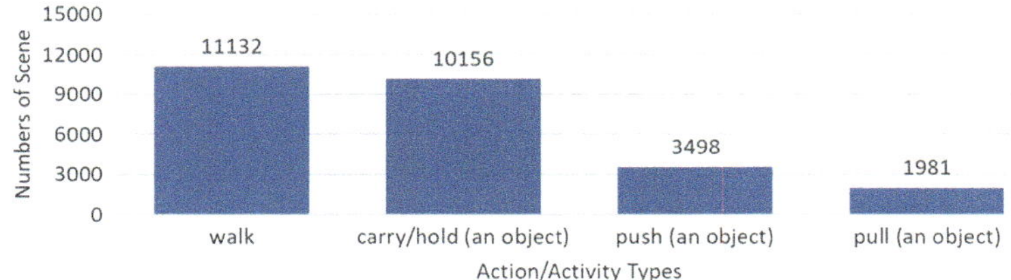

Fig. 4 Summary of the annotated activity scenes according to the general activity classes

5 Conclusion

Our proposed AI-based approach for contextualizing and predicting long-term human activities and walking paths in an industrial context aims to optimally and predictively control processes in production plants while ensuring human safety in human–robot interactions. To implement and evaluate this approach, we made the following contributions:

1. We designed and implemented a multi-layered structure using various AI methods to predict long-term human intentions and trajectories in an industrial context. By fusing the results from individual sensors in 2D/3D space, the classifiers' hypotheses are

verified, ensuring continuous tracking of human activity across the entire workspace and duration.

2. The empirical experiments with over 60 subjects, designed to generate datasets of human actions and activities in an industrial environment, provided an adequate basis for training the end-to-end multitask learning system and verifying the framework. The diverse scenarios provided a sufficient number of action sequences.
3. Our work focused on training the end-to-end multitask learning system to predict long-term human activities and walking paths in an industrial context. This approach considers not only human activities but also movements and interactions with the environment, including the prediction of human walking paths. We aimed to create a dedicated training dataset that maps specific activities within an industrial context. Building on this preliminary work, we can now train the end-to-end multitask learning system on new scenarios in an industrial setting.

Empirical tests showed that complex activities consist of numerous simple actions, such as walking or holding an object, while more specialised actions, such as pushing or pulling an object, make up only a small proportion. This discrepancy makes it much more difficult to train models for the specialised action classes, as only a limited number of sequences are available.

This paper aims to stimulate the development of new methods for activity prediction in an industrial context and foster innovations in vision-based process control. Implementing these methods in industrial applications will create new possibilities for the autonomous control of machines and robots in production processes. In addition to reducing space requirements through hybrid workstations for human–robot cooperation, action prediction approaches minimize cycle time interruptions and enhance machine utilization.

References

1. Vogel, C., Fritzsche, M., Elkmann, N.: Safe human-robot cooperation with high-payload robots in industrial applications. In: 2016 11th ACM/IEEE International Conference on Human-Robot Interaction (HRI), pp. 529–530 (2016)
2. Mao, W., Liu, M., Salzmann, M.: Generating smooth pose sequences for diverse human motion prediction. http://arxiv.org/pdf/2108.08422v3 (2021)
3. Dang, L., Nie, Y., Long, C., Zhang, Q., Li, G.: MSR-GCN: multi-scale residual graph convolution networks for human motion prediction. http://arxiv.org/pdf/2108.07152v2 (2021)
4. Hou, S., Bdiwi, M., Rashid, A., Krusche, S., Ihlenfeldt, S.: A data-driven approach for motion planning of industrial robots controlled by high-level motion commands. Front. Robot. AI **9**, 1030668 (2022). https://doi.org/10.3389/frobt.2022.1030668
5. Bdiwi, M., Rashid, A., Putz, M.: Autonomous disassembly of electric vehicle motors based on robot cognition. In: 2016 IEEE International Conference on Robotics and Automation (ICRA), Stockholm, Sweden, pp. 2500–2505. (2016)

6. Ruiz, A.H., Gall, J., Moreno-Noguer, F.: Human motion prediction via spatio-temporal inpainting. http://arxiv.org/pdf/1812.05478v2. (2018)
7. Feichtenhofer, C., Pinz, A., Zisserman, A.: Convolutional two-stream network fusion for video action recognition. http://arxiv.org/pdf/1604.06573v2. (2016)
8. Hua, G., Jégou, H.: Human interaction prediction using deep temporal features. Springer International Publishing, Cham (2016)
9. Kong, Y., Gao, S., Sun, B., Fu, Y.: Action prediction from videos via memorizing hard-to-predict samples, p. 8. (2018)
10. Sun, L., Jia, K., Chen, K., Yeung, D.Y., Shi, B.E. and Savarese, S.: Lattice long short-term memory for human action recognition. (2017)
11. Gammulle, H., Denman, S., Sridharan, S. and Fookes, C.: Predicting the future: a jointly learnt model for action anticipation: we present an action anticipation model that enables the prediction of plausible future actions by forecast- ing both the visual and temporal future. Kein code. (2019)
12. Mahmud, T., Hasan, M. and Roy-Chowdhury, A.K.: Joint prediction of activity labels and starting times in untrimmed videos. (2017)
13. Zhang, S., Liu, S., Gao, F.: 3D human motion prediction via activity-driven attention-MLP Association: activity predicition kein code This paper presents a novel method to predict future hu man activities from partially observed RGB-D videos. In: 2023 IEEE International Conference on Image Processing (ICIP), Kuala Lumpur, Malaysia, pp. 960–964 (2023)
14. Yadav, G.K., Nandi, G.C.: Development of adaptive sampling based strategy for human activity predictions using sequential networks: predict human activities dynamically kein code this paper presents an innovative idea of developing an adaptive sampling-based strategy to recognize and predict human activities dynamically. In: 2020 IEEE 4th Conference on Information & Communication Technology (CICT), Chennai, India, pp. 1–6 (2020)
15. Ghosh, P., Song, J., Aksan, E., Hilliges, O.: Learning Human motion models for long-term predictions: horizons1 over long-time kein code. In: 2017 International Conference on 3D Vision (3DV), Qingdao, pp. 458–466 (2017)
16. Liu, J., Shahroudy, A., Wang, G., Duan, L.-Y., Kot, A.C.: Skeleton-based online action prediction using scale selection network. IEEE Trans. Pattern Anal. Mach. Intell. **42**(6), 1453–1467 (2020). https://doi.org/10.1109/TPAMI.2019.2898954
17. Liu, J., Shahroudy, A., Wang, G., Duan, L.-Y., Kot, A.C.: SSNet: Scale selection network for online 3D action prediction: no code. In 2018 IEEE/CVF Conference on Computer Vision and Pattern Recognition, pp. 8349–8358. https://ieeexplore.ieee.org/document/8578969/ (2018)
18. Morais, R., Le Vuong, T.T., Venkatesh, S.: Learning to abstract and predict human actions. http://arxiv.org/pdf/2008.09234v1 (2020)
19. Chen, J., Bao, W., Kong, Y.: Group activity prediction with sequential relational anticipation model: to predict group activities code vorhanden In this paper, we propose a novel approach to predict group activities given the beginning frames with incomplete activity execu- tions. (2020)
20. Liang, J., Jiang, L., Niebles, J.C., Hauptmann, A., Fei-Fei, L.: Peeking into the future: predicting future person activities and locations in videos. http://arxiv.org/pdf/1902.03748v3 (2019)
21. Cao, Z., Hidalgo, G., Simon, T., Wei, S.E. and Sheikh, Y.: OpenPose: realtime multi-person 2D pose estimation using part affinity fields. CoRR, abs/1812.08008 (2018)
22. Li, J., Wang, C., Zhu, H., Mao, Y., Fang, H.-S., Lu, C.: CrowdPose: efficient crowded scenes pose estimation and a new benchmark. arXiv
23. Krusche, S., Al Naser, I., Bdiwi, M., Ihlenfeldt, S.: A novel approach for automatic annotation of human actions in 3D point clouds for flexible collaborative tasks with industrial robots. Front. Robot. AI **10**, 1028329 (2023). https://doi.org/10.3389/frobt.2023.1028329
24. Kitani, K.M., Ziebart, B.D., Bagnell, J.A., Hebert, M.: Activity forecasting. (2012)

25. Gupta, A., Johnson, J., Fei-Fei, L., Savarese, S., Alahi, A.: Social GAN: socially acceptable trajectories with generative adversarial networks. In: 2018 IEEE/CVF Conference on Computer Vision and Pattern Recognition, 2018, pp. 2255–2264. https://ieeexplore.ieee.org/document/8578338/. Last accessed May 19 2021
26. Alahi, A., Goel, K., Ramanathan, V., Robicquet, A., Fei-Fei, L., Savarese, S.: Social LSTM: human trajectory prediction in crowded spaces, pp. 961–971. http://ieeexplore.ieee.org/document/7780479/. Last accessed May 19 2021
27. He, K., Gkioxari, G., Dollár, P., Girshick, R.: Mask R-CNN. http://arxiv.org/pdf/1703.06870v3 (2017)
28. Fang, H.-S., Xie, S., Tai, Y.-W., Lu, C.: RMPE: regional multi-person pose estimation. In: 2017 IEEE International Conference on Computer Vision (ICCV), Venice, pp. 2353–2362 (2017)
29. Chen, L.-C., Zhu, Y., Papandreou, G., Schroff, F., Adam, H.: Encoder-decoder with atrous separable convolution for semantic image segmentation. http://arxiv.org/pdf/1802.02611v3 (2018)
30. Liang, J., Jiang, L., Cao, L., Li, L.J., Hauptmann, A.G.: Focal visual-text attention for visual question answering

Collaborative Screw Fastening Using Behavior Trees and Computer Vision

David Kötter, Manuel Belke, Oliver Petrovic, and Christian Brecher

Abstract

Collaborative robotics are one possible solution for businesses to increase efficiency, safety, and productivity while also offering cost-effective and flexible solutions for a wide range of tasks. As demographic changes lead to a shrinking workforce, the implementation of robots in various industries emerges as a powerful solution to counteract labor shortages. This paper presents a novel approach to automate the task of fastening screws using collaborative robotics (cobots) and behavior trees alongside artificial neural networks. The system uses a camera and the YOLOv8 nano network architecture to detect screws and their position in the workspace of the cobot. Therefore, the artificial neural network trains with an augmented dataset of 200 images, showing different screw heads with varying background and level of corrosion. The proposed system utilizes the flexibility and adaptability of cobots to perform the physical task of screw fastening, while behavior trees provide a high-level, modular control framework for cobots. To validate our approach, we perform a sample implementation and the experimental results demonstrate the effectiveness and flexibility of the proposed approach,

D. Kötter (✉) · M. Belke · O. Petrovic · C. Brecher
Laboratory for Machine Tools and Production Engineering, RWTH Aachen University, Aachen, Germany
e-mail: d.koetter@wzl.rwth-aachen.de

M. Belke
e-mail: m.belke@wzl.rwth-aachen.de

O. Petrovic
e-mail: o.petrovic@wzl.rwth-aachen.de

C. Brecher
e-mail: c.brecher@wzl.rwth-aachen.de

M.-C. Wanner et al. (eds.), *Annals of Scientific Society for Assembly, Handling and Industrial Robotics 2024*, https://doi.org/10.1007/978-3-031-91463-8_3

although the non-precise calibration of the camera results in a small deviation between screw head and screwdriver, depending on the angle between camera and assembled product.

Keywords

Cobots • Manufacturing • Screw fastening • Computer vision

1 Introduction and Background

In today's dynamic business environment, it is essential for the manufacturing industry to leverage flexible automation solutions such as robotic systems to remain competitive and to uphold the delivery of high-quality products to customer. Furthermore, through demographic change in industrialized countries like Germany, workforce decreases [1]. Therefore, we need solutions to counteract the shortcoming of labor force. One possible part of the solution is the integration of robots into the manufacturing process. Robots exhibit high precision and repeatability, operating at consistently fast speeds, resulting in high quality and fast manufacturing [2]. Conversely, humans possess the ability to comprehend complexities, assess problems, and react appropriately, as well as the capability to move freely in space [2]. The goal of the human–robot collaboration is to harness the unique strengths of both humans and robots, allowing them to operate together in a shared workspace [3]. In order to use the advantages of cobots in the manufacturing process, the cobot needs to be programmed easily, with behavior trees emerging as a possible solution due to their ability to simplify the programming process and create more intuitive and flexible robotic behaviors.

Behavior trees are hierarchical structures that organize the behavior of a system using nodes activated by ticks [4]. They provide modularity, allowing for easy addition or removal of nodes or subtrees. Condition nodes verify if a specific condition is met before an action node is executed, influencing the reactivity of the trees. Nodes execute in sequence or parallel, and they indicate their state returning RUNNING, SUCCESS, or FAILURE. Selector nodes activate their child nodes until one returns RUNNING or SUCCESS. Additionally, decorators are nodes, customizable by the user. Behavior trees are intuitive and understandable for humans, making them accessible to non-experts. Furthermore, several extensions exists, like a human action node [5], which incorporates the human subtasks in the tree to reduce idling time or whole subtrees, like a safety subtree [6] to ensure the integrity of the worker during human–robot collaboration. In order to detect screws in the workspace, computer vision is utilized.

Artificial neural networks [7] have emerged as a powerful framework within the field of computer vision and artificial intelligence mimicking the structure and functionality of the human brain. With their ability to learn from vast amounts of data, adapt to new information, and make accurate predictions, artificial neural networks have revolutionized various domains such as computer vision, natural language processing, and speech recognition.

The contribution of this paper is the following:

- Development of a system to recognize screws in the workspace of the cobot
- Integration of the screw fastening process into a behavior tree, controlling the cobot in a modular, reusable and flexible war
- Evaluation of the approach using heat maps.

The remainder of the document is structures as follows. Section 2 provides an overview of the state of the art regarding screw fastening with cobots. The system design, containing the communication with the hardware, the design of the behavior tree and the underlying algorithm is described in Sect. 3. In Sect. 4, the system is evaluated and Sect. 5 provides a summary giving an outlook on future lines of investigation.

2 Collaborative Screw Fastening

This subsection provides state of the art approaches to incorporate screw fastening into the skillset of a collaborative robotic system. Therefore, even hard-to-automate processes, like manufacturing textiles or cables, can be automated by combining the high precision and accuracy of the cobot as well as the high-reasoning ability of the worker [8]. The topic of collaborative screwdriving also includes the development of several tools to optimize the screwdriving process [8]. In scenarios where the exact position of the screw is not known to the cobot, there are three distinct methods for fastening (or unfastening) screws using a collaborative robot:

- The cobot moves first and the force sensors are utilized to detect the screw in the workspace of the cobot [9, 10].
- Before the cobot moves, a sensor detects the position of the screw. The cobot moves to the respective coordinates of the screw afterwards [11].
- A human being detects the position of the screw position and moves the cobot afterwards to the respective position of the screw [12].

Li et al. use an electric nutrunner spindle attached to a cobot as well as the internal force sensors of the robot to unfasten screws. Their spiral search motion and control strategy is able to unfasten screws, although the position of the screws is only known approximately [9]. Furthermore, Huang et al. present a method of automated screw unfastening for robotic disassembly by combining torque control, position control and active compliance. To hit the screws in the center, the cobot performs a spiral search motion, similar to the approach from Li et al. [10].

To detect the screw before the robot moves, a suitable sensor like a camera is required to detect the position of the screw in the workspace of the cobot. Peralta et al. use a camera, a tunable liquid crystal lense and active near-infrared reflectance alongside segmentation artificial neural networks to detect screw fasteners and their position in the workspace relatively to the cobot. They work with different fastener sizes and an embedded force sensor to verify the success of the fastener localization in the real world [11].

In contrast to using artificial sensors to handle the uncertainty of the screw, Villa et al. develop a framework for robotic manipulation tasks, integrating two adaptive controllers to modulate robot compliance in contact with the environment along constrained directions and to enable human guidance through touch, if human guidance needed. The cobot system is able to identify the position of screws through an online human pose-tracking system, and performs the screwing using the proposed controller [12].

The existing state of the art indicates a lack of approaches integrating screw tightening into the high-level control structure of a cobot, such as a behavior tree, to facilitate the easy reuse in order to maximize scalability and efficiency. The contribution of this paper tackles the problem of the demographic change in industrialized countries, enabling workers without programming knowledge to integrate screwdriving into the skillset of a robot, to counteract the shortcoming of labor force in industrialized countries.

3 System Design

3.1 Communication with Hardware

In order to move the cobot with the attached screwdriver to the screws, the system must recognize the screws, their position in the workspace and send the movement signal to the cobot. Figure 1 shows the architecture of the system. The BT as high-level control architecture of the cobot takes an RGB image stream as input and queries a depth-image, if a screw is detected in the workspace of the cobot. If screws are detected, they are localized and the signal to move the cobot is sent to the controller of the cobot (see Fig. 1).

3.2 Detection Algorithm and Training

In order to detect screws in the workspace, a detection model is required. Therefore, the YOLO (You Only Look Once) model [13] is chosen. YOLO models are regularly

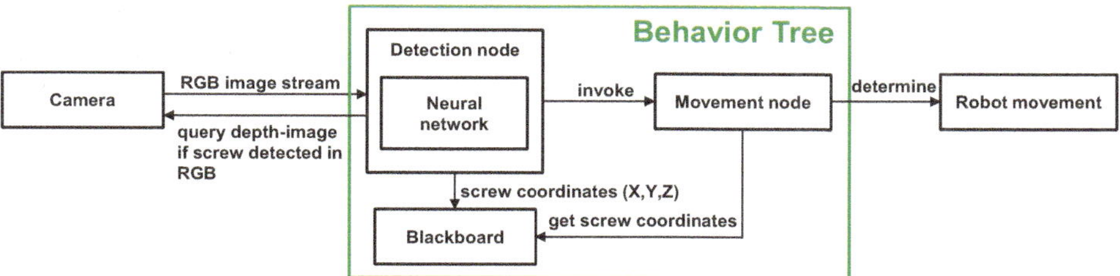

Fig. 1 Concept of the system to fasten screws with a cobot using a behavior tree

Fig. 2 Example of three pictures from the dataset used for training the YOLO-algorithm. The bounding boxes are highlighted in red. From left to right: calculator, smartwatch, hinge

chosen for screw detection [14, 15] due to their real-time processing capabilities and high accuracy in object detection tasks. The single-stage architecture of YOLO models allows for faster inference times compared to other object detection models, making them ideal for applications where speed is crucial, such as industrial automation. Additionally, YOLO models have shown impressive performance in detecting small objects like screws [14, 15], thanks to their ability to capture fine details and accurately localize objects in images. For the detection node, the YOLOv8n (nano) model is used, due to the small training dataset, consisting of 20 pictures with different screw heads of the birds-eye view. The dataset is labeled using bounding boxes and augmented [16] to upscale the dataset to 200 pictures. Figure 2 shows an example of three images used for training the model. The images depict a calculator, a smartwatch and a hinge. The bounding boxes are shown in red.

3.3 Transformation to the Cobot Coordinate Frame

In order to move the cobot to the respective position of the screw, the coordinates of the screws are transformed from the image coordinate frame to the coordinate system of the cobot. Figure 3 shows the graphical representation and relations of the image, cobot and camera coordinate system. Since the camera (Intel RealSense D435i [17]) outputs the

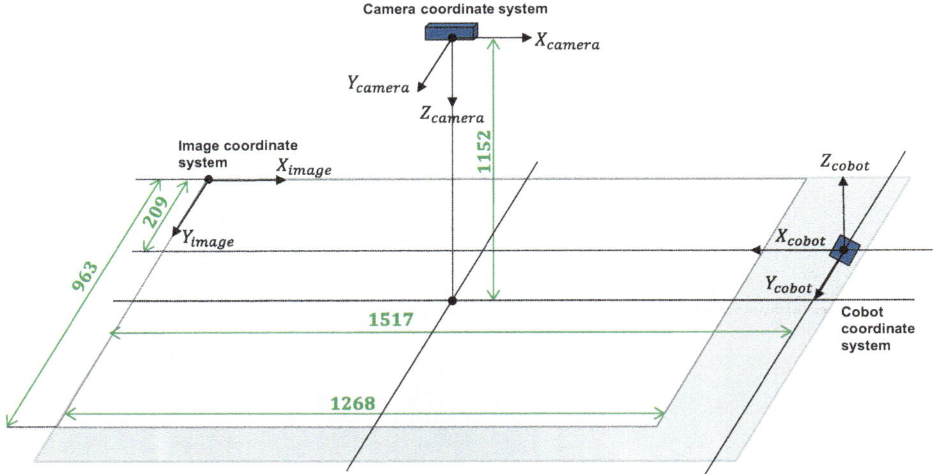

Fig. 3 Spatial representation of the image, cobot and camera coordinate system

coordinates of the screw in pixels (image coordinate system in Fig. 3), the coordinates are transformed from pixels to millimeters using Eqs. (1) and (2). Herein, the output of the camera is divided by the number of pixels in the respective direction and multiplied by the total length of the camera-sight in the same direction.

$$X_{\text{image}_{\text{mm}}} = \frac{X_{\text{image}}}{1960} * 1268 \text{ mm} \tag{1}$$

$$Y_{\text{image}_{\text{mm}}} = \frac{Y_{\text{image}}}{1080} * 963 \text{ mm} \tag{2}$$

To transform the coordinates from the image to the cobot coordinate frame, Eqs. (3) and (4) are utilized.

$$X_{\text{cobot}} = 1517 \text{ mm} - X_{\text{image}_{\text{mm}}} \tag{3}$$

$$Y_{\text{cobot}} = Y_{\text{image}_{\text{mm}}} - 209 \text{ mm} \tag{4}$$

Furthermore, Eq. (5) is used to calculate the distance in z direction. Since the output for the third direction is in mm, there is no need to transform the value from pixels to millimeters.

$$Z_{\text{cobot}} = 1152 \text{ mm} - Z_{\text{camera}} \tag{5}$$

3.4 Design of the Behavior Tree and Algorithm

Figure 4 shows the architecture of the behavior tree after the integration of the task subtree to detect screws in the workspace of the cobot and move the screwdriver to the corresponding positions to tighten the screws. The left subtree holds a list to write the coordinates of the detected screws to the blackboard, *ScrewCoords*. In order to detect the screws in the condition node *Screw Detected*, an Intel RealSense D435 camera [17] along with the YOLOv8-nano model [18] is used. If a screw is detected, the condition node *Screw Detected* returns SUCCESS (FAILURE otherwise), and the action node *Move to Screw* is invoked. The movement is realized using the open-source manipulation platform MoveIt [19]. The system is implemented utilizing the Robot Operating System framework [20].

The screw detection and motion planning interact according to the following scheme:

1. Use the RGB image to detect screws in the two-dimensional space to get the X, Y coordinates
2. Evaluate the depth image at the X, Y coordinates to get the Z coordinate
3. Write a list with all screw positions to the blackboard
4. Get the list from the blackboard and transform the coordinates from the camera to the cobot coordinate frame
5. Move to the X, Y, Z position to hit the screw head via an intermediate position, ten centimeters above the screw

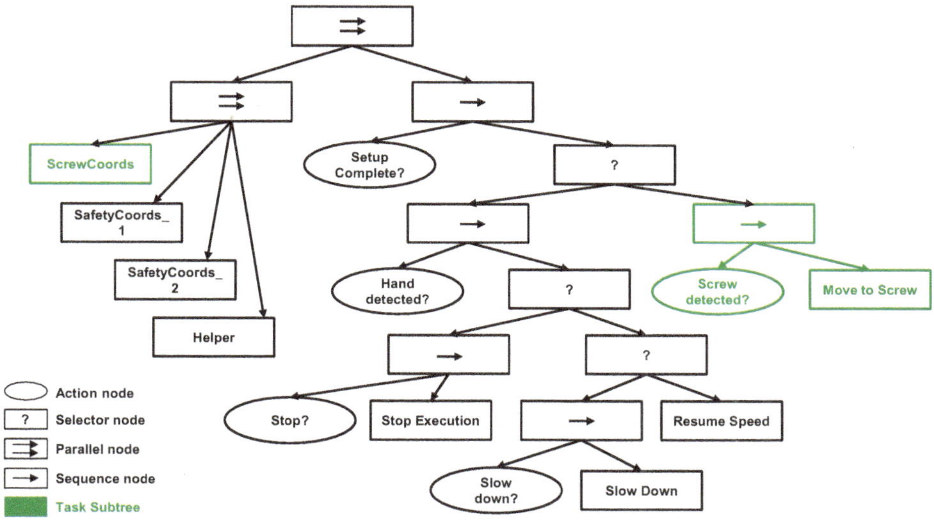

Fig. 4 Graphical representation of the behavior tree with the integrated task subtree

Algorithm 1: Screw detection and movement

1: **SDAM** $(RGB\_image, depth\_image)$
2: $X_k, Y_k \leftarrow find\_screw\_center(RGB\_image)$
$with\ k \in [1, n] \ \wedge \ n \leftarrow number\ screws\ in\ image$
3: $Z_k \leftarrow get\_z\ (X_k, Y_k, depth\_image)$
4: $write\ (X_k, Y_k, Z_k)\ to\ blackboard$
5: **for** $k\ in\ range\ 1 \ldots n$
6: $(X_{k\_cobot}, Y_{k\_cobot}, Z_{k\_cobot}) \leftarrow transform\ (X_k, Y_k, Z_k)$
7: $move\_to\ (X_{k\_cobot}, Y_{k\_cobot}, Z_{k\_cobot} + 100)$
8: $move\_to\ (X_{k\_cobot}, Y_{k\_cobot}, Z_{k\_cobot})$

Fig. 5 Screw detection and movement algorithm

6. Repeat step five until all all screws from the list are approached.

Afterwards, if the screwdriver approached all screw heads successfully, the action node Move to Screw returns SUCCESS. The action node returns FAILURE otherwise. Figure 5 shows the pseudocode for the screw detection and movement algorithm.

4 Evaluation

4.1 Experimental Setup

The approach is evaluated in the Smart Automation Lab [21] at the RWTH Aachen University. Figure 6 shows the collaborative working cell on the left side, the product in its support frame in the middle and the assembled product on the right. The worker has to place four parts of the lamp and three nuts into the support frame, as well as three screws to connect the parts. The cobot tightens the screws afterwards. The cobot is a Doosan M1013 from Doosan Robotics [22]. At the tool-center point, the ERS12-M20 screwdriver from Desoutter [23] is attached.

4.2 Results

The evaluation is split into two different parts. First, the angle between the assembled part and the camera is set to zero degrees, which means the cobot moves in its x-direction (see Fig. 3) to hit the three screws. The results are seen in the middle of Fig. 7. Afterwards, the assembled part is turned by 90°, so the cobot has to move in its y-direction to hit the screws. The results are shown in the right picture of Fig. 7. The aggregated results are shown on the left picture of Fig. 7. The center of each screw is at position (0, 0).

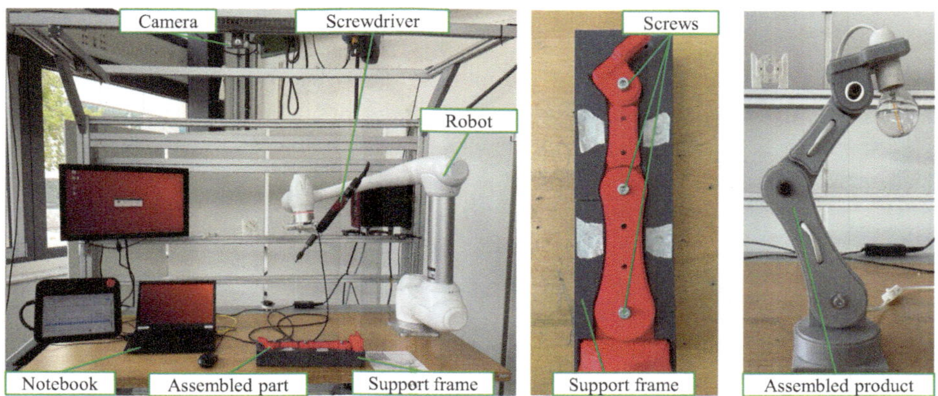

Fig. 6 Left side: workspace of the cobot. Middle: assembled product in red with the support structure in grey. Right side: assembled product

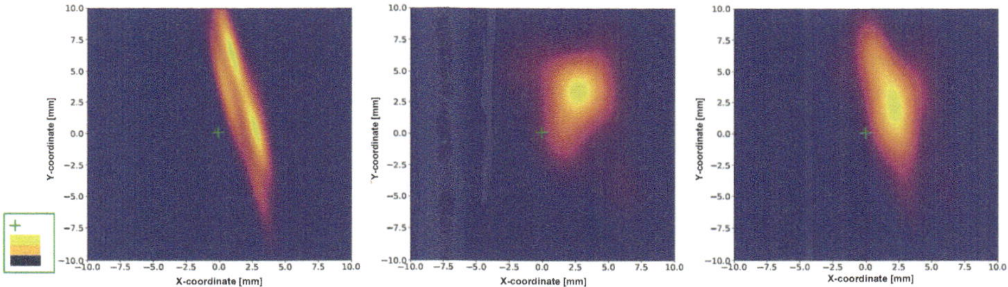

Fig. 7 Heat maps of the deviation between screwdriver and screw head. Left side: aggregated deviation. Middle: camera and assembled part parallel. Right side: camera and assembled part with right degree angle

For both rotations, twelve experiments are conducted, which leads to a total recognition of 72 screw heads. The screw heads have diameters of 12.5 mm. The height between assembled part and camera is constant at 1100 mm. This leads to an average deviation of 3.4 mm (with a standard deviation of 1.8 mm) between screwdriver and screw head. In summary, 72/72 screws are detected correctly from the YOLOv8n model, without false positive detections.

Overall, every screw head is detected correctly by the artificial neural network. Nevertheless, the calibration of the camera in respect to the coordinate system of the cobot is done via hand-eye calibration, which reduces the precision of the system. Furthermore, the higher deviation in y-direction could be explained by a stronger distortion of the camera image in this direction. However, since the cobot attempts to reach every detected screw, the logic of the behavior tree works well.

5 Conclusion

This paper presents a novel approach to integrate screw fastening into the behavior tree, used as high-level control structure for a cobot. In order to detect the screws in the workspace, an Intel RealSense D435i camera along with a YOLOv8n algorithm is used, and trained on an augmented dataset of 200 screw heads. The list with the coordinates of the detected screws is transformed from the image and camera to the cobot coordinate frame and written on the blackboard. The framework MoveIt is used to carry out the movement of the cobot to the position of the screw heads. The evaluation shows, that 72 out of 72 screws are detected correctly. Nevertheless, the insufficiently accurate calibration of the camera leads to an average deviation between the screw head and the screwdriver of 3.4 mm, with a standard deviation of 1.8 mm.

For future research, the calibration of the camera should be studied more in-depth to hit the screw heads more precisely. Futhermore, the evaluation should be carried out with different cobots and use cases to show the generalization of our approach.

Acknowledgements Funded by the Deutsche Forschungsgemeinschaft (DFG, German Research Foundation) under Germany's Excellence Strategy—EXC-2023 Internet of Production—390621612 and funded as IGF-project 22648 N/2 (ROOKIE) of the research association FVP via the AiF within the funding program "Industrielle Gemeinschaftsforschung und -entwicklung (IGF)" by the Federal Ministry for Economic Affairs and Climate Action (BMWK) due to a decision of the German Parliament.

References

1. Burstedde, A., Hickmann, H., Schirner, S., Werner, D.: Ohne Zuwanderung sinkt das Arbeitskräftepotential schon heute. In: IW-Report 25/2021 (2021)
2. Buxbaum, H.-J., Kleutges, M.: Evolution oder revolution? Die Mensch-Roboter-Kollaboration. In: Mensch-Roboter-Kollaboration, pp. 15–33. Springer Gabler, Wiesbaden (2020)
3. Hietanen, A., Pieters, R., Lanz, M., Latokartano, J., Kämäräinen, J.-K.: AR-based interaction for human-robot collaborative manufacturing. In: Robotics and Computer-Integrated Manufacturing, vol. 63 (2020)
4. Colledanchise, M., Ögren, P.: Behavior Trees in Robotics and AI. CRC Press (2018)
5. Behery, M., Deutsch, J., Trinh, M., Kötter, D., Brecher, C., Lakemeyer, G.: Human robot collaborative assembly using behavior trees and dynamic tree dispatching. In: 56th International Symposium on Robotics, Stuttgart (2023)
6. Trinh, M., Kötter, D., Chu, A., Behery, M., Lakemeyer, G., Petrovic, O., Brecher, C.: Safe and flexible collaborative assembly processes using behavior trees and computer vision. In: Intelligent Human Systems Integration (2023)
7. Müller, B., Reinhardt, J., Strickland, M.: Neural Networks: An Introduction. Springer (1995)
8. Han, S., Choi, M.-S., Shin, Y.-W., Jang, G.-R., Lee, D.-H., Cho, J., Park, J.-H., Bae, J.-H.: Screw driving gripper that mimics human two-handed assembly tasks. Robotics **11**(1) (2022)

9. Li, R., Pham, D., Huang, J., Tan, Y., Qu, M., Wang, Y., Kerin, M., Jiang, K., Su, S., Ji, C., Liu, Q., Zhou, Z.: Unfastening of hexagonal headed screws by a collaborative robot. IEEE Trans. Autom. Sci. Eng. **17**(3) (2020)
10. Huang, J., Pham, D., Li, R., Jiang, K., Qu, M., Wang, Y., Su, S., Ji, C.: A screw unfastening method for robotic disassembly. In: Industry 4.0—Shaping the Future of the Digital World (2021)
11. Peralta, P., Ferre, M., Sánchez-Urán, M.: Robust fastener detection based on force and vision algorithms in robotic (un)screwing applications. Sensors **2023** (2023)
12. Villa, N., Mobedi, E., Ajoudani, A.: A contact-adaptive control framework for co-manipulation tasks with application to collaborative screwing. In: 31st IEEE International Conference on Robot and Human Interactive Communication (2022)
13. YOLO from Ultralytics. https://github.com/ultralytics/ultralytics. Last accessed 28 Feb 2024
14. Ambaye, G., Krishnan, K., Boldsaikhan, E.: Detection of small screws using machine learning. In: International Conference on Information and Communication Technology for Development for Africa (2023)
15. Mangold, S., Steiner, C., Friedmann, M., Fleischer, J.: Vision-based screw head detection for automated disassembly for remanufacturing. In: 29th CIRP Life Cycle Engineering Conference (2022)
16. Xu, M., Yoon, S., Fuentes, A., Park, D.: A comprehensive survey of image augmentation techniques for deep learning. Pattern Recognit. (2022)
17. Intel RealSense D435 Camera Website. https://www.intelrealsense.com/depth-camera-d435/. Last accessed 24 Feb 2024
18. Jocher, G., Chaurasia, A., Qiu, J.: YOLO by Ultralytics. https://github.com/ultralytics/ultralytics. Last accessed 31 Jan 2024
19. Chitta, S., Sucan, I., Cousins, S.: MoveIt! IEEE Robot. Autom. Mag. (2012, March)
20. Koubaa, A.: Robot Operating System (ROS). Springer (2017)
21. SAL Homepage: https://smartautomationlab.de/cms/~mrsju/smart-automation-lab/. Last accessed 30 Jan 2024
22. Doosan M1013 Collaborative Robot. https://www.doosanrobotics.com/de/product/Products/LineupOptions/M1013. Last accessed 20 July 2024
23. Desoutter ERS12-M20 Screwdriver. https://webshop.desouttertools.com/de-de/products/Elektrische-Schraubendreher/ERS12-M20-sku6151656930-cat31011.32042. Last accessed 20 July 2024

Gesture-Based Robot Programming Enabling Non-sequential Control Flow

Fabian Mikula, Sascha Sucker and Dominik Henrich

Abstract

Medium-sized companies have the potential to automate numerous tasks. However, this often faces challenges due to small batch sizes with costs outweighing the benefits. One solution is enabling non-experts to program robots with an intuitive interface. Prior research hardly focused on gestures for such an interface. In this paper, we contribute a concept for a unimodal control structure-based robot programming system. To this end, we present a concept and prototype that records the user's gestures, classifies them and converts them into a robot program with control structures. Further, we present methods for error detection. We assessed the usability of our prototype with a user study resulting in a median SUS score of 81.25. We found that users can program both simple and complex tasks in less than 2 min on average.

Keywords

Gesture programming • Intelligent robots • Intuitive robot programming • Human-robot interaction

1 Introduction

Automating tasks presents a significant value for companies [21]. Nevertheless, robot programming demands expertise, incurring high costs [20, 21]. A common solution lies in having a robot capable of intuitive and flexible reprogramming by non-experts enabling seamless adaptation to evolving task requirements [20, 21]. Research is currently underway

F. Mikula (✉) · S. Sucker · D. Henrich
Chair for Robotics and Embedded Systems, University of Bayreuth, Bayreuth, Germany
e-mail: Fabian.Mikula@uni-bayreuth.de

M.-C. Wanner et al. (eds.), *Annals of Scientific Society for Assembly, Handling and Industrial Robotics 2024*, https://doi.org/10.1007/978-3-031-91463-8_4

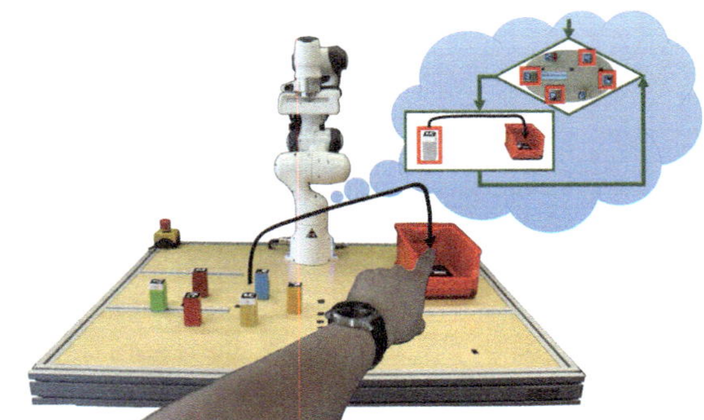

Fig. 1 We envision robot gesture programming to instruct complex tasks. Here, we instruct a for-each loop to place the selected objects into the box

to increase programming accessibility using natural language [22, 23], kinesthetic guidance [25], augmented reality [11, 18, 19], or gestures [14, 15]. *Gestures* are body movements conveying ideas or meanings [28]. Further, they are applicable in loud environments and can express complex ideas (e.g., sign language). Thus, gestures offer a promising interface for intuitive human-robot interaction [26, 27].

This paper targets intuitive robot programming of complex tasks using only gestures (see Fig. 1). In robot programming, gestures often supplement a main input modality (e.g., voice commands in [14]). Here, we specifically focus on gestures as unimodal programming input modality. *Programming* refers to complex control flow and parametrization of robot skills differing from simple motion primitives – thus, allowing the expression of complex tasks. Such control flow encompasses various control structures, including sequences, branches, and loops (for, while, and for-each).

As our contribution, we present intuitive robot programming of complex tasks using only gestures (Sect. 3). This includes a dynamic world representation, gesture recognition (Sect. 3.1), and program construction including syntax (Sect. 3.2) and semantic interpretation (Sect. 3.3). Based on a user study, we examine the usability of robot programming with gestures only, given our prototype (Sect. 4).

2 Related Work

Research in robot programming using gestures focuses significantly on teleoperations. In this method, users often direct the robot to move in increments in a specific direction through gestures [1–5]. Another form of teleoperation involves extracting the skeletal model of the user from a video stream with depth data and transferring the model's joint positions to a humanoid robot [6] or robot arm [7] in real-time. Alternative approaches specify the position of a robot, using arm movements or predefined hand gestures that trigger a linked robot program [8–11, 16]. However, these programming methods cannot handle more complex tasks as no branches are supported. Furthermore, programming techniques that connect robot programs with gestures would necessitate expert knowledge to program the robot's movements.

To program more complex tasks, we must extract control structures from gestures. For instance, there are methods for programming pick and place operations using hand gestures [12–15, 17]. Individual objects [12, 13, 15, 17] or groups [14] can be selected using pointing gestures and placed at the target location. Some of these approaches are multimodal [14][17], allowing for input in multiple forms. For example, voice commands and gestures can be used for pick and place operations [14, 17]. However, these implementations only include control structures such as sequences for the successive execution of operations (all works) or the for-each loop [14] for moving entire selected object groups. Other control structures need to be included in these approaches, thus not allowing complex programs, which is the central focus of this work.

Additionally, prior research explored gesture programming in augmented reality, which specifies the target position of an object more precisely. For instance, augmented reality allows the user to grasp and move virtual objects, enabling the parameterization of both the object orientation and target position [18, 19]. Another approach uses object relations to define target positions through gesticulation [15]. However, we aim to limit ourselves to the most necessary hardware.

Regarding programming with gestures, prior approaches focus on control structures such as sequences (all works), for-each loops [14], and for loops [16]. However, they lack the integration of branches and while-loops, which are essential for the control flow of complex programs. Therefore, this work fills the gap by focusing on programming control structures using gestures.

3 Robot Programming with Gestures

This work aims to realize the programming of the control structures using gestures. The first step in the programming process with gestures is to record a video stream with depth information (Recording, Fig. 2). For gesture recognition, a hand skeleton model, consisting of markers with a fixed position on the hand, is extracted from the video stream. A neural network classifies these markers into gesture tokens. After the classification, the tokens are filtered to avoid noise (unintended gestures) contributing to the program's correctness. The filtered gesture tokens are analyzed syntactically to further check correctness utilizing a context-free grammar. After this, a semantic parsing is conducted to generate a control structure based robot program consisting of semantic gesture information (Programming, Fig. 2). Before execution, the semantic gesture information is analyzed and checked for semantic correctness in the relevant context. For instance, in a pick and place operation, it is verified whether the selected objects exist or the target position is reachable. Finally, the robot executes the program.

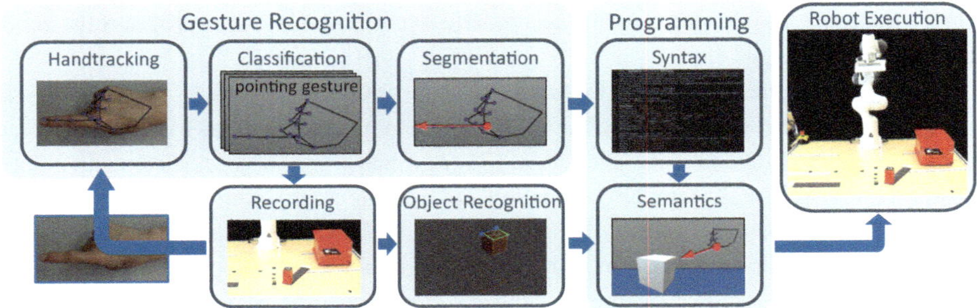

Fig. 2 This is an overview of the programming process with gestures. Gesture and Object Recognition are done parallelly. The programming step interprets both results to create the program for robot execution

3.1 Object and Gesture Recognition

During *object recognition* we detect and localize objects in the robot's workspace creating a dynamic world model [30, 31]. For this, we require a video stream with depth information from a calibrated sensor encompassing the workspace. Concurrently, in *gesture recognition*, we extract a sequence of gesture tokens from image data of the same video stream. Gesture recognition can be achieved using a neural network, typically through CNN-based approaches [32, 33]. However, the programming system can be complex to train due to the varying characteristics of many individuals, such as different skin colors. Moreover, incorporating the orientation of the gesture into the training data presents a challenge. Therefore, we have implemented gesture recognition based on hand pose estimation. Hand pose estimation can be approached through vision-based or sensor-based methods [36]. Vision-based approaches often use neural networks to determine a *hand skeleton model* M from an input image I containing individual hand skeleton *markers* $m_i \in M \subseteq I$ [36]. Sensor-based methods often use a data glove to measure the bending angle and the level of adduction of each finger captured by embedded sensors [36]. We opted for the vision-based neural network approach as publicly available software (e.g., media pipe [35]) can generate 21 markers M. Furthermore, there is no need for additional hardware on the body. To recognize the gestures, we use a neuronal network, which receives the hand skeleton markers M as input and delivers the gesture class as output. There is no pre-trained neural network that classifies gestures for programming control structures. Thus, a custom neural network is required (e.g., on the basis of [33]). The neural network's training data can be generated by recording a gesture over several frames and storing the extracted labeled data. However, the training data would be required to cover all orientations and positions of a gesture for accurate classification. To remedy this, we normalize the training data for translation and rotation invariance. We achieve translation invariance of the hand skeleton model by specifying all markers m_i relative to the base marker m_0. We obtain rotational invariance by aligning the hand skeleton model with the y-axis. Accordingly, we can use this normalization to record fewer training samples.

Natural behavior, such as deliberation intervals or transitional movements between relevant gestures, may occur during the programming process. It is essential to acknowledge this as it can impact the correctness of the program. To prevent a false robot program, we have to filter out gestures irrelevant to the programming task. For example, suppose a user extends his index finger as he thinks about which object to point next. In that case, the inadvertent extension of the index finger is also accidentally recognized as a gesture. It is feasible to solve this problem in several ways. One possibility would be to analyze the posture data and the direction of the programmer's gaze and deduce whether the user is currently programming or engaging in natural behavior such as thinking. However, this requires further research and needs to be considered in another paper. A more straightforward solution would be to enforce a specific dwell time for the gestures before transmitting them to the robot controller. This would result in minor uncertainties and hand movements being ignored. We implement this type of filtering by looking at the probability of a gesture occurring in a time window. If this probability exceeds a certain threshold, the gesture is then added to the program's gesture token stream.

3.2 Program Construction with Control Structures

This work aims to realize the programming of the control structures using a *context-free grammar* $G = (N, T, P, S)$. The production rules P map the non-terminal symbols N into sequences of terminal symbols T, where a terminal symbol $t \in T$ corresponds to a single gesture. From the start symbol S, all words w of the language $L(G)$ are derived from the grammar G. Using context-free grammar implies that the set of all terminal symbols is interchangeable without losing the functionality of the language. Furthermore, the assignment of the words $w \in L(G)$ can be checked for syntactic correctness, allowing us to decide whether a gesture sequence w makes sense and is part of the defined language, which is essential for the generation of a correct robot program. In addition, semantic parsing enriches the grammar to ensure executable code.

The program is constructed by first sending the filtered gestures to the robot programming instance, which checks the transmitted gesture sequence for affiliation to the language $L(G)$ by syntactic analysis. The framework of each control structure can only be constructed as part of the program using syntactic analysis if the program of the gesture is syntactically correct. Therefore, each control structure has a grammar G', a subset of the grammar G from which we derive the control structures. When recognizing a word $w \in L(G')$, we apply semantic parsing to generate a control structure using the semantic information from the gestures. The terminal symbols of the grammar result from a preliminary study with three participants. First, the participants are gesticulating an instruction for a task and then executing a gestured instruction. However, the choice of gestures requires a further, more elaborate user study with more participants. All loops start and end with an initial gesture. Following this, we provide loop-specific instructions, such as a condition (thumbs up, down, or an interval with index finger and thumbs) or an object selection with the flat of the hand over an object group. Now, we can define the loop body, which can contain further instructions. Loops conclude with an end gesture corresponding to the start gesture. The only exception is the for loop with a fixed number of repetitions, particularly if the number 5 is gesticulated. Therefore, we define the number of repetitions of the for loop at the end. This is important to avoid confusion when selecting objects with a flat hand in a for-each loop. The number n of repetitions can be defined in different ways. One possibility is to define n by gestures from the most significant digit x_p to the least significant digit x_0. This method is independent of the base b of n and can be calculated with $n = \sum_{i=0}^{p} x_i \cdot b^i$. For humans, the decimal system ($b = 10$) is suitable because of the ten fingers. Another possibility is an additive method, calculating n as the sum of the numbers x_i shown in the numerical gestures. The last possibility is a hybrid approach. For calculating $n = c \cdot 10 + d$, we first add the number c of tens and then the digits d. This was commonly used in our preliminary study leading us to choose this approach.

Branching is initiated with a branching gesture, followed by a set of conditions represented by gestures such as thumb down for negative, thumb up for positive, and an interval. To define the interval, we use the distance between the thumb and index finger. Now, we can define the body of the branch and terminate the branch with a branch gesture. In the case of multiple conditions in a single branch, the logical 'and' is applied to them. Several branches can be instructed in succession to implement the logical 'or'.

3.3 Semantic Interpretation

Following the program setup step, the control structures and operations provide an understanding of the context of individual gestures. However, the deictic gestures, such as pointing gestures and object selection with a flat hand in the for-each loop, require interpretation and semantic verification to ensure an executable program. For instance, verifying if the objects have been selected is necessary during the pick and place operation. Additionally, we indicate an accessible point on the table when selecting a target by a pointing gesture. For interpreting the pointing gestures, we calculate a line

$$g = \hat{m}_5 + \lambda \cdot (\hat{m}_7 - \hat{m}_5); \text{ with } \lambda \in \mathbf{R}, \hat{m}_i \in \hat{M} \quad (1)$$

from the hand skeleton model, where $\hat{m}_i \in \hat{M}$ represents the world coordinates of the marker i from the hand skeleton model M. The program checks whether the straight line g intersects with one of all oriented bounding boxes of the world model. If it is an object selection and the intersection test is positive, the corresponding object is noted in the pick and place operation. If the cut test of an object selection is negative, an error is output, and the program is not executed. When selecting a target, a positive intersection test results in stacking, and a negative intersection test results in calculating the intersection point between the worktop and the straight line g. We model the worktop as a plane D, resulting in the following intersection test:

$$E = u \cdot \begin{pmatrix} 1 \\ 0 \\ 0 \end{pmatrix} + v \cdot \begin{pmatrix} 0 \\ 1 \\ 0 \end{pmatrix} = g; \text{ with } u, v \in \mathbf{R}. \tag{2}$$

For the group selection, we project the palm's center point

$$p = \frac{1}{n} \cdot \sum_{i=1}^{n} \hat{m_i}; \text{ with } n = |\hat{M}|, \hat{m_i} \in \hat{M} \tag{3}$$

onto the plane E of the robot's work surface. Next, we perform a K-Means clustering for all objects in the world model. Then, we select the object group closest to the projected point p of the palm when p lays within a radius from the nearest cluster's center. An error message will be displayed if no cluster is found near the palm. Otherwise, the program can be executed.

4 Evaluation

The following user study aims to answer the scientific question of the extent to which robots can be programmed to use gestures. To this end, the study consists of four tasks, each requiring different types of control structures, which are used to analyze the program's usability with the SUS score and the programming time. The first task involves programming a sequence of pick and place operations, while the second task requires a for loop to program a pick and place operation. Task three involves a for-each loop to move a set of objects using a pick and place operation. The fourth task requires programming a for-each loop that evaluates a condition within the loop body. We also set up a monitor for feedback about the hand skeleton model, the recognized gesture, and the filtered gesture. A suspected confounding factor is unreliable hand tracking. As the hand skeleton model is not displayed correctly on the monitor, it may affect the overall usability.

The user study commences with a written briefing for each participant and addresses any questions. Next, we give the participants a brief period to acquaint themselves with the robot programming system. Subsequently, we present individual tasks to the participants in paper form, which they can solve one after the other. Afterwards they complete a SUS and a demographic questionnaire. The questionnaire covers age, gender, highest degree attained, programming experience, experience with robots, and experience with gesture recognition systems.

The study involved twelve participants, all from convenience sampling, primarily students and employees of the University of Bayreuth. The age range of the participants was between 22 and 57 years, with a median age of 25 years, including eleven male and one female

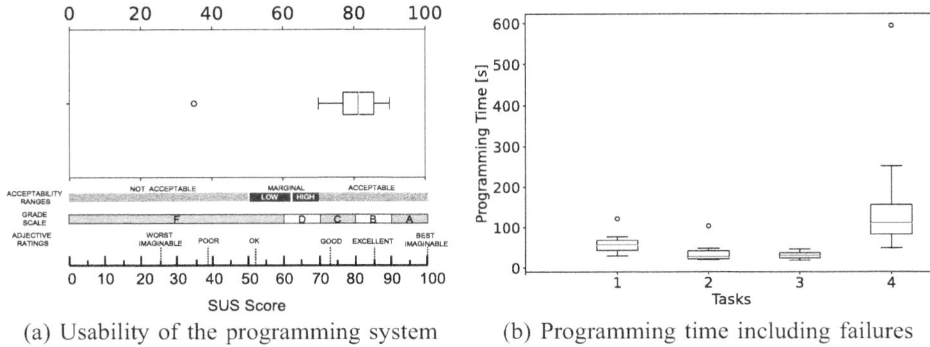

(a) Usability of the programming system (b) Programming time including failures

Fig. 3 The program's usability is 'good' [29], with a median SUS over 81.25 across all tasks. We suspect a familiarization effect at the beginning reducing the programming time of Task 2 and 3. With increasing complexity in Task 4, however, the programming time increases due to multiple failed attempts

participants. Half of the participants had prior experience with handling robots, and 28.8% of all participants had previously worked with gesture recognition systems, all of whom had prior experience with robots. The analysis of the effects of previous knowledge on handling the programming system was not possible due to the small number of participants. Additionally, all participants have a programming background, confounding the questions about the accurate assessment of programming intuitiveness given their prior experience. However, this could be investigated in further research with a larger user group including individuals without programming experience.

The data shows that the median SUS score awarded is approximately 81.25, with an average SUS score of 77.7 (Fig. 3a). These values are higher than the average SUS values in the hardware (=71.8) and GUI (=76.2) categories and can be classified as good [29]. The outlier with a SUS Score of 35 is due to hand tracking. In particular, the hand skeleton model is not always detected or is lost during movement making programming the robot more complex and requiring additional effort. This was the case with the female subject, possibly due underrepresented training data in Media Pipe. However, in general, the user study evaluation by the SUS score demonstrates the convenience of programming complex tasks with gestures.

When evaluating programming time, it is measured until a syntactically correct program is generated, including the times of failed programming attempts. As tasks get more complex, programming time increases, and more failed attempts occur before a correct program is achieved. Task four took twice as long as tasks two and three combined, with an average completion time of 111 s over two attempts and an outlier with 594 s due to hand tracking. Task one had a longer completion time of 57 s due to unfamiliarity with the programming system (Fig. 3b). The evaluation of the programming time in the user study showed that simple and complex tasks can be programmed quite quickly using gestures.

Most of the feedback from all participants relates to hand tracking. The hand skeleton model is not found during movements, which causes irritations. Additionally, participants suggested adding meta gestures that would influence the programming process to improve user-friendliness. For example, meta-gestures would allow users to undo gestures sent incorrectly rather than having to cancel programming and repeat programming the whole task. The study participants suggested including acoustic feedback to characterize the transmission of the gesture to the programming system, in addition to displaying text on the monitor. Another suggestion was to recognize natural behavior and segment it out.

5 Conclusions and Future Work

This work contributes a gesture-based robot programming system with good usability. In this case, gesture-based programming means the robot is programmed unimodally, and control structure-based with gestures. The programming system has proven to be good, as the median SUS score from the user study is above 71.4. For user-friendly programming with gestures, it is crucial to distinguish relevant gestures from natural behavior, which is achieved through gesture recognition and a sliding window filtering method. A context-free grammar is presented to check the syntactic correctness of a gesture input sequence and build up the context of the instructions using control structures. To make the program executable, the gestures in the control structures must be semantically interpreted. The programming system's usability was evaluated as part of a user study. The average SUS score was 77.7, which is considered good. The evaluation of programming time to generate a syntactically correct program showed a increase for more complex tasks. This was due to additional failed attempts caused by incorrect hand tracking during programming.

Therefore, our prototype needs further improvements. For example, quicker and more robust hand tracking could improve the system's usability. Additionally, reducing the filtering time of the gestures enhances user-friendliness and the programming time. Moreover, a customized feedback could improve the user's understanding and, thus, the overall usability. Also specifying the object position more precise would increase the number of industrial assembly tasks. In conclusion, this concept allows users to program a robot with good usability using only gestures. With this, even complex tasks can be programmed quickly.

References

1. Tsarouchi, et al.: High level robot programming using body and hand gestures. Procedia CIRP **55**, 1–5 (2016)

2. Neto, P. et al.: Accelerometer-based control of an industrial robotic arm. RO-MAN, 1192–1197 (2009)
3. Coronado, E., et al.: Gesture-based robot control: Design challenges and evaluation with humans. ICRA **2017**, 2761–2767 (2017)
4. Shin, S., Tafreshi, R., Langari, R.: EMG and IMU based real-time HCI using dynamic hand gestures for a multiple-DoF robot arm. J. Intell. Fuzzy Syst. **35**(1), 861–876 (2018)
5. Mashood, A., et al.: A gesture based kinect for quadrotor control. ICTRC **2015**, 298–301 (2015)
6. Yavşan, E., Uçar, A.: Gesture imitation and recognition using Kinect sensor and extreme learning machines. Measurement **94**, 852–861 (2016)
7. Fang, B. et al.: A novel data glove using inertial and magnetic sensors for motion capture and robotic arm-hand teleoperation. Ind. Rob.: An Int. J. 155–165 (2017)
8. Sylari, A., Ferrer, B.R., Lastra, J.L.M.: Hand gesture-based on-line programming of industrial robot manipulators. INDIN **1**, 827–834 (2019)
9. Rudd, G. et al.: Intuitive gesture-based control system with collision avoidance for robotic manipulators. Ind. Rob.: The Int. J. 243–251 (2020)
10. Devine, S. et al.: Real time robotic arm control using hand gestures with multiple end effectors. UKACC (CONTROL), 1–5 (2016)
11. Peppoloni, L. et al.: Immersive ROS-integrated framework for robot teleoperation. 3DUI, 177–178 (2015)
12. Quintero, C.P. et al.: Visual pointing gestures for bi-directional human robot interaction in a pick-and-place task. RO-MAN, 349–354 (2015)
13. Becker, M., et al.: GripSee: a gesture-controlled robot for object perception and manipulation. Auton. Robot. **6**(2), 203–221 (1999)
14. Han, J. et al.: Structuring Human-Robot Interactions via Interaction Conventions. RO-MAN, 341–348 (2020)
15. Gäbert, C. et al.: Gesture Based Symbiotic Robot Programming for Agile Production. CIVEMSA, 1–6 (2022)
16. Nuzzi, C. et al.: MEGURU: A gesture-based robot program builder for Meta-Collaborative workstations. Rob. and Computer-Integrated Manufacturing (2021)
17. Van Delden, S., et al.: Pick-and-place application development using voice and visual commands. Ind. Robot: An Int. J. **39**(6), 592–600 (2012)
18. Lambrecht, J., Krüger, J.: Spatial programming for industrial robots based on gestures and Augmented Reality. IEEE/RSJ Int. Conf. Intell. Robots and Syst. **2012**, 466–472 (2012)
19. Lambrecht, J., Krüger, J.: Spatial programming for industrial robots: efficient, effective and user-optimised through natural communication and augmented reality. Adv. Mater. Res. **1018**, 39–46 (2014)
20. Dietz, T. et al.: Programming system for efficient use of industrial robots for deburring in SME environments. In: ROBOTIK; 7th German Conference on Rob, pp. 1–6 (2012)
21. Goel, R., Gupta, P.: Robotics and Industry 4.0. In: Nayyar, A., Kumar, A. (eds.) A Roadmap to Industry 4.0: Smart Production, Sharp Business and Sustainable Development (pp. 157–169). Springer International Publishing (2020)
22. Weigelt, S.: Eine agentenbasierte Architektur für Programmierung mit gesprochener Sprache (2022)
23. Sucker, S., Henrich, D.: A layered Pipeline for Natural Language Robot Programming with Control Structures. WG MHI (to appear) (2023)
24. Sauer, L., Henrich, D.: Synchronisation in Extended Robot State Automata. IRC, 360–363 (2022)
25. Riedl, M.: Intuitive sensorbasierte, editierbare, kinästhetische Programmierung von Pick-and-Place-Aufgaben für Mehrrobotersysteme [Dissertation] (2022)
26. Xia, Z.et al.: Vision-Based Hand Gesture Recognition for Human-Robot Collaboration: A Survey. ICCAR, 198–205 (2019)

27. Martin, A.: Exploring 2D and 3D gesture interaction with an assistive system for a workplace for impaired workers [Diploma Thesis] (2014)
28. Gesture: Oxford Reference. Retrieved 19 Mar. 2024, from https://www.oxfordreference.com/view/10.1093/oi/authority.20110803095850220
29. Bangor, A., Kortum, P., Miller, J.T.: Determining what individual SUS scores mean: Adding an adjective rating scale. J. Usability Stud. Arch. (2009)
30. Holz, D. et al.: Real-time object detection, localization and verification for fast robotic depalletizing. IROS, 1459–1466 (2015)
31. Farag, M. et al.: Real-Time Robotic Grasping and Localization Using Deep Learning-Based Object Detection Technique. I2CACIS (2019)
32. Shi, Y., et al.: Review of dynamic gesture recognition. Virtual Reality Intell. Hardware **3**(3), 183–206 (2021)
33. Jain, R., Karsh, R.K., Barbhuiya, A.A.: Literature review of vision-based dynamic gesture recognition using deep learning techniques. Concurr. Comput.: Pract. Exp. **34**(22), e7159 (2022)
34. Mueller, F. et al.: GANerated Hands For Real-time 3D Hand Tracking From Monocular RGB. In: IEEE/CVF Conference on Computer Vision and Pattern Recognition, pp. 49–59 (2018)
35. Zhang, F. et al.: MediaPipe Hands: On-device Real-time Hand Tracking (2020). arXiv:2006.10214
36. Chen, W. et al.: A survey on hand pose estimation with wearable sensors and computer-vision-based methods. Sensors **20**(4), Article 4 (2020)

Human Takeover: Overriding Task Assignments Mid-Execution in Human-Robot Interaction

David Kostolani, Bernd Brenter, and Sebastian Schlund

Abstract

Collaborative robots promise added value to manufacturing due to their flexibility and the possibility of working alongside human workers. Although the goal of human-robot collaboration is to create coordinated teams with common goals, current task allocation models define distinct roles that hinder interactions and enforce task boundaries. We challenge the notion of clear-cut boundaries between the agents by introducing the concept of task takeovers in human-robot collaboration. We demonstrate that creating flexible boundaries, to empower users to interrupt and take over tasks previously assigned to the robot, can be easily achieved by transferring the principles of context-aware computing into industrial human-robot collaboration. Therefore, we ground the need for task takeovers on literature regarding human-automation collaboration and discuss practical applications and future research directions towards making such collaboration more human-centred.

Keywords

Human-robot collaboration • Human-centred automation • Context-aware computing

D. Kostolani (✉) · B. Brenter · S. Schlund
Human-Machine Interaction, TU Wien, Vienna, Austria
e-mail: david.kostolani@tuwien.ac.at

M.-C. Wanner et al. (eds.), *Annals of Scientific Society for Assembly, Handling and Industrial Robotics 2024*, https://doi.org/10.1007/978-3-031-91463-8_5

1 Introduction

Collaborative robots are lightweight robots equipped with collision sensors that enable them to operate in direct proximity to humans. The integration of collaborative robots (cobots) in manufacturing and assembly has prompted a new line of research, where humans work together with robots to achieve a common goal, referred to as human-robot collaboration (HRC). HRC in the industry aims to leverage the strengths of the human operator (e.g., cognition and reasoning) and the robot (e.g., precision and durability). Through this combination, HRC applications promise productivity gains and flexibility in the manufacturing process [4, 15].

A key challenge in human-robot collaboration is task allocation, which describes the division of labour between the agents. Current research mostly focuses on pre-planning the division of tasks, which simplifies transitions and makes allocation more transparent [6, 8]. However, this also leads to task fragmentation and issues when unforeseen errors arise [16]. If a human operator wants to assume control of a task assigned to the robot, they must either interrupt the robot's program or wait until the task's completion before re-negotiating the task allocation. Moreover, task assignment continues to pose a challenge to decision-making autonomy. Prior works have shown that users are prompted to assign authority to the robot, thereby ceding their control over task allocation [5]. This stands in contrast with the goal of human-centred manufacturing, where automation supports the operators by enhancing their decision-making skills to increase production flexibility.

We believe that effective human-robot cooperation hinges on empowering the operator. This also includes the possibility of the human taking over the control of the robot's task at any time, with such task transactions needing to be seamless and efficient. In our research, we introduce task takeovers in cooperative robotics. Our approach blurs the boundary that pre-programmed task allocation models impose, allowing the human user to override robot actions, and thereby reassert control. By applying the principles of context-aware computing, we demonstrate that current AI models can easily enable such seamless transitions. We explore the implications of human-robot interaction and discuss how applications of HRC can benefit from this approach. The contributions of our research are twofold. Firstly, we (1) propose a novel approach for human takeovers, which is generalisable for different settings, and aims to tackle the challenges related to the division of labour between robots and humans. Secondly, we (2) conceptualise criteria for implementing this approach in practical applications and demonstrate how takeovers can be effectively deployed in realistic settings.

2 Background

2.1 Human-Robot Interaction and Task Allocation

Compared to other industrial sectors, assembly processes are to a large extent still performed manually [10]. Industrial human-robot interaction aims to aid manual operations by augmenting the workspace with a cobot assisting the assembly operator. Cobots are typically equipped with easy-to-use interfaces that encourage re-programming, which makes them suitable for flexible tasks. Although the types of interactions have not been fully standardised, prior works commonly refer to two different levels of teaming that involve joint effort, i.e., cooperation and collaboration [1]. While both involve working towards a common goal, collaboration generally involves a greater degree of interdependence between the human and the robot agent in terms of task execution. The task interdependence can be modelled based on (a) time constraints, (b) process constraints, or (c) workpiece constraints [3]. Hence, tasks are either performed sequentially or simultaneously by the human and the robot [12]. In settings requiring joint effort, agents need to plan the division of labour. In engineering and historical contexts, a common paradigm involves the MABA approach ("machines-are-better-at") which distributes tasks based on metrics of efficiency or cost [9], leaving the residual tasks for human completion. This approach has been criticised in prior works due to job fragmentation and the monotonous nature of such task division [16].

Modern task allocation models promote a human-centred approach to the process of dividing tasks. Task planning is generally performed a priori and can be adapted during the collaborative process. In [16], an intuitive interface for a priori role planning was introduced, allowing human operators to determine a task division between the robot and themselves based on ad-hoc recommendations. During the process, the operator can re-assign the tasks, which can enhance the flexibility of the allocation. In contrast to that, Messeri et al. opted for a robot-driven adaptation of the task allocation. In their approach, the robot proposes adjustments to the task allocation based on video-based ergonomic evaluation of the operator, who then has the option to accept or reject these proposals [11]. A different approach presents the leader-follower paradigm. Zhang et al. used neural networks to detect the actions of the leader, i.e., the human, and to predict the optimal timing for the robot's involvement in the collaboration [22]. The UHTP algorithm, introduced in a study by [13], also leverages a neural network to detect human actions. This detection is then used to structure the task allocation hierarchically based on the actions performed by one of the agents. Mahadevan et al. investigated an alternative approach to task allocation in a virtual setup that relies on either explicit or implicit human delegation [8]. In addition to explicit techniques such as voice commands or pre-defined territories, they propose proxemic relationships between the agents to implicitly delegate a task to the robot by placing it closer to its base. While users have the ability to dynamically re-schedule the task assignment at present, the option to take over or interrupt a task that a robot has already commenced, to the best of our knowledge, has not been addressed in task allocation discussions.

2.2 Authority and Agency in Task Allocation

Current engineering paradigms in task allocation have been shaped by the concept of transparent role assignment [7]. However, the approaches to user involvement differ when it comes to adapting the task division during the assembly process. One approach is to involve the operator by proposing suggestions, which can be approved or rejected [11]. Other works argue for a communication-free allocation strategy to reduce the load imposed on the operator [14].

Empirical works that investigated workers' preferences regarding authority within the allocation process remain inconclusive. In a study performed by Gombolay et al., human operators preferred the robot to perform the task assignment, as participants found it more efficient [5]. In contrast to that, Schmidbauer et al. evaluated their system for self-organised a priori planning with industrial workers, and their results demonstrated that workers prefer to have control over the task allocation [17]. This view is supported by a large study presented in [18], which found that participants were happier when given agency in the task allocation process. Moreover, having control over the allocation also resulted in participants feeling a greater sense of autonomy. These constructs have also been discussed as ways to increase trust and self-efficacy. However, the execution of control can also increase mental effort, as discussed in another study by [19]. Giving robot agency in task allocation also affects their acceptance. Zafari and Koeszegi have found that a low perceived level of control can result in negative attitudes towards robots, which was more pronounced in women and participants with limited prior experience with robots [21]. Therefore, we argue to strengthen the user control over HRC tasks by enabling them to override robot actions.

3 Task Takeovers: Overriding Robot Actions

By allotting roles to the agents, task allocation defines distinct boundaries between them, without considering the possibility of mid-task takeovers. To our knowledge, the concept of overriding or taking over tasks started by the robot has not been included in existing task allocation models. Lending from other domains researching human-autonomy collaboration such as autonomous driving, takeovers denote passing the control of the task from one agent to another at any given moment. In autonomous driving, takeovers are bidirectional, meaning the human can delegate the control to the vehicle or vice versa, depending on the task context [20]. We argue for a similar concept in HRC to extend task allocation and override robot actions. This involves blurring the boundary between the human and the robot tasks, enabling the human to take over. Figure 1 depicts an example of a takeover scenario. The reasons for takeovers in HRC can vary significantly. These include preventing errors by manually assuming control of a task, or for efficiency purposes in sequential collaborations, such when the human participant can complete the task faster and chooses to do so. Moreover,

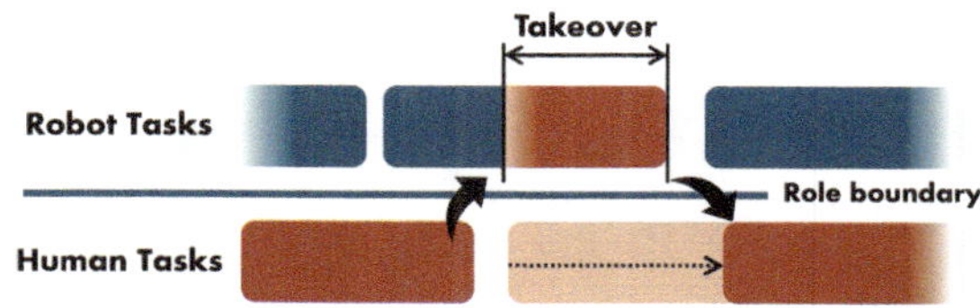

Fig. 1 Conceptual takeover scenario

we argue that the possibility to intervene and override tasks mid-way increases the level of human control over the task, regardless of whether the human operator delegated the task to the robot or not.

We conceptualise takeovers in HRC as unidirectional requests initialised from the human to the robot who has to adapt accordingly. The takeover ends after the human accomplishes the task or delegates the control back to the robot. A challenge to this is that interrupting tasks mid-way results in uncertainty regarding how the robot should proceed with the next task. To tackle this, we propose to make robotic applications context-aware to understand the actions undertaken by the human and continue where the human has left off. Building upon previous research on context awareness [2], we define three levels of context relevant for takeovers:

- **Human intent**: Recognising human intent to take over a task is crucial for a seamless execution of task transitions. Similar to the interaction modes proposed in [8], we argue that takeovers can be explicitly communicated, e.g., through a tactile signal redirecting the robot away from its current trajectory, or implicitly. Implicit means can include cues like proximity, such as when a human reaches for an object while a robot is simultaneously performing a picking operation on the very same object.
- **Status**: In our model, status refers to the degree of completion of the task that has been interrupted mid-way and instead performed by the human. In other words, the robot has to comprehend whether the human aimed to correct the robot's task or to fully take over.
- **Location**: As takeovers require the robot to re-plan the path, location refers to detecting the next target destination for the robot. Generally, this involves determining whether the original destination has been altered by a human or remains available for the robot to use.

4 Experiment: Overriding Pick-and-Place Operations

In our experiment, we consider a pick-and-place task due to its generalisability to other robotic applications. Our experiment aims to demonstrate that takeovers can be easily realised using off-the-shelf machine learning algorithms, thereby, enhancing current engineering practices in collaborative human-robot settings. The application consists of a 7-axis collaborative robot from Franka Robotics, which performs a pick-and-place task in an end-

Fig. 2 Experimental setup and example takeover scenario

less loop to enable takeovers at any given time. When a takeover request is detected, the robot is instructed to halt its current operation, and assume a static position away from the task space to free space for human manipulation. The robot application was programmed using the ROS framework utilizing the robot vendor's specific libraries. Motion plans are computed and executed via the MoveIt motion planning framework. Figure 2 depicts the setup of the application, including takeover scenarios.

To achieve context-awareness, we opted for a visual approach and integrated a ZED2 stereo camera from Stereo Labs to monitor the task space. To detect intent to take over, we implemented explicit communication via haptic feedback utilizing the cobot's torque sensors, and an implicit measure triggered when the human remains idle in the proximity of the pick-and-place locations and actively reaches out towards these areas. This was achieved by pose estimation performed on RGB images, and lifted in the depth map provided by the stereo camera. In particular, we estimate the middle of the hand palm by extrapolating the vector defined by the wrist and elbow keypoints. We also defined regions of interest fixed in the robot coordinate system that corresponded with the location of the two bins for the pick-and-place task. Additionally, we defined a collaborative space in the proximity of the robot. Figure 3 depicts a point cloud visualization, with the regions of interest highlighted: (1) green—collaborative space, (2) red & yellow—pick-and-place bins, (3) turquoise—human skeleton. We use this 3D information to monitor the movement of the human body through trunk keypoints, and define the idling state when the human stops in the proximity of the robot. This is employed to differentiate between the intent to take over and passing by, aiming to prevent misclassification of the takeover intent. Respectively, we monitor the bin locations and initialise a takeover when the palm of the hand enters the proximity of one bin location.

In Addition to recognising the intent to take over, the robot must keep an internal representation of the task and of any elements that have been manipulated by the human. We implement this through the concept of a digital twin, i.e., a state machine that mirrors the context in the physical space, as visualised in Fig. 4. The state machine includes the condition of the binning locations, in our case, 8 possible pick-and-place positions, which are either free or occupied. Furthermore, six robot states are modelled to accommodate various scenarios.

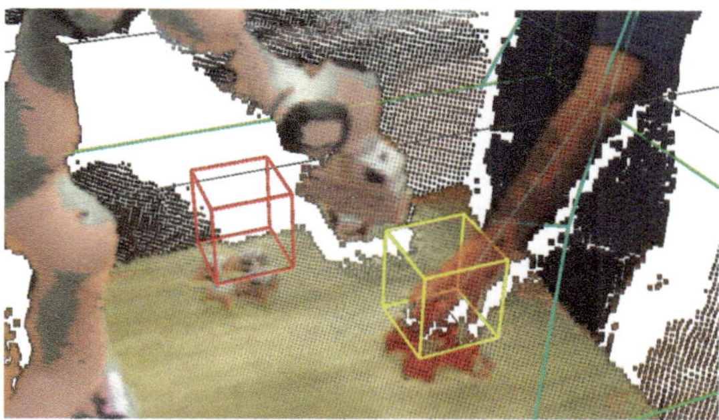

Fig. 3 Point cloud visualisation of the monitored regions of interests and their interaction with the hand keypoint estimated using pose detection

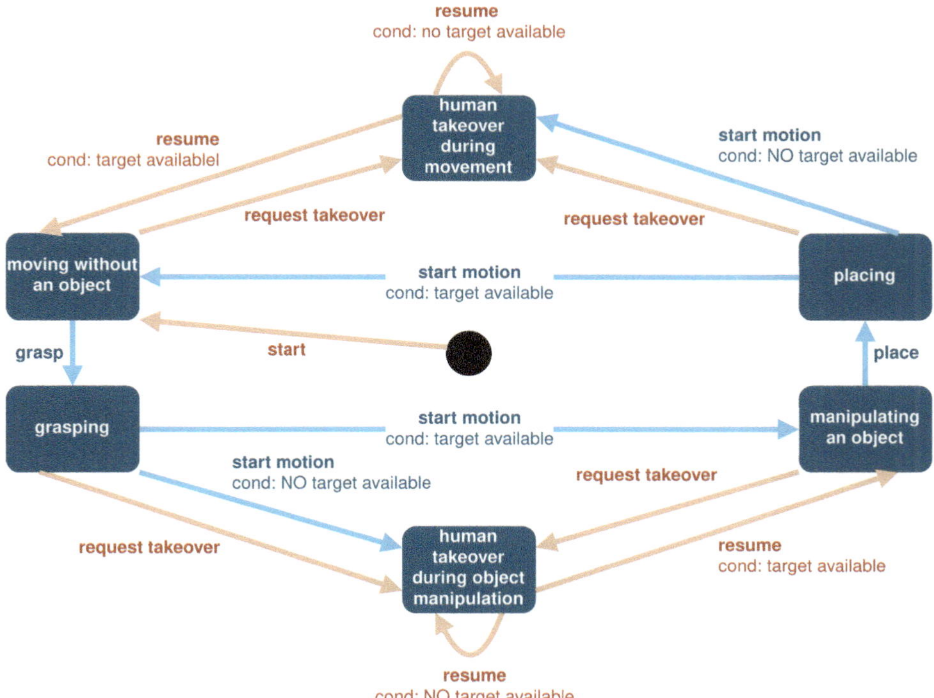

Fig. 4 Contextual representation of the twin modelled as a state machine. Blue components depict state changes of the robot, while orange components visualise state transitions that are triggered by human actions

These states include grasping and placing, moving without or manipulating an object, and human takeover during movement or during object manipulation. This enables the robot to proceed with the prior task after the interruption. Internally, the number of manipulation objects in the task space can be adapted. The virtual representation of the application communicates with the physical application through the visual detection system. Additionally, it infers the internal states that are triggered by the robot's program, enabling an interaction between the virtual and physical components of the system. The 8 different pick-and-place locations are included in the twin, and the recognition is performed using the defined region of interest of the binning locations and structural similarity index measure (SSIM). By comparing frames before and after the takeover, SSIM returns information on whether a bin slot was manipulated by the human during the task. Therefore, the robot can react by re-planning the operation with the next free bin slot.

5 Discussion

Our work argues for including the possibility of taking over robot tasks mid-execution to increase the level of human control over collaborative and cooperative robot tasks. We demonstrate that such a concept can be easily integrated into human-robot interactions by representing the states as a digital twin and recognising context in the workspace using off-the-shelf algorithms.

Besides human-robot cooperation and collaboration, one possible application of this concept is the supervision of robot tasks, e.g., with multiple robots, when the human wants to override the robot's actions. This includes tasks that the robot has not executed correctly or is on the verge of executing incorrectly. In such scenarios, the human can perform the task manually, without interrupting the robot's program, as the robot understands the interaction context and can proceed with the next task.

Open challenges to human-robot task takeovers include the communication of human intent to take over. In our work, we have tested both explicit and simple implicit interaction modes, e.g. via haptic feedback and proximal relationships of the human body and the task. In general, while explicit techniques offer a greater sense of control over the task, implicit means allow users to multitask or engage in parallel activities [8]. In the case of takeovers, implicit communication can be embodied by actions, such as actively reaching out. Other modalities could include gaze patterns to actively infer visual attention and predict the takeover request. One challenge to the implicit approach is the recognition delay until the robot stops the execution of the task and assumes a safe pose outside of the task space. In our case, the reason for this delay is the lower sampling rate of cameras, compared to robot torque sensors. This could potentially impact the usability and efficiency of takeovers, as well as how safety is perceived within the context of these interactions. A possible remedy to this could be the use of more sophisticated ML-enabled action and trajectory prediction algorithms to infer probabilities of future takeovers based on current and past human actions.

Respectively, another open challenge arises in the communication of the robot's intent, which could inform the operator about the current and future tasks the robot is performing, allowing the operator to optimally time their intervention for a takeover. Although prior works refer to communication-free approaches, we argue that successful human-robot collaboration is also hinged on purposeful communication between both agents. Especially when it comes to human-initiated task takeovers, we see benefits in having the robot communicate its understanding of the takeover request back to the human. Without this feedback, the human worker might not be able to tell whether the robot recognised the intent to take over. Future work should explore how this communication of understanding can influence the success of the shared task.

Future work can look to evaluate the effect of task takeovers on the perception of control over the application. In particular, we speculate that applications where task assignment is not performed by the human operator could benefit from our concept. This includes scenarios such as ad hoc task planning or situations where delegating task allocation to the human is limited due to constraints. Moreover, in prior works on task delegation [8], explicit measures were shown to be more efficient and provided a greater level of control. However, this evaluation has been performed through the Wizard-of-Oz concept which did not simulate the overhead time for the recognition, and the delegation did not include the possibility of taking over the robot's task. Therefore, we aim to further evaluate the usability of explicit and implicit mechanisms for indicating takeovers and derive design implications for their implementation.

6 Conclusion

In conclusion, our paper challenges conventional task allocation models in industrial human-robot interaction by introducing the concept of task takeovers. We advocate for empowering human operators to seamlessly interrupt and assume control of tasks designated to robots and argue that this fosters greater flexibility and efficiency in assembly operations. Through the application of context-aware computing principles, we demonstrate the feasibility of smooth transitions mid-task between human and robot agents. Our proposed approach aims to instil a more human-centric perspective in robot-assisted assembly, enabling adaptability to dynamic work environments. Further research directions include exploring practical applications and implications of task takeovers in industrial contexts, with the overarching objective of enhancing collaboration and productivity.

Acknowledgements This work was supported by the Austrian Research Promotion Agency as a part of the A2P project (FFG-891196).

References

1. Aaltonen, I., Salmi, T., Marstio, I.: Refining levels of collaboration to support the design and evaluation of human-robot interaction in the manufacturing industry. Procedia CIRP **72**, 93–98 (2018). https://doi.org/10.1016/j.procir.2018.03.214
2. Dey, A.K., Abowd, G.D., Salber, D.: A conceptual framework and a toolkit for supporting the rapid prototyping of context-aware applications. Hum.– Comput. Interact. **16**(2–4), 97–166 (2001). https://doi.org/10.1207/S15327051HCI16234_02
3. El Zaatari, S., Marei, M., Li, W., Usman, Z.: Cobot programming for collaborative industrial tasks: an overview. Robot. Auton. Syst. **116**, 162–180, (2019). https://doi.org/10.1016/j.robot.2019.03.003
4. Fast-Berglund, Å., Palmkvist, F., Nyqvist, P., Ekered, S., Åkerman, M.: Evaluating cobots for final assembly. Procedia CIRP **44**, 175–180 (2016). https://doi.org/10.1016/j.procir.2016.02.114
5. Gombolay, M.C., Gutierrez, R.A., Clarke, S.G., Sturla, G.F., Shah, J.A.: Decision-making authority, team efficiency and human worker satisfaction in mixed human–robot teams. Auton. Robot. **39**, 293–312 (2015). https://doi.org/10.1007/s10514-015-9457-9
6. Herr, S., Gross, T., Gradmann, M., Henrich, D.: Exploring interaction modalities and task allocation for household robotic arms. In: Proceedings of the 2016 CHI Conference Extended Abstracts on Human Factors in Computing Systems, pp. 2844–2850 (2016). https://doi.org/10.1145/2851581.2892456
7. Kiyokawa, T., Shirakura, N., Wang, Z., Yamanobe, N., Ramirez-Alpizar, I. G., Wan, W., Harada, K.: Difficulty and complexity definitions for assembly task allocation and assignment in human–robot collaborations: a review. Robot. Comput.-Integr. Manuf. (Print) **84,** 102598 (2023) https://doi.org/10.1016/j.rcim.2023.102598
8. Mahadevan, K., Sousa, M., Tang, A., Grossman, T.: "grip-that-there": an investigation of explicit and implicit task allocation techniques for human-robot collaboration. In: Proceedings of the 2021 CHI Conference on Human Factors in Computing Systems, pp. 1–14 (2021). https://doi.org/10.1145/3411764.3445355
9. Malik, A.A., Bilberg, A.: Collaborative robots in assembly: a practical approach for tasks distribution. Procedia Cirp **81**, 665–670 (2019). https://doi.org/10.1016/j.procir.2019.03.173
10. Meissner, A., Trübswetter, A., Conti-Kufner, A.S., Schmidtler, J.: Friend or foe? understanding assembly workers' acceptance of human-robot collaboration. ACM Trans. Hum.-Robot. Interact. (THRI) **10**(1), 1–30. (2020). https://doi.org/10.1145/3399433
11. Messeri, C., Bicchi, A., Zanchettin, A.M., Rocco, P.: A dynamic task allocation strategy to mitigate the human physical fatigue in collaborative robotics. IEEE Robot. Autom. Lett. **7**(2), 2178–2185 (2022). https://doi.org/10.1109/LRA.2022.3143520
12. Petzoldt, C., Harms, M., Freitag, M.: Review of task allocation for human-robot collaboration in assembly. Int. J. Comput. Integr. Manuf. **36**(11), 1675-1-715 (2023). https://doi.org/10.1080/0951192X.2023.2204467
13. Ramachandruni, K., Kent, C., Chernova, S.: UHTP: a user-aware hierarchical task planning framework for communication-free, mutually-adaptive human-robot collaboration. ACM Trans. Hum.-Robot. Interact. (2023). https://doi.org/10.1145/3623387
14. Roncone, A., Mangin, O., Scassellati, B.: Transparent role assignment and task allocation in human robot collaboration. In: 2017 IEEE International Conference on Robotics and Automation (ICRA), pp. 1014–1021. IEEE (2017). https://doi.org/10.1109/ICRA.2017.7989122
15. Sherwani, F., Asad, M.M., Ibrahim, B.S.K.K.: Collaborative robots and industrial revolution 4.0 (IR 4.0). In: 2020 International Conference on Emerging Trends in Smart Technologies (ICETST), pp. 1–5. IEEE. (2020). https://doi.org/10.1109/ICETST49965.2020.9080724

16. Schmidbauer, C., Schlund, S., Ionescu, T.B., Hader, B.: Adaptive task sharing in human-robot interaction in assembly. In: 2020 IEEE International Conference on Iindustrial Engineering and Engineering Management (IEEM), pp. 546–550. IEEE.(2020). https://doi.org/10.1109/IEEM45057.2020.9309971
17. Schmidbauer, C., Zafari, S., Hader, B., Schlund, S.: An empirical study on workers' preferences in human–robot task assignment in industrial assembly systems. IEEE Trans. Hum.-Mach. Syst. **53**(2), 293–302 (2023). https://doi.org/10.1109/THMS.2022.3230667
18. Tausch, A., Kluge, A.: The best task allocation process is to decide on one's own: effects of the allocation agent in human–robot interaction on perceived work characteristics and satisfaction. Cogn., Technol. Work **24**(1), 39–55 (2022). https://doi.org/10.1007/s10111-020-00656-7
19. Tausch, A., Kluge, A., Adolph, L.: Psychological effects of the allocation process in human–robot interaction–a model for research on ad hoc task allocation. Front. Psychol. **11**, 564672 (2020). https://doi.org/10.3389/fpsyg.2020.564672
20. Xing, Y., Lv, C., Cao, D., Hang, P.: Toward human-vehicle collaboration: review and perspectives on human-centered collaborative automated driving. Transp. Res. Part C: Emerg. Technol. **128**, 103199 (2021). https://doi.org/10.1016/j.trc.2021.103199
21. Zafari, S., Koeszegi, S.T.: Attitudes toward attributed agency: role of perceived control. Int. J. Soc. Robot. **13**(8), 2071–2080 (2021). https://doi.org/10.1007/s12369-020-00672-7
22. Zhang, R., Li, J., Zheng, P., Lu, Y., Bao, J., Sun, X.: A fusion-based spiking neural network approach for predicting collaboration request in human-robot collaboration. Robot. Comput.-Integr. Manuf. **78**, 102383 (2022). https://doi.org/10.1016/j.rcim.2022.102383

A System Architecture for Real-Time Trajectory Adaptation in Human-Robot Interaction Using Depth Cameras

Timo Habersang, Paul Glogowski, Alfred Hypki, and Bernd Kuhlenkötter

Abstract

In human-robot interaction (HRI), robots equipped with appropriate safety mechanisms work together with humans within a shared workspace. To ensure the safety of human co-workers, robots must move slowly or come to a complete stop if the co-worker enters their workspace. However, this circumstance reduces the efficiency and cost-effectiveness of HRI applications and is an obstacle to the widespread use of HRI robots. One solution could be that robots act to changes by adapting their path, thereby ensuring the co-worker's safety and ultimately raising the robot's moving speed. For example, a robot aware of the human pose can adapt its real-time trajectory to changing conditions. In this regard, depth cameras can perform human pose estimation reliably, and the robot can use this information for collision avoidance. In this article, the potential of adapting the trajectory of a robot in real-time, utilizing human pose data, is shown, and an initial solution for a system architecture is provided.

Keywords

Human-robot interaction • Human pose estimation • Assembly • Trajectory adaptation

T. Habersang (✉) · A. Hypki · B. Kuhlenkötter
Ruhr University Bochum, Chair of Production Systems, Bochum, Germany
e-mail: habersang@lps.ruhr-uni-bochum.de

P. Glogowski · B. Kuhlenkötter
Ruhr University Bochum, Research Center ZESS, Bochum, Germany

M.-C. Wanner et al. (eds.), *Annals of Scientific Society for Assembly, Handling and Industrial Robotics 2024*, https://doi.org/10.1007/978-3-031-91463-8_6

1 Introduction

The landscape of industrial manufacturing is transforming as collaborative robots, emerge as key players in augmenting human capabilities within assembly processes. In an era where efficiency, precision, and ergonomic considerations are paramount, the seamless integration of humans and robots holds immense promise. According to the ISO/TS 15066 [1] guidelines, four permissible operations exist for implementing and certifying collaborative production systems in human-robot interaction (HRI). One of these operations is *speed and separation monitoring* (SSM). A dynamic separation distance must be maintained, which depends on the cartesian speeds of the robot and the human. If this distance is not maintained, the robot slows until it reaches a standstill. When reactively planning and executing a trajectory, the standstill time of the robot can be reduced. Any reduction in the time the robot does not move leads to increased productivity and efficiency of an assembly task.

This paper proposes an addon for the SSM. When a robot and a human perform a task with active SSM, the robot adapts its speed but not the path on which it moves. If a human blocks the path, the robot will start to get slower and stop when it is too close to the human. This paper addresses that problem and proposes a software architecture that reactively plans and executes a collision-free trajectory. This architecture is then evaluated in a real-time capable simulation, where a pick-and-place task is performed, with captured human motion data. First, an overview of trajectory adaptation and workspace monitoring is given, followed by a presentation of the simulation setup.

1.1 Related Work

The trajectory adaptation for robot movement has been the subject of various research and development activities for several years. An overview is given by Miro et al. [2]. Liu et al. [3] defined a risk index determined by the robot's speed. If the risk index exceeds a certain limit, the robot starts evasive motion or reduces its speed. Bobka et al. [4] measure the shortest distance between the point clouds of the robot and the human and adapt the robot's speed according to this information. Glogowski et al. [5] proposed a method to determine the dynamic separation distance between a human and a robot precisely and thus calculate the robot's speed according to ensure the requirements given in SSM. In [6], Glogowski presented a simulation tool to evaluate SSM in assembly tasks. This tool serves as the foundation for the enhancement presented in this work.

Another form of trajectory adaptation is path adaptation. Werner et al. [7] provided an architecture for risk-minimal path planning in HRI. In [8], an overview of different ways to adapt the path is provided. A distinction is made between local and global path adaptation. In local path adaptation, the robot utilizes information about a small segment of workspace occupancy. Local path planning is frequently performed with virtual potential fields, where obstacles are modeled as hills in the robot's joint space and collision-free configurations as valleys. Like water, the robot's path tends to follow the valleys and so stays collision-free. Khatib [9] first investigated this approach. The problem with local path planning is that an unfortunate path to the goal can be taken because of the lack of knowledge of the whole workspace. In contrast, a global path planning algorithm considers the whole workspace environment. *Rapidly-exploring Random Trees* (RRTs), where the configuration space is explored rapidly by expanding a tree of random samples toward unexplored regions, are commonly used for global path planning. It was proposed by Lavalle [10], and variants that further improved the algorithm were elaborated by Kuffner [11] and Karaman [12]. An overview of more planning algorithms is given by Latombe [13].

Workspace monitoring is necessary to reactively plan a global path in the context of HRI. Therefore, it must be possible to estimate the pose of a human at all times reliably. Numerous depth cameras on the market suit this task [14]. Many of these cameras contain a software integration of the OpenPose library [15], which is a powerful tool for reliably estimating the pose of moving humans by utilizing a deep neural network trained on the COCO dataset [16].

1.2 System Setup of the Simulation Environment

The Robot Operating System (ROS) [17] stands out for its ability to seamlessly integrate a camera, process human pose data, and simulate a robot manipulator within a single framework. This efficiency is a testament to its modularity, where sensors, actuators, and simulation are harmoniously combined in one software system. The simulation, which is based on *RViz*,[1] enables the modeling of the working environment as well as the planning and visualization of robot trajectories. For path planning, which is realized with *MoveIt*,[2] the robot has to be equipped with the knowledge of which workspace areas are blocked to ensure smooth trajectory planning. This is where the *planning scene*[3] comes in, translating the obstacles visualized in *RViz* into blocked zones for joint configuration in the planning algorithm. The versatility of this system is further demonstrated by its ability to model both stationary and moving objects, making it ideal for a pick-and-place task that is only occasionally interrupted by a human. In this setup, stationary objects like the workstation are modeled and loaded in advance. The motion data of the person in the form of the joint positions, captured by a passive RGB stereovision-based ZED camera [18], are published as *tf-frames*[4] in *RViz*. To ensure accurate trajectory planning, enveloping spheres are placed around the person's joints and added to the *planning scene*. Initially, only enveloping spheres are used, as more

[1] http://wiki.ros.org/rviz.

[2] https://moveit.ros.org/.

[3] The planning scene provides the environment information to the planning algorithm.

[4] http://wiki.ros.org/tf.

complex human geometries necessitate more intricate human-to-robot distance calculations. The robot in use, the 7-DoF LBR KUKA iiwa14 with a gripper, also has enveloping spheres placed around its joints for distance calculation.

2 Reactive Trajectory Planning

In this environment, reactive trajectory replanning should be investigated. If there is an obstacle in the form of human body parts in the robot's path and the robot comes too close to the human, the robot will stop. Then, with the knowledge of the current planning scene, a new path should be planned and executed, which avoids the human. In contrast to [7], this paper investigates the performance of a pick-and-place task and also adds the safety mechanism of SSM.

2.1 Application of Reactive Trajectory Planning in a Shared Human-Robot Environment

To address the demands of practical and interdisciplinary research endeavors, a versatile production facility called *COssembly* [19] was engineered and installed at the Learning and Research Factory at the Chair of Production Systems at the Ruhr University Bochum. On this production line, humans and robots work in a shared environment. In Fig. 1, the *COssembly* and its stations are depicted. For this paper, the only station of interest is station 3, whose task is simplified to a pick-and-place task.

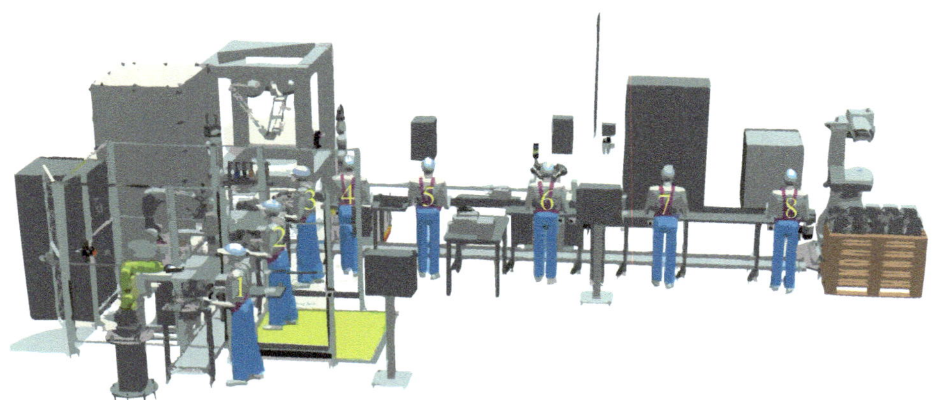

Fig. 1 3D animation of the COssembly

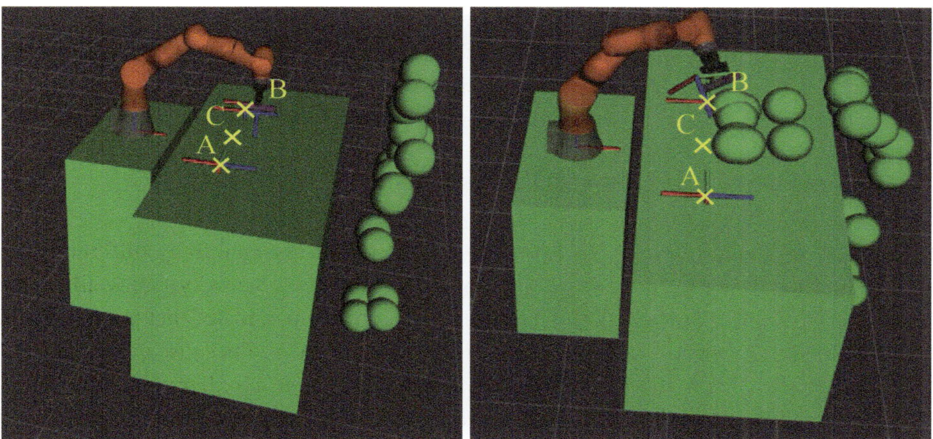

Fig. 2 Simulation of a LBR KUKA iiwa14 performing a pick-and-place task, which may get interrupted by a human

In ROS, the *move group's*[5] method **plan**() is used to plan trajectories for the manipulator to a desired goal configuration or a desired pose and orientation of the end effector. In this work, the trajectories are planned with the RRT-connect algorithm [11]. is us A pick task, as well as a place task, consists of four stages to pick up or release an object correctly. These stages are explained using the example of a pick task, with the place task following a similar process. In the first stage, the robot moves towards the object and stops at a predefined point above it with its gripper open. In the second stage, the robot moves straight towards the object so that it is between the gripper jaws. In the third stage, the gripper closes around the object. In the final stage, the robot lifts the object to a predefined point off the ground. Imagine a scenario as depicted in Fig. 2 where a robot picks up an object at location A and transports it to location B. The path between these two locations is always the same when the task is executed repeatedly. At some point in time, a human decides to perform a task at location C. When the robot attempts to repeat the task, it will stop when it reaches a certain distance, as determined by the SSM's active safety mechanism. The robot will then keep its position until the human moves out of close proximity. Reactive trajectory planning can offer significant benefits in such a case by allowing the robot to replan its trajectory and execute it more efficiently, reducing stillstand time.

[5] The *move group* is the class that controls the robot in ROS.

2.2 Design of a Reactive Trajectory Planning Simulation Module

The present section outlines the architecture of a simple reactive trajectory planning and execution environment in ROS. This simulation environment is an extension of the architecture proposed in Glogowski et al. [6] for SSM.

As depicted in Fig. 3, the architecture comprises two software blocks. The block enclosed within the dashed box executes the SSM function (see [6]) and can work independently. With SSM, the robot follows a predefined path, whereby the speed changes depending on the distance and direction of the body motions. If a human blocks the robot's path, the robot has to stop to avoid a collision. The robot waits to resume its movement until the human does not block the predetermined path. The grey box in Fig. 3 proposes an architecture that

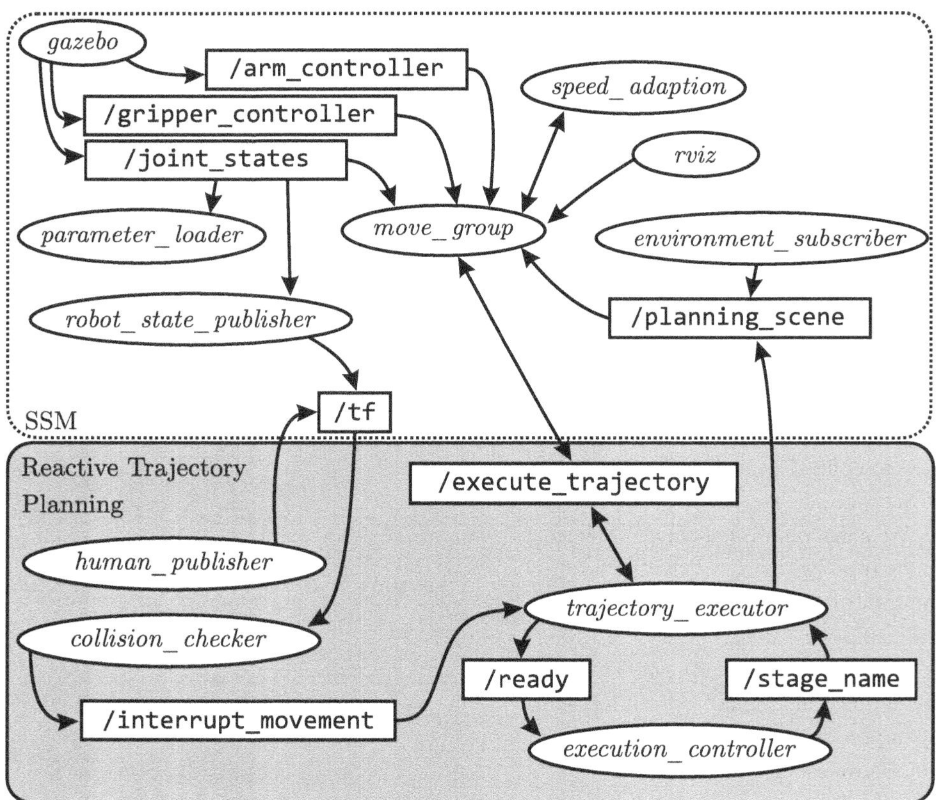

Fig. 3 Architecture of the ROS simulation environment. The nodes are depicted with ellipses, the topics with rectangles, and the communication between the nodes with arrows

extends the SSM by integrating a reactive trajectory planning and execution module. If this software module is activated, the robot can replan its path to avoid colliding with a human. The functionality of this module is elaborated in the following.

The extended architecture includes four new nodes and four new topics. The *human_publisher* publishes the joint positions and orientations captured by the ZED camera as *tf-frames* and applies the human body as a collision object to the planning scene. The collision_checker monitors the distance between the robot's joints and the captured human joints. This node also publishes a command to stop the robot on the `/interrupt_movement` topic if the distance falls below a certain threshold. The *execution_controller* controls the pick-and-place task loop. When a stage of the pick-and-place task concludes, indicated by the `/ready` topic, it publishes the command to plan the trajectory of the next stage on the `/stage_name` topic. The *trajectory_executor* plans and publishes the trajectory of the current stage on the `/execute_trajectory` topic. The trajectory must be executed asynchronously. Otherwise, it would not be possible to interrupt its execution in *MoveIt*.

In ROS, executing a planned trajectory is commonly performed synchronously by using the **`execute`**`()` command. However, this command has a limitation; new commands cannot be given until the program finishes executing the **`execute`**`()` command. In an HRI scenario, the execution of a trajectory must be flexible and adaptable. During a stage's trajectory planning, the robot is only aware of the current human position. While the stage is executed, the human will change his position. Thus, a mechanism must be implemented to change the trajectory correctly when the human moves closer to the robot. Figure 4 presents the flowchart of such a mechanism, which is essentially the logic of the *trajectory_executor*. Upon receiving the `/stage_name` message, the trajectory_executor begins to plan and execute a trajectory for the given target joint values of the stage. The **`asynchExecute`**`()` allows the asynchronous execution of code. That means code in the trajectory_executor can be executed while the *move group* executes the trajectory. So, a loop is implemented to check whether the movement is interrupted or the robot has reached its target joint configuration. The frequency of this loop is determined by **`wait`**`()` and is set to 1 kHz. The robot is allowed to move as long as the interrupt variable has the value `false`. When the collision_checker publishes `true` on the `/interrupt_movement` topic, `interrupt` is set to `true`, and the robot stops its movement. After that, a new trajectory with the current joint states as start states is planned, and `interrupt` is set to `false` so that the robot can start to execute the trajectory. If the attempt to find a collision-free trajectory by **`plan`**`()` fails, it is retried in the next time stamp. While executing the trajectory, the *planning scene* is refreshed by the *human_publisher* to ensure that the human pose data stays current. This way, the robot remains aware of the human position at the moment when the **`plan`**`()` method is called.

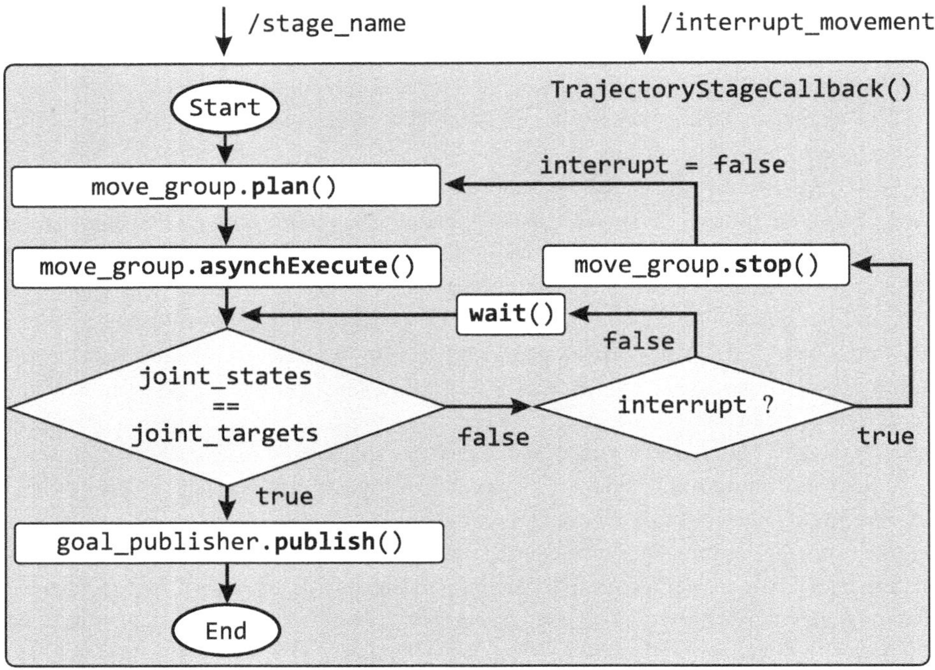

Fig. 4 Program flow chart of the *trajectory_executor*

3 Simulation

In this section, a simulation study is presented that demonstrates the potential of reactive trajectory planning and execution to reduce the time required for a pick-and-place task compared to using solely SSM. The pick-and-place task involves picking up an object from location A and placing it at location B before returning to location A (see Fig. 2). The cycle time is defined as the time required to complete this entire process. The movement of the human was captured with the ZED camera and the movement of the robot is merely simulated.

Table 1 compares the cycle times and productivity of both methods. For this simulation, it is assumed that a human worker reaches location C once an hour and spends about three minutes performing tasks such as grasping, cleaning, or any other task where the workspace is shared long enough. If the human is not blocking the robot's path, only SSM is active. The average cycle time is then 21.84 s. As soon as the human blocks the path of the robot, the robot stops as the distance to the human becomes too small. The robot remains stationary for 3 min while the human performs its task at location C. If reactive trajectory planning is used, the robot can continue its task, although the human blocks the original path (see Fig. 3). The standstill time is reduced by 162.42 s because the robot can execute a new path, which

Table 1 Comparison of productivity of reactive trajectory planning and execution with SSM

	SSM with path adaptation	SSM
Standstil time per h	17.58 s	180.00 s
Cycle time without interruption	21.84 s	21.84 s
Cycle time with interruption	25.44 s	/
Cycles per h	162.97	156.59
Shift duration	8 h	8 h
Cycles per shift	1303.79	1252.72

avoids the human. During this time, the execution of a cycle takes an average of 25.44 s. This cycle time is slightly longer than when the human is not performing a task at location C because the replanned trajectory takes longer to execute. It can be seen that the standstill time is reduced drastically when reactive trajectory planning and execution are used. This reduction leads to an increase in the number of cycles per hour. The amount of reduced standstill time using reactive trajectory planning and execution depends on the frequency and duration of these human interruptions. The more frequent and longer the interruptions, the more significant the potential time savings.

4 Conclusion and Next Steps

The paper presents an initial solution for a system architecture that allows robots to plan and execute a collision-free trajectory reactively. This reduces the robot's standstill time and increases productivity and efficiency in assembly tasks. The proposed architecture is evaluated in a real-time simulation, where a pick-and-place task is performed with captured human motion data. The results show that trajectory adaptation can significantly improve the efficiency of this HRI application.

The next step is to add a physical robot controller to verify the simulated task in the real world. Future work aims to make path planning even more intelligent. Currently, the robot reacts only to changes in the environment. This means that if a new trajectory is planned, it can be interrupted in the next moment because of the movement of the human. However, a question of investigation is how the prediction of changes in the environment can be used to control the robot.

Acknowledgements This research work is partially funded by the German Research Foundation (DFG) within the research project 'High-speed motion tracking and coupling for human-robot collaborative assembly tasks' (KU 1543/43-1). The authors thank the DFG for promoting and facilitating the research.

References

1. ISO/TS 15066:2016.: Robots and robotic devices - collaborative robots (2016)
2. Miro, M., Simulation Technology and Application of Safe Collaborative Operations in Human-Robot Interaction, ISR Europe, et al.: 54th International Symposium on Robotics, vol. 2022, pp. 1–9. Munich, Germany (2022)
3. Liu, Z., et al.: Dynamic risk assessment and active response strategy for industrial human-robot collaboration. Comput. Ind. Eng. **141**, 106302 (2020). https://doi.org/10.1016/j.cie.2020.106302
4. Bobka, P., Germann, T., Heyn, J.K., Gerbers, R., Dietrich, F., Dröder, K.: Simulation platform to investigate safe operation of human-robot collaboration systems. In: 6th CIRP Conference on Assembly Technologies and Systems, pp. 187–192. Göteborg, Schweden (2016)
5. Glogowski, P., Böhmer, A., Hypki, A., Kuhlenkötter, B.: Robot speed adaption in multiple trajectory planning and integration in a simulation tool for human-robot interaction. J. Intell. Robot. Syst. **102**(1), Article 25 (2021). https://doi.org/10.1007/s10846-020-01309-7
6. Glogowski, P.: Entwicklung und Simulation einer adaptiven Bewegungsplanung für Roboter in der Mensch-Roboter-Interaktion, Dissertation, Ruhr-Universität Bochum, Bochum (2022). https://doi.org/10.13154/294-9046
7. Werner, T., Riedelbauch, D., Henrich, D.: Design and evaluation of a multi-agent software architecture for risk-minimized path planning in human-robot workcells, Tagungsband des 2. Kongresses Montage Handhabung Industrieroboter (2017)
8. Müller, R., Franke, J., Henrich, D., Kuhlenkötter, B., Raatz, A., Verl, A. (eds.): Handbuch Mensch-Roboter-Kollaboration. Carl Hanser Verlag München, Munich, Germany (2019)
9. Khatib, O.: Real-time obstacle avoidance for manipulators and mobile robots. Int. J. Robot. Res. **5**(1), 90–98 (1986)
10. Lavalle, S.: Rapidly-exploring random trees: a new tool for path planning. Technical Report, Iowa State University (1998)
11. Kuffner, L.: RRT-connect: an efficient approach to singlequery path planning. In: IEEE International Conference on Robotics and Automation (2000)
12. Karaman, S., Frazzoli, E.: Sampling-based algorithms for optimal motion planning. Int. J. Robot. Res. (2011)
13. Latombe, J.-C.: Robot Motion Planning, vol. 124. Springer Science & Business Media (1991)
14. Ramasubramanian, A., Kazasidis, M., Fay, B., Papakostas, N.: On the evaluation of diverse vision systems towards detecting human pose in collaborative robot applications. Sensors **24**, 578 (2024). https://doi.org/10.3390/s24020578
15. Cao, Z., Hidalgo, G., Simon, T., Wei, S., Sheikh, Y.: Openpose: realtime multi-person 2D pose estimation using part affinity fields. IEEE Trans. Pattern Anal. Mach. Intell. **43**(1), 172–186 (2019)
16. Lin, T., Maire, M., Belongie, S., Hays, J., Perona, P., Ramanan, D., Dollár, P., Zitnick, C.: Microsoft coco: common objects in context. In: European Conference on Computer Vision, pp. 740–755. Springer, Berlin (2014)
17. Quigley, M., Conley, K., Gerkey, B., Faust, J., Foote, T., Leibs, J., Wheeler, R.: ROS: an open-source robot operating system. In: ICRA Workshop on Open Source Software (2009)
18. Ortiz, L., Cabrera, E., Gonçalves, L.: Depth data error modeling of the ZED 3D vision sensor from Stereolabs. Electron. Lett. Comput. Vis. Image Anal. **17**, 1–15 (2018)
19. Christ, L., Milloch, E., Boshoff, M., Hypki, A., Kuhlenkötter, B.: Implementation of digital twin and real production system to address actual and future challenges in assembly technology. Automation **4**, 345–358 (2023). https://doi.org/10.3390/automation4040020

Optimizing Force Signals from Human Demonstrations of In-Contact Motions

Johannes Hartwig, Fabian Viessmann, and Dominik Henrich

Abstract

For non-robot-programming experts, kinesthetic guiding can be an intuitive input method, as robot programming of in-contact tasks is becoming more prominent. However, imprecise and noisy input signals from human demonstrations pose problems when reproducing motions directly or using the signal as input for machine learning methods. This paper explores optimizing force signals to correspond better to the human intention of the demonstrated signal. We compare different signal filtering methods and propose a peak detection method for dealing with first-contact deviations in the signal. The evaluation of these methods considers a specialized error criterion between the input and the human-intended signal. In addition, we analyze the critical parameters' influence on the filtering methods. The quality for an individual motion could be increased by up to 20 % concerning the error criterion. The proposed contribution can improve the usability of robot programming and the interaction between humans and robots.

Keywords

Programming by demonstration • Kinesthetic teaching • Hybrid motion/force control • Signal filtering • Error minimization

J. Hartwig and F. Viessmann authors contributed equally.

J. Hartwig (✉) · F. Viessmann · D. Henrich
Chair for Applied Computer Science III (Robotics and Embedded Systems), University of Bayreuth, Bayreuth, Germany
e-mail: johannes.hartwig@uni-bayreuth.de

F. Viessmann
e-mail: fabian.viessmann@uni-bayreuth.de

M.-C. Wanner et al. (eds.), *Annals of Scientific Society for Assembly, Handling and Industrial Robotics 2024*, https://doi.org/10.1007/978-3-031-91463-8_7

1 Introduction and Related Work

Due to current market demands, raising the flexibility and agility in automation using cobots is an increasing trend in robotics [1]. As small and medium-sized enterprises (SMEs) lack robot programming experts for economic reasons, enabling the available task experts to program and reconfigure robot systems easily is critical [2, 3]. This allows them to contribute their task knowledge directly. Therefore, a future-oriented goal is to improve the usability of robot programming and the interaction between humans and robots [4]. Utilizing the programming by demonstration (PbD) paradigm, specifically with kinesthetic guiding, can enable humans to program robot trajectories without prior robot programming knowledge [5]. Thus, a human can program robot trajectories by guiding the robot directly and fulfilling a given task only with his domain knowledge.

Nevertheless, specific tasks can only be adequately addressed partially through trajectory programming, such as when a robot task consists of polishing [6], engraving [7], or grinding complex shapes [8]. Here, continuous contact between the tool and the object becomes imperative, and a method to specify the contact forces to be applied over time and space is needed. There are possibilities to achieve this simultaneous input of position and force. For example, in assembly tasks, the contact forces can be measured by a force/torque sensor located underneath a mounting bracket [9], or the needed robot stiffness is estimated by an electromyographic wristband from the muscle activation of the user [10]. The most prominent approach we will also use here is an additional force/torque sensor attached between the user's hand-guiding contact point and the robot's tool [11–13]. In this way, we can measure the robot's trajectory and the forces exerted on the environment during the demonstration by hand guiding.

However, the force profiles generated by humans during the demonstration are often noisy or inaccurate (see Fig. 1a). A prominent approach here is to learn from multiple demonstrations, often combining methods such as dynamic time warping (DTW), Gaussian mixture models (GMMs), or dynamical systems. A categorizing overview for robot manipulation in

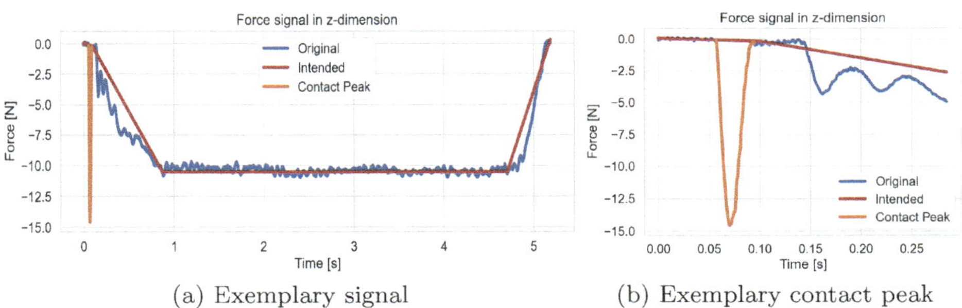

(a) Exemplary signal (b) Exemplary contact peak

Fig. 1 Exemplary signal of a human demonstration (blue), an ideal signal (red) and a contact peak (orange) in Fig. 1a. Contact peak is enlarged in Fig. 1b

contact can be found here [14]. Otherwise, when programming the force profiles directly, these inaccuracies can be visualized and corrected by the user like in our previous work [15, 16], or one can enhance them automatically using an error correction or signal filtering technique. Most of these methods employ either frequency-domain filtering (e.g., using Fourier Analysis [17]), averaging over filter windows (e.g., Savitzky-Golay filter [18]), or fit primitives (e.g., using RANSAC [19]). Each method has different advantages and drawbacks. We will explore these methods, applying their fundamental approach to our inaccurate signal.

Therefore, this paper examines the question of to what extent measured force signals can be optimized so that they correspond better to the human intention of the demonstrated signal. This can be used to minimize the error when reproducing the motion directly as in playback programming or as a signal pre-processing for machine learning methods to improve the signal-to-noise ratio. We evaluate three different filtering techniques for force profiles and aim to contribute insights on what to consider in such a case for practitioners.

First, Sect. 2 formalizes our input signal and error criterion before describing different signal filtering methods. Subsequently, Sect. 3 discusses their evaluation regarding the error criterion between the input and the human-intended signal. The critical parameters of the filtering methods are especially considered here. Section 4 summarizes the paper and discusses future work.

2 Methodology

When recording an in-contact motion through kinesthetic guiding, the robot configurations $\mathcal{Q} = (q_1, \ldots, q_n)$ are recorded with $q_i \in \mathbb{R}^d$ and d joints of the robot. The motion's sampling frequency f_S and the duration T determines the number of demonstrated configurations by $n = f_S \cdot T$. In addition, we record the force/torque value $w_i \in \mathbb{R}^6$ for each q_i using a force/torque sensor located between the user's hand-guiding contact point and the end-effector. This sensor measures the wrench $\mathcal{F} = (w_1, \ldots, w_n)$ exerted by the user on the robot's tool as it interacts with the environment. Each w_i contains three force and three torque values, one along each force/torque sensor coordinate axis. For this paper, we only consider one dimension at a time so that we can define a one-dimensional force/torque signal $\mathcal{F}_j = (w_{1,j}, \ldots, w_{n,j})$ with $w_{i,j} \in \mathbb{R}$ representing the force/torque value in dimension j at time i. We will refer to the signal $\mathcal{F}_j$ as $\mathcal{S} = (k_1, \ldots, k_n)$.

Our approach aims to optimize the force signals obtained from human demonstrations by kinesthetic guiding. The goal is to minimize the error between a force signal $\mathcal{S}$ from a human demonstration and an ideal signal $\hat{\mathcal{S}} = (\hat{k}_1, \ldots, \hat{k}_n)$ that meets the intentions of the user (see Fig. 1a). The user's intended signal is usually predefined by the given task. The deviations from the ideal signal occur due to various effects. Firstly, there are inaccuracies due to the unfamiliar guidance of the tool via the robot arm. Also, there are inaccuracies due to human muscle activation and sensor noise. As no ideal force signal is usually available for a human demonstration, we define the ideal force signals for different scenarios (see Sect. 3). We can define an optimization criterion based on these ideal signals and the recorded ones (see Sect. 2.1).

To minimize the error criterion, we apply different filtering/smoothing algorithms to the force signal to generate an optimized signal $\tilde{S}$. Each smoothing algorithm has a critical parameter that strongly influences the optimization result. Therefore, we must find the critical parameter that minimizes the error criteria for each method. Since force signals obtained from human demonstrations typically contain significant peaks due to the first contact with the environment, the force signals can not be used directly as input for the smoothing algorithms (see Fig. 1a). Large peaks in the signal distort the results. Thus, our approach consists of two steps. The first step is identifying and removing unwanted peaks (see Sect. 2.2). The second step is applying different smoothing algorithms (see Sect. 2.3): *Sinc-in-time filter*, *Savitzky-Golay filter*, and *RANSAC line approximation*.

2.1 Error Criterion

Regarding the error criterion, we first have to define how we want to weight the error components between the given signal S and the optimal signal $\hat{S}$ at a given time t. We can either use an L1 norm based on the absolute error value, like mean absolute error (MAE), or an L2 norm squaring the result, like mean squared error (MSE). As we assume that outliers, i.e., a single sub-optimal demonstration by a user, should not be weighted more, we consider an L1 error criterion more suitable here.

However, as seen in Fig. 1a, the force signal of a human demonstration often deviates strongly from the expectation at the start and end. In order to obtain information about the error in the central area of the signal, we need to weight the error at the start and end lower. There is a similar line of thought in the error weighting of control loops. Here, the error in the step response should not dominate the measure, and the oscillations later should also be considered.

For control loop problems, one of the prominent criterion functions is the integral of the time-multiplied absolute value of error (ITAE) criterion [20]. This criterion tries to solve the described problem by weighting the absolute error regarding the time t:

$$J_{\text{ITAE}} = \int_0^{\infty} |e(t)| \cdot t \, dt \tag{1}$$

Since only the start is weighted less using the ITAE criterion, we modify it by using a bidirectional integral weighted from start and end for our finite signal of duration T:

$$J_{\text{bidirectional-ITAE}} = \int_0^{\frac{T}{2}} |e(t)| \cdot t \, dt + \int_{\frac{T}{2}}^{T} |e(t)| \cdot (T - t) \, dt \tag{2}$$

We now adapt this criterion to our discrete case with the recorded signal S and ideal $\hat{S}$:

$$\tilde{J}(\mathcal{S}, \hat{\mathcal{S}}) = \sum_{i=1}^{\lfloor \frac{n}{2} \rfloor} i \cdot |k_i - \hat{k}_i| + \sum_{i=\lfloor \frac{n}{2} \rfloor + 1}^{n} (n + 1 - i) \cdot |k_i - \hat{k}_i| \quad (3)$$

In order to make signals of different lengths comparable, we also normalize the error regarding the sum of time weights:

$$J(\mathcal{S}, \hat{\mathcal{S}}) = \frac{\sum_{i=1}^{\lfloor \frac{n}{2} \rfloor} i \cdot |k_i - \hat{k}_i| + \sum_{i=\lfloor \frac{n}{2} \rfloor + 1}^{n} (n + 1 - i) \cdot |k_i - \hat{k}_i|}{\sum_{i=1}^{\lfloor \frac{n}{2} \rfloor} i + \sum_{i=\lfloor \frac{n}{2} \rfloor + 1}^{n} (n + 1 - i)} \quad (4)$$

Although we normalize the criterion by the sum of time weights, only similar signals should be compared because a fair comparison requires the same number of significant changes (resp. step responses in the controller analogy). In addition, the ratio of significant changes to oscillation phases should be similar. Otherwise, they would be weighted differently using this criterion. This means only signals of similar individual motions should be compared using this criterion, not signals of entire demonstrated tasks.

2.2 Contact Peak Removal

When analyzing the signals from human demonstrations, we observe that, typically, force signals contain so-called contact peaks. These represent a large deflection in the force signal, which is very short in time. The peak occurs during initial contact between the tool and the environment, as humans typically establish contact with the environment too quickly and lose it again immediately. Figure 1b depicts such a contact peak.

Therefore formally, a peak in our force signal $\mathcal{S}$ is defined as a sequence $\mathcal{P} = (k_u, k_{u+1}, \dots, k_{v-1}, k_v, k_{v+1}, \dots, k_{w-1}, k_w)$ at times i between peak start u and end w with $1 \leq u < v < w \leq n$ and $\text{argmin}_{u \leq i \leq w} k_i = v$. Let now $\mathcal{S}' = (k'_1, \dots, k'_n)$ be the derivation of the force signal $\mathcal{S}$. Since $\mathcal{S}$ is discrete, we must approximate the derivative using finite differences. Then the following must hold: $k'_i < 0$ with $u \leq i < v$, $k'_v = 0$ and $k'_i > 0$ with $v < i \leq w$.

Since this basic definition also applies to every force signal down- an upswing, we have to define further requirements:

1. When dealing with these unwanted short peaks, it must generally hold that the force k_u at the end of the peak u corresponds approximately to the force k_w at the beginning of the peak w. For this purpose, we use a relative threshold to set the difference in magnitude between the forces k_u and k_w to the total peak height $|k_u - k_v|$. This relative threshold should be less than or equal to a given parameter δ.

$$\frac{|k_u - k_w|}{|k_u - k_v|} \leq \delta \tag{5}$$

2. When considering the derivative, there is a strong deflection for the decreasing and the increasing part of the contact peak. This means the value changes significantly during the contact peak's time interval in relation to the rest of the signal. Therefore, we can identify these as outliers utilizing the so-called z-score. If the z-score deviates from the mean by more than μ standard deviations, we consider this value an outlier. Therefore, let be $\bar{S}'$ the mean value and σ be the standard deviation of the signal S'. Then the following must hold:

$$\exists u \leq i < v : \frac{k_i' - \bar{S}'}{\sigma} > \mu \wedge \exists v < l \leq w : \frac{k_l' - \bar{S}'}{\sigma} > \mu \tag{6}$$

3. After the contact peak, additional smaller peaks may occur in the signal during the human step response. However, we only intend to filter the first peak as it represents the beginning of a motion. Hence, we must take into account the temporal distance between potential peaks. Let $\mathcal{P}_1$ and $\mathcal{P}_2$ be two peaks, Then the following must hold:

$$u_2 - w_1 \geq \tau \tag{7}$$

After using these conditions to identify all contact peaks, we remove the found peaks by setting $k_i = k_u$ for all i in the peak interval $[u, w]$.

2.3 Application of Filtering Techniques

After identifying and removing the signal's contact peaks, we can apply different filtering approaches: Sinc-in-time filter using Fourier Transform, Savitzky-Golay filter, and RANSAC line approximation. Each algorithm has a critical parameter that strongly influences the smoothing result. These three algorithms are considered because each represents the basic idea of either frequency-domain filtering, averaging over a filter window, or fitting primitives. We want to consider all of these approaches for our optimization problem. For an example of the application and the smoothed results, see Fig. 2.

Sinc-in-time filter: For this, we determine the frequency components of the force signal first. Since our signal is discrete, we use the Fast Fourier Transform [17]. To smooth our force signal, we remove all frequencies above a predetermined cutoff frequency f_c and obtain a filtered signal by executing the Inverse Fourier Transform. Therefore, this is an optimal low-pass filter, and the critical parameter of this algorithm is the cutoff frequency f_c. The lower the cutoff frequency is, the greater the smoothing, and vice versa.

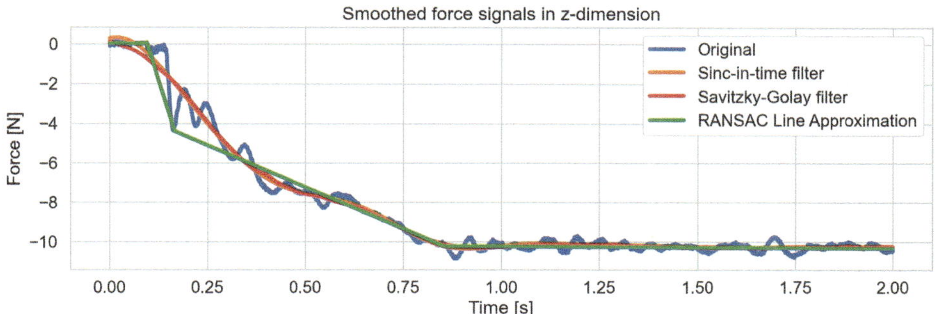

Fig. 2 Application of the filtering techniques of an exemplary signal. Used parameters are $f_c = 3\,\text{Hz}$ for the *Sinc-in-time filter*, $h = 0.25\,\text{s}$ for the *Savitzky-Golay filter* and $l = 1\,\text{s}$ for *RANSAC line approximation*

Savitzky-Golay filter [18]: This filter performs a weighted moving average using coefficients to perform a polynomial regression in a moving window with a low-degree polynomial. The polynomial is then evaluated in the center of the window, and its value is used as the new signal value at this position. Therefore, the critical parameter of the Savitzky-Golay filter is the window width h; The wider the window is, the greater the smoothing, and vice versa.

Random Sample Consensus (RANSAC) [19] *Line Approximation*: Here, we fit a model into the signal. The RANSAC method first selects the number of data points required to instantiate the model to be fitted. The supporting data points of this model are then counted, and the model with the most supporters is selected. The model is then fitted according to the least squares among these supporters so that outliers do not distort the fitting process. In this work, we fit linear models into the force signals. Therefore, we have to execute the RANSAC algorithm multiple times to find multiple linear model approximations explaining the whole signal. There is an additional parameter to determine the granularity of the linear models, which is the maximum segment length l. This length limits the time interval in which the respective model applies. Therefore, the critical parameter of this smoothing algorithm is the maximum segment length l; The greater the length l is, the greater the smoothing, and vice versa.

3 Evaluation

To determine the critical parameters of the filtering techniques and subsequently evaluate the signal optimization, we gathered demonstrations by 10 different users. The participants were tasked to demonstrate 10 different in-contact motions by kinesthetically guiding a Franka Emika Panda robot with an attached Schunk Gamma 6-axis force/torque sensor (z-axis measuring range $\pm 100\,\text{N}$, resolution $0.013\,\text{N}$). After getting familiar with the robot's hand guidance, the participants repeated every motion five times. We provided them with visual feedback on the current force value using a bar representation of the current deviation and intuition about the motion (e.g., pressing on a wood dowel, or applying and pressing on an adhesive bond). A 3D-printed stump was attached to the robot's gripper with which the forces were to be exerted.

The demonstrated motions can be divided into the following categories: Firstly, point with constant force in z-direction with 7.5 and 15 N, or linearly increasing force with 7.5 and 15 N; Secondly, a part-wise linear path along a triangle with constant force in z-direction with 0, 7.5 and 15 N; Lastly, a circular path with constant force in z-direction with 0, 7.5 and 15 N.

3.1 Contact Peak Detection

To evaluate the peak detection (see Sect. 2.2), we use 25 randomly selected demonstrations of each motion group. We further split the point demonstrations into subgroups with constant or linear increasing force. This differentiation makes sense here as the needed jump responses demonstrated by the users differ significantly, which may influence the peak detection results. All contact peaks were labeled manually for the evaluation and then compared with the automatically detected peaks. We distinguish between correctly recognized peaks (true positives), unrecognized peaks (false negatives), and additional wrongly recognized peaks (false positives). Peak detection was performed with empirically determined parameters $\delta = 0.15$, $\mu = 3$, and $\tau = 600$. Figure 3 shows the results.

When analyzing the results of the point demonstrations with a constant force, it is evident that 84 % of the peaks are correctly identified. With the chosen parameter set, peak detection works well for point demonstrations, as the results with increasing force were similar. For the linear motions, 92 % of the peaks are correctly detected. However, three false positive

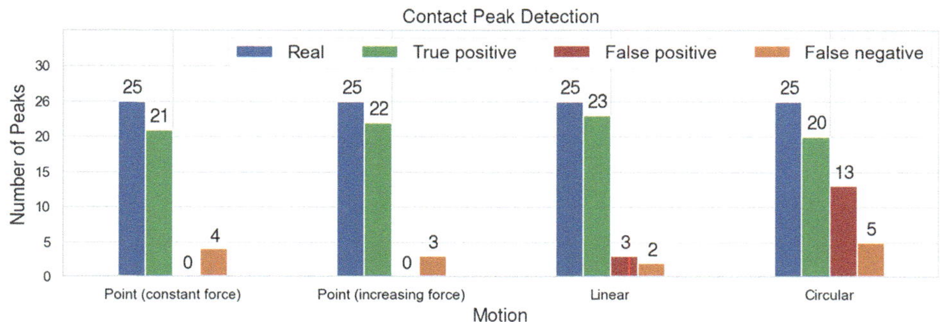

Fig. 3 Evaluation of the automatically detected contact peaks compared to the manually labeled ones

peaks originating from the signal's noise are detected. For circular motions, slightly fewer peaks are correctly detected at 80 %. A total of 13 false positive peaks are identified. Here, whether these peaks belong in the false positive class is unclear. We observed that it is challenging to exert a force while the user has to move the robot. The robot may lose contact with the environment during the demonstration, representing a contact peak when it regains contact with the environment. Also, the force applied may jump through stick-slip effects.

Overall, peak detection gives good results, with an average sensitivity of 86 %. However, false positives cannot be ruled out, especially when the human demonstrator simultaneously applies force and moves the robot.

3.2 Filtering Techniques

As a final step, we have to determine the critical parameter for each filtering technique that minimizes the error to the optimal motion on average for all demonstrations. To achieve this, we sampled the parameter space for each approach and smoothed each signal with the sampled critical parameter. The individual error curves are then averaged (Fig. 4). In addition to the mentioned demonstrations, we also consider point demonstrations with varying force profiles (step- and parabola-shaped) to analyze the influence of force complexity on optimization. These were generated with a different user group.

With the Sinc-in-time filter, the minimum for the point demonstration is $f_c = 6.5$ Hz, and for the point demonstration with varying force, it is 1.0 Hz. Accordingly, the difficulty of the force profile to be applied influences the smoothing, and the more complex the force profile, the greater the smoothing must be. In comparison, the cutoff frequency for linear demonstrations is significantly lower and is $f_c = 0.15$ Hz. For circular demonstrations, this is even $f_c = 0.1$ Hz. This leads to the conclusion that the smoothing must be increased if the human moves the robot in addition to exerting force. Furthermore, the minimum error for linear and circular demonstrations is still much higher than for point demonstrations. From this, we conclude that the difficulty of the positional motion should be weighted more heavily than the difficulty of force profile.

We observe similar results for the Savitzky-Golay filter and RANSAC line approximation (point: $h = 0.8$ s, $l = 0.9$ s; point with varying force: $h = 1.2$ s, $l = 0.9$ s; linear: $h = 12.8$ s, $l = 9.9$ s; circular: $h = 14.4$ s, $l = 10.5$ s). When comparing the relative improvement between the input signal and filtered signal for linear demonstrations, the Sinc-in-time filter (18.47 %) performs similarly to the Savitzky-Golay filter (15.96 %). RANSAC line approximation (−5.82 %) results in a worsening due to the inaccurate approximation of the signal by lines. Therefore, a piece-wise linear approximation with our line model using RANSAC is not useful considering the given input signals, although many of our predefined test motions are linear. Overall, we can improve the error measure with Sinc-in-time and Savitzky-Golay filters, but the best parameter values differ heavily (see Fig. 4). This becomes

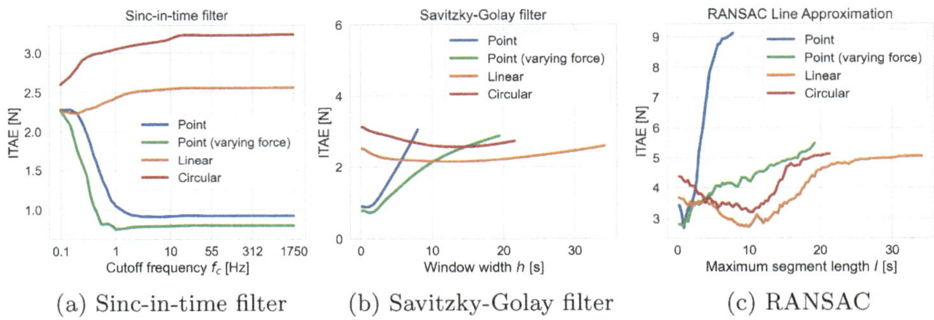

(a) Sinc-in-time filter (b) Savitzky-Golay filter (c) RANSAC

Fig. 4 The error criterion averaged over all motions of each category plotted against the critical parameter of each method

particularly clear if, for example, we use the Savitzky-Golay filter with the point motion parameter for the circular motion (4.69 %) instead of the best parameter for the circular motion (20.45 %). Thus, additional information about the motion must be considered, or user assistance is required.

4 Conclusion

This paper compares different methods to optimize force signals from kinesthetic guiding. We proposed a method to remove contact peaks in force signals and showed that it is possible to filter signals so that the smoothed signal corresponds more closely to the human intention. Two different approaches improve the error measure and might be helpful for varying use cases. Although significant improvement is achieved, a high degree of filtering is required, particularly for motions in which the human demonstrator exerts a force and moves the robot simultaneously. This phenomenon is mainly observed in the different critical parameters for the motions in Sect. 3.2. Therefore, future work should address the modeling of critical parameters dependent on the positional motion. In addition, it should be investigated to what extent such filtering offers advantages when learning robot motions utilizing machine learning methods.

References

1. Dietz, T., et al.: Programming system for efficient use of industrial robots for deburring in SME environments. In: 7th German Conference on Robotics. VDE (2012)

2. Riedelbauch, D., Hartwig, J., Henrich, D.: Enabling end-users to deploy flexible human-robot teams to factories of the future. In: IROS 2019 Workshop Factory of the Future. IEEE (2019)
3. Perzylo, A., et al.: SMErobotics: smart robots for flexible manufacturing. IEEE Robot. Autom. Mag. **26**(1) (2019)
4. Kildal, J., et al.: Potential users' key concerns and expectations for the adoption of cobots. Procedia CIRP **72** (2018)
5. Villani, V., et al.: Survey on human–robot collaboration in industrial settings: safety, intuitive interfaces and applications. Mechatronics **55** (2018)
6. Amanhoud, W., et al.: A dynamical system approach to motion and force generation in contact tasks. In: Robotics: Science and Systems (2019)
7. Koropouli, V., et al.: Learning interaction control policies by demonstration. In: 2011 IEEE/RSJ International Conference on Intelligent Robots and Systems (2011)
8. Jinno, M., et al.: Development of a force controlled robot for grinding, chamfering and polishing. In: Proceedings of 1995 IEEE International Conference on Robotics and Automation. IEEE (1995)
9. Bargmann, D., et al.: Unobstructed programming-by-demonstration for force-based assembly utilizing external force-torque sensors. In: 2021 IEEE International Conference on Systems, Man, and Cybernetics. IEEE (2021)
10. Yang, C., et al.: A learning framework of adaptive manipulative skills from human to robot. IEEE Trans. Ind. Inform. **15**(2) (2018)
11. Montebelli, A., et al.: On handing down our tools to robots: Single-phase kinesthetic teaching for dynamic in-contact tasks. In: 2015 IEEE International Conference on Robotics and Automation. IEEE (2015)
12. Steinmetz, F., et al.: Simultaneous kinesthetic teaching of positional and force requirements for sequential in-contact tasks. In: 2015 IEEE-RAS 15th International Conference on Humanoid Robots. IEEE (2015)
13. Conkey, A., Hermans, T.: Learning task constraints from demonstration for hybrid force/position control. In: 2019 IEEE-RAS 19th International Conference on Humanoid Robots. IEEE (2019)
14. Suomalainen, M., et al.: A survey of robot manipulation in contact. Robot. Auton. Syst. **156** (2022)
15. Hartwig, J., et al.: Visualization of forces and torques for robot-programming of in-contact tasks. In: Annals of Scientific Society for Assembly, Handling and Industrial Robotics 2023 (to appear) (2023)
16. Hartwig, J., et al.: Input and editing of force profiles of in-contact robot motions via a touch graphical user interface. In: 2023 IEEE International Conference on Robotic Computing (to appear). IEEE (2023)
17. Cooley, J.W., Tukey, J.W.: An algorithm for the machine calculation of complex Fourier series. Math. Comput. **19**(90) (1965)
18. Savitzky, A., et al.: Smoothing and differentiation of data by simplified least squares procedures. Anal. Chem. **36** (1964)
19. Fischler, M., et al.: Random sample consensus: a paradigm for model fitting with applications to image analysis and automated cartography. Commun. ACM **24** (1981)
20. Joseph, S.B., et al.: Metaheuristic algorithms for PID controller parameters tuning: review, approaches and open problems. Heliyon **8**(5) (2022)

Modelling and Simulation

Concepts and Requirements for Flexible Dismantling Systems for End-of-Life Vehicles

Lukas Gründel, Marc-André Weismüller, Oliver Petrovic, and Christian Brecher

Abstract

The demand for critical raw materials (CRMs) and resilient supply chains for important technologies like electric vehicles necessitates efficient end-of-life vehicle (ELV) dismantling systems to enable a circular automotive industry. As labor shortages continue to rise and manual dismantling processes prove increasingly time-consuming, automated systems become indispensable. This paper outlines a methodology for developing flexible and automated ELV dismantling systems in line with VDI 2206 and VDI 2221 guidelines. Therefore, the requirements for automation solutions to meet environmental, economic, and regulatory requirements are analyzed. Furthermore, key enabling technologies such as the Digital Product Passport (DPP), a Dynamic Target Part List (DTP) and a Dynamic Disassembly Map (DDM) are proposed to facilitate efficient dismantling operations while meeting the aforementioned requirements. The concept is developed following a functional analysis for the relevant dismantling steps and a morphological box to select potential automation solutions. Based on this concept a layout consisting of dismantling areas for ELVs and sub-components is presented, followed by an exemplary setup of a flexible and automated dismantling cell. This approach emphasizes automation while incorporating human–robot collaboration (HRC) to handle complex dismantling tasks. The proposed concept serves as a blueprint to orient and drive forthcoming research efforts focused on the advancement of dismantling systems.

Lukas Gründel and Marc-André Weismüller are contributed equally to this work.

L. Gründel (✉) · M.-A. Weismüller · O. Petrovic · C. Brecher
RWTH Aachen University, Aachen, Germany
e-mail: l.gruendel@wzl.rwth-aachen.de

M.-C. Wanner et al. (eds.), *Annals of Scientific Society for Assembly, Handling and Industrial Robotics 2024*, https://doi.org/10.1007/978-3-031-91463-8_8

Keywords

End-of-life vehicles • Dismantling systems • Automation

1 Introduction

Particularly technologies relevant for future challenges, such as electric vehicles, solar power plants, and wind turbines, are driving an increased demand for CRMs [1]. These materials, as defined by the European Union, encompass substances of significant economic importance to the EU, characterized by a high risk of supply disruption [2].

To ensure the security and sustainability of CRM supply chains, the European Commission proposed the European Critical Raw Materials Act in March 2023, including the goal to cover 15% of the EU's annual CRM consumption via recycling [3]. According to the EU, the automotive industry is one of the most resource-intensive sectors of the European economy, accounting for 19% of the total steel consumption, 10% of plastics, 42% of aluminum, 6% of copper and 65% of rubber consumption within the EU [4]. Therefore, ELVs represent significant reservoirs of CRMs and other important resources. In 2020, according to the German Federal Statistical Office's waste statistics, only 19.3% of the unladen weight of ELVs were dismantled for the recovery of spare parts or recyclable materials in Germany. Especially electronics are rarely dismantled (0.8 kg per ELV) [5]. The dismantling of car parts is the key enabler for all subsequent R-strategies like reuse, repurpose, remanufacture or recycle. The low dismantling rate therefore prevents the reuse or remanufacturing of car parts and therefore leads to a loss of CRMs and a degradation of expensive alloys within the subsequent recycling processes. As presented by Gradin et al. the complete dismantling of ELVs is significantly better in terms of environmental and resource impacts than the conventional recycling scenario via shredding [6].

In Germany a significant proportion of ELVs are currently dismantled in non-certified facilities and subsequently exported or returned to the legal material flows. The other fraction is deregistered in collection centers and then dismantled in certified facilities. The extracted fluids and dismantled parts are then treated and reconditioned for reuse. The remaining car body is usually shredded and the resulting fractions are either used for energy recovery or recycled. The remaining dusts and other residues are then disposed of in landfills. Proff et al. found that only 2% by weight undergo reuse or remanufacturing, while this already includes the reutilization of fuel and other fluids post-depollution [7]. This observation underscores the contemporary state of low-quality ELV recycling in Germany, leading to adverse ramifications for both the environment and industrial supply chains. This is fundamentally due to the low fraction of dismantled components before shredding and the low dismantling depth. Only 19.3% of the vehicle weight is dismantled according to [5]. The industrial context currently lacks any discernible fraction of finely dismantled sub-components for material sorting purposes.

In both Germany and the EU, dismantling enterprises predominantly consist of small and medium-sized enterprises, which face considerable challenges, particularly stemming from labour shortages. Presently, dismantling processes rely on manual procedures, which according to Tian and Chen would lead to a total dismantling time of about 6.2 h with three workers for an entire ELV [8]. Therefore, the automation of dismantling processes is one of the major enablers for economic, high-quality recycling processes.

This paper presents a concept for an automated ELV dismantling cell. The following chapters are structured as follows: In Sect. 2 the requirements for such a dismantling system are discussed followed by the State of the Art. In Sect. 4 the concept is developed, which is applied to a flexible dismantling setup presented in Sect. 5. The paper ends with a conclusion and an outlook for further research activities.

2 Requirement Analysis

The following requirements for a flexible dismantling system result from a review of the current legal, environmental and safety situation by examining laws and regulations, disposal guidelines and resources for compliance with occupational health and safety. Further, reference was made to the new EU proposal for regulation for management of ELVs [4].

This work focuses on ELVs of category M1, i.e. vehicles used for the transport of passengers with no more than eight seats. The maximum permissible dimensions of M1 vehicles are $2.55 \times 4 \times 12$ m and the weight limit is 3.5 t total weight, in most EU countries [9]. Storage facilities for such ELVs, including those at collection centers prior to treatment, as well as their components, parts and materials, shall have impermeable surfaces with spillage collection and segregation facilities, degreasing detergents and water treatment facilities, including rainwater. Further, suitable containers for the storage of batteries and storage tanks for fluids from ELVs shall be provided. Fuel, engine oil, transmission oil, transmission fluid, hydraulic oil, coolant, antifreeze, brake fluid, battery acids and air conditioning system fluids in the ELV must be removed and the removal of pollutants must be documented.

Besides components that are already being dismantled today, the proposed EU ELV Directive defines several new components. Most significant change is the required removal of all printed circuit boards larger than 10 cm^2, photovoltaic modules larger than 0.2 m^2 and metal or plastic mono material parts with a total mass of more than 10 kg. In addition, glass must be directly identified and sorted during dismantling into container glass, fibre glass or glass of equivalent quality [4].

All mentioned parts in [4] must either be reused, remanufactured or recycled depending on their condition and intended R-strategy. Parts intended for reuse must be checked for functionality. Components intended for remanufacturing must be checked for completeness and subjected to an assessment to determine any damage, reduced functionality and necessary repairs. Despite the intention to increase the recycled content, most of the safety-related components (e.g. airbags) and parts of the waste gas treatment system are not allowed to be reused.

Considering the safety requirements of the dismantling system, most operations involve the risk of falling parts, overexertion and back pain from handling heavy objects, oil and liquid splashes, excessive noise, burns from hot parts and sparks, inhalation of dust, and especially contact with parts containing mercury [10]. In addition, some processes involve the risk of explosion and fire due to gases and fuels. Therefore, appropriate safety measures, following local regulations must be taken (e.g. ventilation, protection, lifting equipment, etc.). With the increased dismantling depth proposed by [4] and the highly non-deterministic nature of the ELV dismantling process, an automated dismantling system must be highly flexible and able to adapt to the broad range of vehicle and component states.

3 State of the Art

The research domain concentrating on machine construction and automation solutions for ELV dismantling remains relatively constrained. El Halabi et al. conducted an analysis of the environmental repercussions of machine-based dismantling as a substitute for conventional shredding and sorting techniques [11]. Their findings demonstrate a favourable effect on CO_2 emissions compared to traditional process chains, assuming a fossil fuel-based energy mix. However, the utilization of fuel-driven modified excavators could lead to increased emissions in comparison to a renewable-driven shredding and sorting process.

In the context of automated dismantling solutions, some publications address the entire ELV, while others focus solely on specific sub-components like the fine dismantling of batteries or drivetrains. Fleischer et al. propose a robot-based setup for a flexible drive train dismantling including two kinematics, machine vision, a clamping device and a transportation system [12]. The requirements are mainly derived by a product analysis neglecting current regulations or environmental boundaries. The current state of the art regarding battery dismantling of electric vehicles is summarized in [13], focusing on the technology fields of artificial intelligence and human–robot collaboration. Yuan et al. propose a HRC-based dismantling of lithium-ion batteries increasing the flexibility in complex tasks performed by humans and robots [14].

Xia et al. presents a process flow for ELV dismantling and also suggests the respective stations of a vehicle disassembly plant. However, the requirements are formulated superficially and the provided approach lacks automation [15]. Zhang and Chen also discuss an approach for ELV recycling in China [16]. To meet the demand of processing 30,000 ELVs annually, a line-based approach with a cycle time of 7 min is adopted. The proposed line prioritizes maximum throughput over flexibility, resulting in complex ELVs being immediately shredded. To handle non-dismantlable components within the cycle time, forcible entry tools such as hydraulic clamps are utilized. Another plant layout is presented in [17]. The authors use a system layout design method to develop a feasible layout scheme for a 30,000 m^2 area in Southern China.

In summary, the referenced publications concerning entire ELVs fail to address automation solutions, resulting in increased dismantling costs, especially in high wage countries like Germany. Conversely, the cited papers focusing on ELV sub-components demonstrate promising automation solutions, which require implementation for the complexity of entire ELVs.

4 Concept Development

Understanding KPIs such as line throughput and car types is essential to the effective design of a dismantling line. Equally important is the identification of the ELV components that will be dismantled, the dismantling order and the tools and equipment that will be required. This determination is complex, considering factors like component value, regulatory compliance, and extraction effort. Finding the optimal components involves balancing value and extraction effort, influenced by ELV type and condition, which may change dynamically. This paper introduces innovative concepts and technologies aimed at facilitating economically viable remanufacturing systems in Europe.

4.1 Enabling Technologies

Numerous emerging technologies show promise for automated or semi-automated dismantling, with particular emphasis on the digital twin concept for automation. Given the substantial uncertainty inherent in the dismantling process, contextual data derived from vehicle usage and information enrichment throughout the dismantling is crucial. The **Digital Product Passport**, advocated notably by the European Commission through initiatives like the Green Deal and the Circular Economy Action Plan, holds significant importance in this regard. The DPP aims to transparently store product information across life cycle phases to facilitate R-strategies such as repair and reuse. While still in a conceptual phase, the DPP's technical implementation and data scope remain under development. The Asset Administration Shell (AAS) emerges as a promising avenue for technical realization. While initiatives such as the digital battery pass exist for specific components like batteries, a comparable effort for ELVs is currently lacking. Thus, the authors advocate for collaborative efforts among automotive stakeholders to devise suitable concepts for ELVs.

One integral component of the DPP is the inclusion of a target part list, crucial for tracking essential components and establishing an optimized dismantling sequence. Recognizing the imperative for the dismantling line to operate economically and in compliance with regulatory standards, this paper introduces the concept of a **Dynamic Target Part List**. The DTP delineates a hierarchical list of parts deemed most pertinent for dismantling, taking into consideration environmental, economic, and regulatory factors, as well as the intended R-strategy approach for each component. Continuously updated throughout the dismantling process to accommodate unforeseen circumstances and evolving task requirements, the DTP incorporates real-time knowledge gained about the condition of the ELV. It encompasses the essential components slated for extraction from the ELV and outlines the requisite methods for their extraction, thereby informing decisions regarding tool selection and dismantling steps. Given the dynamic nature of dismantling operations, wherein parts may undergo changes in status (such as damage occurring during dismantling), it is imperative to update the DTP continuously throughout the process.

In addition to the DTP, the authors propose integrating it with a Disassembly Map to enhance part dismantlability and product repairability. The DM presented in [18] illustrates the product's architecture alongside specific dismantling information, marking target parts to assess their accessibility and potential design adaptations. Combining the DM with the DTP yields a **Dynamic Disassembly Map**, a pivotal advancement towards adaptable sequence control within automated dismantling cells. Leveraging real-time data from the DTP, the DDM facilitates reconfiguration of dismantling sequences, enabling agile adjustments to tooling, path planning, and gripping processes to ensure optimal efficiency and effectiveness in the dismantling process.

4.2 Development Method

To start with the system design the sequence of dismantling steps must be known. These may vary depending on local regulations and company specific procedures. With future development towards higher sustainability as proposed by [4], the remanufacturing rate will increase which requires systems that can achieve a higher dismantling depth. For this reason, a process sequence for both ELV and its sub-components has been developed.

The single steps that must be carried out for the complete dismantling of an ELV are repetitive and mostly involve similar functions and tools. Figure 1 shows the most common functions and in which dismantling process they are required. These functions can be categorized into digital and physical functions.

Digital functions access, process and display information relevant to the dismantling process, such as the parts to be disassembled, the dismantling sequence and the tools required. Physical functions interact directly with the ELV or its components. This could be, for example, unscrewing a joint or handling a dismantled component away from the ELV. There are numerous solutions available for each function.

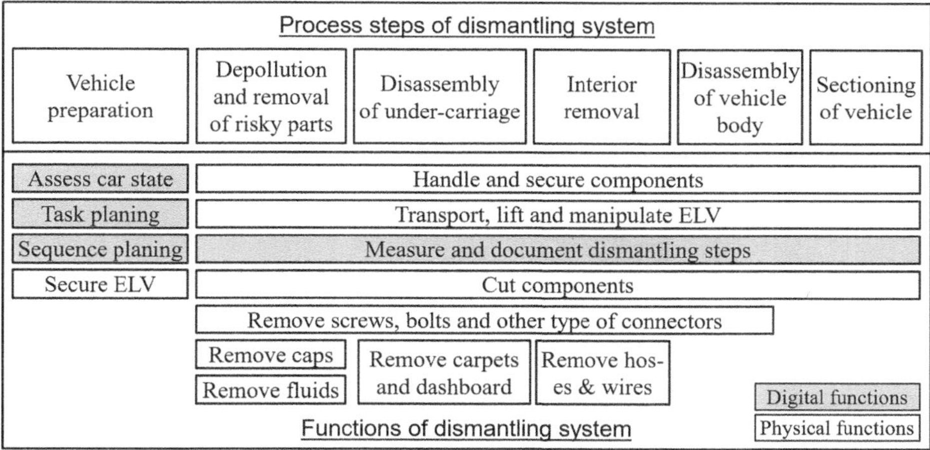

Fig. 1 Functions of the ELV dismantling process

For each of the functions, a comprehensive list of suitable automated and non-automated solutions is generated and structured in a morphological box according to VDI 2206 [19]. An excerpt of the morphological box is shown in Table 1. From this list, system designers can derive their own design of an ELV dismantling system tailored to their needs.

Table 1 Excerpt of morphological box for a dismantling system design

Functions	Solutions			
Handle and secure components	Linear axes with jigs	Robot with univ. gripper	Balancer with custom jigs	Manual handled jigs
Transport, lift and manipulate ELV	Tilt and lift system	Tilting system	Single column lift	Vehicle trolley
Cut components	Milling cutter	Angle grinder	Blow torch	Cutting plier
Remove screws, bolts and connectors	Automated adv. tools	Smart wrench	Plier	Drill unit
Measure and document dismantling	AI and CV system	Integrated sensor	Barcode scanner	Manual document
Assess car state	AI and CV system	Digital data passport	Data from digital twin	Manual inspection
Task- and sequence planning	DTP and DDM	Automated algorithms	Manual planning	Fixes tasks and sequences

5 A Flexible Dismantling Setup

To illustrate the process of system design using the developed method in Sect. 4.2, an example setup of a flexible dismantling system is presented, focusing on a highly economic and flexible dismantling system that can adapt to dynamic market trends, changing regulations and the fluctuating supply of ELVs.

The system developed uses the enabling technologies discussed in Sect. 4.1, like DDP, DTP and DDM, for task and sequence planning and initial ELV condition assessment. This results in a dynamic disassembly process that is unique to each ELV. This dynamic process requires a highly flexible and reconfigurable system, which cannot be solved by a classic unidirectional line concept, as most production systems are designed today. Therefore, a modular line concept is chosen that is capable of handling the highly dynamic process. Figure 2 shows the layout of the developed modular dismantling system and a 3D CAD model of an example dismantling platform.

The dismantling system consists of two areas, the main area covering the dismantling of ELVs and the second area dealing with the inspection, assessment and dismantling of sub-components. Dismantled ELVs in the main area results in sorted residual fractions that are recycled and reused as raw materials. The extracted ELV sub-components from the “ELV dismantling area” are further processed in a second stage. This step is performed in the “sub-component area”, which outputs processed sub-components according to the local legislation and economic viability. These components can be either directly reused, remanufactured or, in the event of damage, recycled. However, the detailed design of the “sub-component area” is outside the scope of this paper and has its own requirements and specific needs.

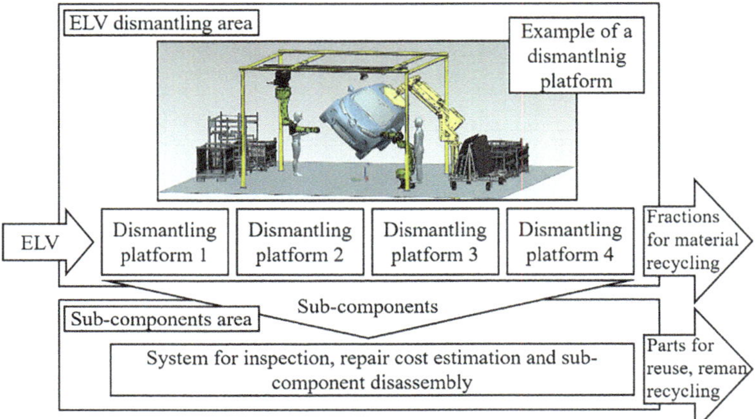

Fig. 2 Exemplary illustration of the outlined dismantling system

In order to achieve the required flexibility, the "ELV dismantling area" of the dismantling system is made up of several so-called dismantling platforms. These are semi-automatic robotic cells that can be used to perform certain dismantling steps. The individual processes carried out on each platform can be varied by equipping it with additional tools and technologies. The throughput and dismantling depth of ELVs can be varied by linking or paralleling them, thus allowing the dismantling system to be adapted to specific market and regulatory requirements.

To make the dismantling platform more flexible, the equipment used is divided into two categories. The first is basic equipment, which can assist in almost all dismantling steps, and the second is additional equipment, which is required to extract specific components. The basic equipment includes a tilt and lift device to lift, transport and manipulate the ELV and a robot with an automatic tool changing system. The robot allows the automation of a wide range of tasks, such as handling heavy components, manipulating tools for disassembly or documentation. Since it is essential to incorporate human intuition regarding complex situations in the highly non-deterministic dismantling process, the selected robot is capable of human–robot collaboration. The additional equipment is task and component specific and can have different levels of automation (LOA). Therefore, the total level of automation (TLOA) of each disassembly platform can vary depending on the selected additional equipment.

As some tasks are more complex for automation, it is highly recommended to divide the tasks performed on the system into two categories to achieve maximum efficiency. The first category includes tasks that are easy to automate, such as destructive disassembly for parts to be recycled (e.g. removal of the exhaust system) and disassembly steps whose counterparts are already well automated in automotive production. These include, for example, the removal of car doors, the removal of tires, and the overall destructive cutting of the vehicle.

The second category is difficult or limited to automate and includes steps involving complex movements, hard-to-reach parts and fragile or flexible parts. These operations have a high probability of failure if fully automated. Therefore, when considering a highly flexible and efficient disassembly system, the most appropriate approach is an HRC-based workflow, where the removal of screws and other hard-to-reach connections is performed by a human, and the handling of the dismantled components away from the ELV is done by a robot. Simultaneously the documentation of the performed steps and an initial assessment of the component is conducted by using the sensors of the tools, the robot, the gripper and a vision-based classification system to register the removal of specific parts. To achieve this workflow the robot needs a flexible clamping and gripping system that can be quickly and easily reconfigured to handle a wide variety of parts and to respond to unpredictable component conditions. To ensure safety, the gripping and clamping system has to meet the safety requirements for HRC.

6 Conclusion

Within this paper a method for developing flexible ELV dismantling systems that can cope with future market trends and regulations is proposed. The paper outlines the overall requirements, system functions, and a potential process sequence for such a system. A morphological box has been devised, containing various system components with varying LOA, which can be chosen based on the individual preferences of the system designer and the operating dismantling facility.

Existing technology concepts, such as the Battery Passport, were adopted and further developed, and their role as enabling technologies for the economic dismantling of ELVs was discussed. A system design of a dismantling system was derived with from the developed morphological box. The developed solution is capable of flexible adjustment to uncertain market and regulatory situation while optimizing economics by making use of the described enabling technologies and by efficient tasks splitting between robots and humans. Thus, enabling the economic dismantling in high-wage countries like Germany and the EU, increasing resilience, sustainability and competitiveness. While bringing efficiency gains, the potential challenges and risks of the developed system must also be considered. Firstly, the solution relies heavily on the development of a DDP for ELVs in order to track and store important data for the dismantling process throughout the life of the ELV. Therefore, the authors again strongly emphasize the need for an initiative between all stakeholders to promote such a passport. The reconfigurable and adaptable gripping system required to handle various parts of the ELV in the described HRC workflow has to be further investigated and the safety aspects of the HRC workflow need to be critically evaluated, as the workflow involves the handling of sharp and heavy components in the presence of humans. Further investigation is also required to provide a detailed categorization of the dismantling steps in terms of automatability.

References

1. European Commission: Communication from the Commission to the European Parliament, Critical Raw Materials Resilience: Charting a Path towards greater Security and Sustainability. https://perma.cc/GL5H-ZB93. Accessed 3 April 2024
2. European Commission et al.: Methodology for Establishing the EU List of Critical Raw Materials: Guidelines (2017). https://perma.cc/G5R4-NNNF. Accessed 03 April 2024
3. European Commission: Proposal for a Regulation of the European Parliament and of the Council Establishing a Framework for Ensuring a Secure and Sustainable Supply of Critical Raw Materials. https://perma.cc/Y6V3-NN7K. Accessed 03 April 2024
4. European Commission: Proposal for a Regulation of the European Parliament and of the Council on Circularity Requirements for Vehicle Design and on Management of End-of-Life Vehicles. https://perma.cc/83CQ-5HTM. Accessed 03 April 2024
5. Umwelt Bundesamt: Beitrag der Demontagebetriebe für Altfahrzeuge zu den Verwertungsquoten. https://perma.cc/9MCV-6DZ2. Accessed 03 April 2024

6. Tasala Gradin, K., Luttropp, C., Björklund, A.: Investigating Improved Vehicle Dismantling and Fragmentation Technology (2013). https://doi.org/10.1016/j.jclepro.2013.05.023
7. Proff, S., Scholz, A., Welter, A.: IN4climate.RR 2023: Die Verwertung von Altfahrzeugen: Status Quo, Herausforderungen und Potentiale im Hinblick auf eine effizientere Kreislaufwirtschaft in Deutschland und dem Rheinischen Revier. https://perma.cc/LYX8-7JKY. Accessed 03 April 2024
8. Tian, J., Chen, M.: Assessing the Economics of Processing End-of-Life Vehicles Through Manual Dismantling (2016). https://doi.org/10.1016/j.wasman.2016.07.046
9. European Commission: Regulation (EU) 2018/858 of the European Parliament and of the Council on the Approval and Market Surveillance of Motor Vehicles. https://perma.cc/RWA6-B5YP. Accessed 03 April 2024
10. ARPAC: Vehicle Dismantling Procedures. https://perma.cc/M6LS-SG22. Accessed 03 April 2024
11. Halabi, E.E., Third, M., Doolan, M.: Machine-Based Dismantling of End of Life Vehicles: A Life Cycle Perspective (2015). https://doi.org/10.1016/j.procir.2015.02.078
12. Fleischer, J., Gerlitz, E., Rieß, S., Coutandin, S., Hofmann, J.: Concepts and Requirements for Flexible Disassembly Systems for Drive Train Components of Electric Vehicles (2021). https://doi.org/10.1016/j.procir.2021.01.154
13. Li, W., Peng, Y., Zhu, Y., Pham, D.T., Nee, A., Ong, S.K.: End-of-Life Electric Vehicle Battery Disassembly Enabled by Intelligent and Human-Robot Collaboration Technologies: A Review (2024). https://doi.org/10.1016/j.rcim.2024.102758
14. Yuan, G., Liu, X., Zhang, C., Pham, D.T., Li, Z.: A New Heuristic Algorithm Based on Multi-Criteria Resilience Assessment of Human–Robot Collaboration Disassembly for Supporting Spent Lithium-Ion Battery Recycling (2023). https://doi.org/10.1016/j.engappai.2023.106878
15. Xia, X. et al.: The Construction and Cost-Benefit Analysis of End-of-Life Vehicle Disassembly Plant: A Typical Case in China (2016). https://doi.org/10.1007/s10098-016-1185-0
16. Zhang, C., Chen, M.: Designing and Verifying a Disassembly Line Approach to Cope with the Upsurge of End-of-Life Vehicles in China (2018). https://doi.org/10.1016/j.wasman.2018.02.031
17. Li, H., Wang, Y., Fan, F., Yu, H., Chu, J.: Sustainable Plant Layout Design for End of Life Vehicle Recycling and Disassembly Industry Based on SLP Method, a Typical Case in China (2021). https://doi.org/10.1109/ACCESS.2021.3086402
18. de Fazio, F., Bakker, C., Flipsen, B., Balkenende, R.: The Disassembly Map: A New Method to Enhance Design for Product Repairability (2021). https://doi.org/10.1016/j.jclepro.2021.128552
19. Gräßler, I., Wiechel, D., Roesmann, D., Thiele, H.: V-Model Based Development of Cyber-Physical Systems and Cyber-Physical Production Systems (2021). https://doi.org/10.1016/j.procir.2021.05.119

Enabling Trustful Data Sharing in Industry 4.0—A Comparison Between Data Spaces, Data Markets and Other Related Concepts

Lennart Klöpper and Jürgen Roßmann

Abstract

Digitalization is an important part of Industry 4.0 in which data sharing is essential to exploit the full potential of data and enable the use of modern technologies like machine learning or predictive maintenance. In addition to trust, other boundary conditions, e.g., incentives, must be taken into account. This work investigates, compares and groups different approaches that can be used to increase data sharing. Designing a data market, which depends on the respective use case, appears to be a suitable high level solution for this. The paper points out that technological approaches to realize a trustful data market can be grouped in data spaces and data intermediaries as superordinated categories. Afterwards, different combinations in order to build a centralized, decentralized or hybrid trustful data market are shown. Finally, initial abstract design decision rules are derived.

Keywords

Trustful data sharing • Industry 4.0 • Data market • Data intermediary • Data space

1 Introduction

In the course of Industry 4.0, data has become an important resource. In industrial robotics, this data comes from a variety of sources, e.g., multiple robots working together or various sensors installed in one robot [13]. Potential data providers and consumers are diverse, as shown in Fig. 1. For example, Ebert et al. state that a bridge dataset, "a large and diverse

L. Klöpper (✉) · J. Roßmann
Institute for Man-Machine Interaction, RWTH Aachen University, Aachen, Germany
e-mail: kloepper@mmi.rwth-aachen.de

M.-C. Wanner et al. (eds.), *Annals of Scientific Society for Assembly, Handling and Industrial Robotics 2024*, https://doi.org/10.1007/978-3-031-91463-8_9

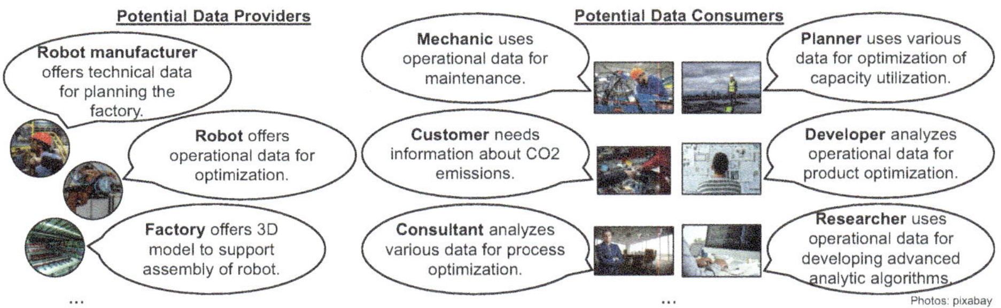

Fig. 1 Excerpt of potential data providers (circles) and consumers (rectangles)

dataset of robotic behaviors collected in a range of settings [...], for a range of different tasks" ([6], p. 2), can be used to increase the generalization in robotic learning. Sharing this dataset can help to train robots without having to repeatedly record huge amounts of data [6]. Additionally, data sharing enables advanced analytics to increase a robot's performance, such as predictice maintenance, anomaly detection, optimization algorithms or real-time decision support [13]. Thus, studies show that in Europe, increasing data sharing in general could result in macroeconomic benefits of 1–2.5% of the European gross domestic product [10].

Nevertheless, nowadays the availability of data is often limited, especially for smaller organizations. Two main reasons for this are data concentration and fragmentation. The former means that data-rich companies possess data on their own platforms and data services, creating data silos that are not available to external parties [14]. In contrast, data fragmentation means that smaller companies are not able to find and bundle sufficiently detailed data from different sources. As a result, data is currently only valuable to organizations which can handle the sharing, discovery and integration of data. Consequently, many innovative companies have no or only difficult access to high-quality data [14].

From the perspective of a data owner, there are various reasons against data sharing. Three main reasons are high transaction costs, missing incentives and the fear to loose competitive advantage. To counteract this, the European Commission gives high priority to the free flow of data. The vision is to create an interconnected network, a so-called data ecosystem, of various organizations that collaboratively use their data [2, 3, 14]. Adapting to innovations like this will help companies to "be well positioned to enhance manufacturing efficiency, competitiveness, and innovation in the dynamic landscape of industrial robotics" ([13], p. 23). This paper therefore investigates and introduces different approaches related to trustful data sharing. The aim is to provide an overview of these approaches, compare them with each other and categorize them. Based on this, initial design decision rules for a data market are derived.

The rest of the paper is structured as follows. Section 2 begins with a general introduction on how to enhance data sharing and introducing different concepts, highlighting three of them. Section 3 then focuses on how the highlighted concepts aim to generate trust.

Afterwards, Sect. 4 shows how data spaces and data intermediaries can be used to realize a trustful data market. Finally, Sect. 5 provides a conclusion and initial design decision rules are derived.

2 How Can Data Sharing Be Enhanced?

Within the European Union (EU), different laws have been adopted that specify which data may not be shared at all or only under which conditions. The General Data Protection Regulation (GDPR)[1] regulates the handling of personal data within the EU. The EU Data Governance Act[2] (DGA) specifies the scope and functionalities of so-called data intermediaries, which are discussed in more detail in the following section. In context of the European Green Deal,[3] the EU has introduced several laws that require global data sharing and can therefore be regarded as legal incentives. For example, the Corporate Sustainability Reporting Directive[4] (CSRD) of the EU requires companies of a certain size to record emissions over the entire product life cycle and make them available to third parties in order to evaluate their sustainability. This applies to all products distributed within the EU. As data sharing not only offers diverse potential for modern technologies, but is also legally required, different approaches to encourage trustful data sharing have emerged [12].

2.1 Introducing Different Concepts

2.1.1 Data Market

In principle, data markets align the providers' supply and the consumers' demand [3]. Their value can be summarized as reducing the cost of searching for data as well as facilitating the combination of data [8]. Currently there is no guideline for the technical implementation. The basic idea is that consumers share their data requirements and markets support their search for suitable data, known as data discovery [3]. However, providers are often not motivated to offer their data in general or in an appropriate way due to a lack of incentives and not knowing how much their data is worth. Pre-processing and standardization as well as combining different data sets, known as data integration, and pricing data are therefore also suitable tasks for a data market [3]. For realizing a data market, the following requirements have to be highlighted here: define a financial value for data, ensure trust between users, counteract strategic buyers and define suitable incentives. Additionally, the type of market (internal, external, for various consortia) has to be considered, as a certain amount of trust

[1] https://gdpr.eu/.

[2] https://www.european-data-governance-act.com/.

[3] https://commission.europa.eu/strategy-and-policy/priorities-2019-2024/european-green-deal_en.

[4] https://eur-lex.europa.eu/legal-content/EN/TXT/HTML/?uri=CELEX:32022L2464#d1e40-15-1.

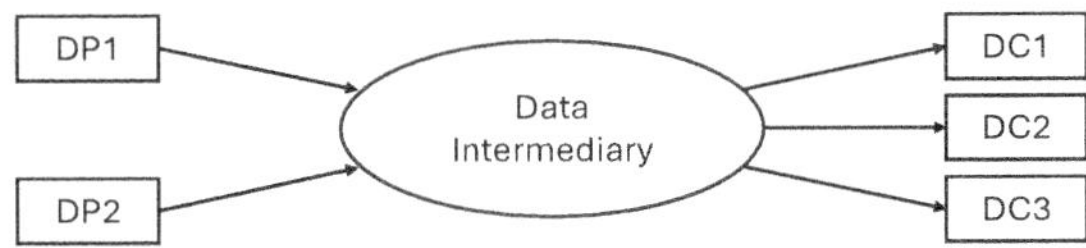

Fig. 2 Data flow between Data Provider (DP) and Data Consumer (DC) using a data intermediary

can be assumed within but not across organizations. These topics have been investigated by various researchers and practitioners in different fields [3].

2.1.2 Data Intermediary

Data intermediaries "are generally defined as third parties which mediate between two parties and support them during the initiation and subsequent execution of transactions" ([14], p. 273). Two or more-sided platforms should be emphasized here. As so-called matchmakers, these enable data sharing between two or more users, see Fig. 2. The aim is to reduce the users' costs, both when searching and sharing data. In addition, they also have the potential to build a certain level of trust by monitoring data sharing, removing sensitive information to ensure this is not shared and at the same time complying with relevant laws, as legislators are often more trusted than a company [14].

2.1.3 Data Space

The idea of leaving data at the source and enabling virtual integration on a semantic level instead of using a centralized data storage was a starting point for the development of data spaces [12]. Nowadays, there are several initiatives that focus on different aspects of data spaces, bundling their knowledge in the European Commission-funded Data Spaces Support Centre (DSSC).[5] Together, they aim to implement data spaces in various domains and establish a trustful environment for interoperable data sharing. To ensure a common understanding, the DSSC provides a glossary [5]. In the second and, at the time of this publication, current version, a data space is defined as a "distributed system defined by a governance framework that enables secure and trustworthy data transactions between participants while supporting trust and data sovereignty" (Chap. 2 in [5]). A data space consists of several components that, e.g., verify a participants identity or support data discovery. Nevertheless, at this point the so-called "Connectors" should be highlighted as they are configurable data space components. As shown in Fig. 3, these provide an endpoint for users to become active participants in a data space, enabling different functionalities like interoperability, authorization and trust services [4, 5]. The International Data Spaces Association gives an overview of various connector implementations in its regularly published Data Connector Report.[6]

[5] https://dssc.eu/.

[6] https://internationaldataspaces.org/data-connector-report/.

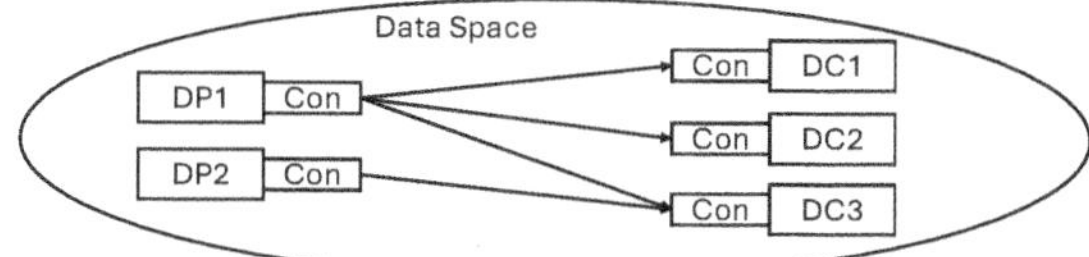

Fig. 3 Data flow between Data Provider (DP) and Data Consumer (DC) inside a data space using Connectors (Con)

2.1.4 Data Stewardship

The term data stewardship is often used regarding the handling of research data. Although there is no uniform definition, it can be said that a main objective is to optimize data management with regard to compliance with the FAIR principles (findability, accessibility, interoperability, reusability). Based on a literature research, Wendelborn et al. [15] suggest a comprehensive view on data stewardship "[...] as the responsibility to optimise research data management and sharing to maximise the potential value of research data [...] and, at the same time, to respect the rights and legitimate interests of all other stakeholders [...]" ([15], p. 4). Concrete technical solutions are less of a focus here. Following this, these ideas can be implemented using data intermediaries or data spaces.

2.1.5 Data Trustee

Data trustees aim to simplify the process of data sharing. However, there is no clear and generally established definition in the current literature [9, 11, 12]. Reiberg et al. suggest a definition seeing data trustees as "[...] institutions that manage data or rights to data on behalf of and in the interest of others. In the course of their activities, trustees obtain control over data [...]" (translated from [11], p. 7). Additionally, a data trustee should handle data securely and transparently in order to generate an environment of trust [9, 11, 12]. As data trustees are centralized components, they can be regarded as data intermediaries.

2.1.6 Data Escrow

A data escrow allows data providers to share their data while retaining full control. Participants send their data to this central infrastructure and define who can access the data under which conditions. At the same time, data consumers share which calculations they would like to execute. As calculations are carried out within the escrow, it is ensured that no raw data leaves the infrastructure without appropriate pre-processing and permission by the provider [2, 16]. Therefore, the basic idea of a data escrow is comparable to a data trustee and considerable as a data intermediary.

2.2 Interim Conclusion

Comparing the concepts presented, it becomes clear that they have more or less in common. In principle, three different approaches can be identified: none, a centralized or a decentralized technical realization. As data markets, data intermediaries and data spaces are more all-encompassing than the other approaches presented, they are considered below as overarching concepts and investigated to identify how these concepts can be used to enable trustful data sharing.

3 How to Establish Trust?

A basic requirement for data sharing is trust [1]. In general, trust is to believe that something is true.[7] However, there is no unique definition of trust in data sharing, as it always refers to the "something" under consideration. Here, the focus is on establishing trust in the infrastructure and that control over sensitive data will not be lost. Both can be achieved with the help of a suitable technical realization. Thereby, the use case has an influence on the required trust level as well, e.g., sharing less sensitive data requires less trust in not loosing control over data than sharing highly sensitive data.

The EU DGA was introduced to create trust in data intermediaries by defining the scope of application and functionalities. The resulting requirements can be divided into three categories. First of all, operators of data intermediaries have to register. Secondly, processing data is only permitted if this is done on behalf of the data provider and to simplify data sharing, e.g., pseudonomization. Lastly, services must be offered in a fair, transparent and non-discriminatory manner, while at the same time security requirements must be satisfied and a certain degree of interoperability ensured. Furthermore, the DGA indicates under which conditions a service is not regarded as a data intermediary, for example if the focus is on sharing copyright protected content.

As mentioned before, a data space is defined by a governance framework, which helps to establish trust in the infrastructure. Part of this is the so-called "trust framework" which is a "[...] composition of policies, rules, standards and procedures designed for trust decisions in data spaces based on verifiable credentials" (Chap. 9 in [5]). Policies contain rules and are linked to data to create a data offering and define what may be done with the data in compliance with the governance framework as well as the data owner [5]. It is essential that policies are both human-readable and machine-readable, e.g., using ODRL,[8] so that they can be easily created by users and automatically enforced by infrastructures. However, it is important to note that the enforcement of policies during use, known as usage control, is not enough to create trust. Even if it is technically possible to control usage within a closed

7 https://www.oxfordlearnersdictionaries.com/definition/english/trust_1.

8 https://www.w3.org/TR/odrl-model/.

system up to a certain point, it is impossible to prevent someone from, e.g., taking a photo of it. For this reason, human-readable contracts must be signed in addition to machine-readable contracts [7].

However, in order to establish trust that sensitive data will not be used without permission, appropriate pre-processing should be performed as close to the data source as possible. For example, anonymization and pseudonymization are suitable for removing personal data, corresponding filters are suitable for removing data that should not be shared and appropriate aggregation with other data sets can prevent conclusions being drawn about the respective origin. The execution of corresponding functions can be specified in policies, for example. In this way, individual offers can be generated for individual data users without the entire raw data being shared [12]. For example, it could be specified that a manufacturer may receive all the operating data to optimize a robot, while a mechanic only receives information about the component to be maintained.

In summary, it can be said that generating trust is no trivial task. Suitable approaches, which are implemented in the concepts presented, are among others pre-processing, consideration of the market type, transparency, data flow monitoring, compliance with relevant laws, exisiting standards and guidelines as well as verifying users and operators. Data markets are initially independent of the technical realization and, in addition to building trust, also take into account which incentives can be created for data sharing. For this reason, data markets represent a suitable approach at the highest level for defining necessary requirements for trustful data sharing. The technical realization can then take place in different ways, which will be discussed in the following section.

4 How to Enable Trustful Data Sharing in Industry 4.0?

Section 2 shows that various concepts to enhance data sharing exist. Section 3 identifies different possibilities to generate trust and data markets as an appropriate high level concept. But how can data markets, data spaces and data intermediaries be combined for enabling trustful data sharing between organizations and helping companies comply with regulations like the CSRD or use advanced analytics? This question is the focus of this section, presenting different possibilities to use data spaces and data intermediaries for realizing a trustful data market. In the following, numbers in brackets refer to the combinations that are shown in Fig. 4.

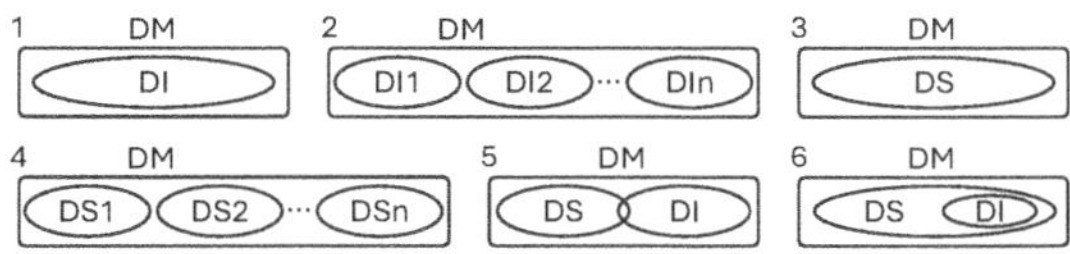

Fig. 4 Different ways to establish a Data Market (DM) using a Data Space (DS) and/or Data Intermediary (DI)

Depending on the chosen implementation, e.g., a data trustee, and the associated functionalities, a data market can be formed based on a single data intermediary (1), see also [2]. However, if a central data intermediary does not meet the requirements of a participant, e.g., due to the storage type of the raw data, the acceptance of a data market is limited. Therefore, it is also suitable for a data market to be built using multiple data intermediaries (2). This allows participants to freely choose which intermediary they trust the most or even develop an own intermediary and integrate it to the data market.

While data intermediaries are technically centralized components, data spaces represent a decentralized solution. In addition to other requirements, most concepts require every participant to have an own Connector to connect to the data space. Current implementations, e.g., Catena-X[9] or Data4Industry-X,[10] focus on specific domains. A robot manufacturer that supplies different domains would therefore have to participate in different data spaces in order to retrieve or share needed data. Consequently, it is reasonable to use one (3) or multiple (4) data spaces for setting up a data market.

Especially for smaller companies with limited resources, setting up a Connector can be challenging. Companies like sovity[11] therefore offer a Connector-as-a-Service. This can significantly reduce the workload of companies. Nevertheless, handling data can be challenging for a company, e.g., due to the combination of different internal data sources or the participation in multiple data spaces, which leads to data fragmentation. Thus, it is suitable for a company to use a data intermediary that manages and combines data on its behalf in order to subsequently share it with a data space, as described in [12]. Consequently, a data market can also be built based on a combination of a data space and a data intermediary (5), resulting in an overlap because the respective Connector is part of both data space and data intermediary. This also has the advantage that data can be shared with third parties who are not part of a specific data space, e.g., members of the data market without a Connector.

Data intermediaries that enable trustful data sharing can also act as building blocks of a data space [12]. The DSSC therefore introduces the term "data space intermediary" as a "[...] data space participant that provides one or more enabling services while not directly participating in the data transactions itself" (Chap. 5 in [5]). Following this, a data market can also be established based on a data space that has a data intermediary as its basis (6). This is especially interesting for companies that actively participate in the design of a data space, as it allows them to integrate their own components.

It should be mentioned at this point that there can also be other combinations to form a data market, while only the basic combinations are shown here. For example, additional data spaces or data intermediaries can be added in (5) or (6).

[9] https://catena-x.net/en/.

[10] https://www.data4industry-x.com/.

[11] https://sovity.de/en/sovity-en/.

This shows that the questions about the role of the concepts presented in enabling trustful data sharing and the best way to build a trustful data market cannot be answered with a general solution. The establishment of a data market already addresses many problems, especially regarding data sharing between companies that do not know or trust each other or have data with an unknown value. To realize a data market, relevant stakeholders must be taken into account, especially the number of domains addressed and the capacities of the companies involved. Can they operate components, which can be technically complex, by themselves or are they able to combine various internal data sources by themselves in the long term? This shows that every solution is dependent on the respective application. For example, a supplier for different sectors could have a data intermediary that brings data into different data spaces and data markets. Conversely, it is also conceivable that a company has a data intermediary that obtains necessary data from data spaces and data markets. Consequently, this data intermediary would have to be a participant in several data markets.

5 Conclusion

In industrial robotics today, data from different sources is available. However, for several reasons, it is often not shared. Different approaches exist to counteract this. Data markets are a suitable overarching concept, as the three main reasons for missing data sharing (high transaction costs, missing incentives, fear to loose competitive advantage) are taken into account. However, there is no single solution for all data markets. Depending on the respective use case, different combinations of data spaces and data intermediaries can be used to implement a centralized, decentralized or hybrid data market. Nevertheless, initial abstract design decision rules for a trustful data market can be derived:

- Requirements analysis: Analyze use case and adapt architecture to potential users, their capabilities and possible benefits.
- Incentives: Use incentives, e.g., money or services, to generate a higher benefit than effort.
- Standardization: Stick to established standards, both for data, e.g., MCAP, and guidelines, e.g., FAIR principles.
- Legislation: Comply with current laws and apply them transparently.
- Processing and integration: Offer suitable algorithms to generate value and protect sensitive data, e.g., anonymization, aggregation, standardization.
- Pricing: Calculate the value of data sets.
- Discovery: Display data with appropriate metadata.
- Transparency: Be transparent regarding data handling and technologies used.

In the future, all-encompassing guidelines or building blocks for the general realization of trustful data sharing or participation in a data market would be desirable. The DSSC is pursuing a similar goal for data spaces. By deriving initial design decision rules, this paper provides starting point in the direction of building blocks for data markets. In addition, it should be investigated in the future which adaptations are necessary to offer services and data in a single market and to what extent data sharing can be automated. Possible approaches here include, for example, the use of Digital Twins. In forestry, research in this area is already part of projects like S3I-X,[12], DTMForst[13] and CO2For-IT.[14]

References

1. Badirova, A., Alangot, B., Dimitrakos, T., Yahyapour, R.: Towards robust trust frameworks for data exchange: a multidisciplinary inquiry. In: Open Identity Summit (2024). https://dl.gi.de/handle/20.500.12116/44089
2. Castro Fernandez, R.: Data-sharing markets: model, protocol, and algorithms to incentivize the formation of data-sharing consortia. Proc. ACM Manag. Data **1**(2) (2023). https://dl.acm.org/doi/10.1145/3589317
3. Castro Fernandez, R., Subramaniam, P., Franklin, M.J.: Data market platforms: trading data assets to solve data problems. Proc. VLDB Endow. **13**(12) (2020). https://doi.org/10.14778/3407790.3407800
4. Data Spaces Support Centre: Building Blocks | Version 0.5 | September 2023 (28.02.2024). https://dssc.eu/space/BBE/178421761/Building+Blocks+%7C+Version+0.5+%7C+September+2023
5. Data Spaces Support Centre: DSSC Glossary | Version 2.0 | September 2023 (28.02.2024). https://dssc.eu/space/Glossary/176553985/DSSC+Glossary+%7C+Version+2.0+%7C+September+2023
6. Ebert, F., Yang, Y., Schmeckpeper, K., Bucher, B., Georgakis, G., Daniilidis, K., Finn, C., Levine, S.: Bridge data: boosting generalization of robotic skills with cross-domain datasets. CoRR (2021). arXiv:2109.13396
7. Eitel, A., Jung, C., Brandstädter, R., Hosseinzadeh, A., Bader, S., Kühnle, C., Birnstill, P., Brost, G., Gall, M., Bruckner, F., Weißenberg, N., Korth, B.: Usage control in the international data spaces. In: Whitepaper, International Data Spaces Association, Version 3.0 (2021). https://zenodo.org/record/5675884
8. Kennedy, J., Subramaniam, P., Galhotra, S., Castro Fernandez, R.: Revisiting online data markets in 2022: a seller and buyer perspective. ACM SIGMOD Record **51**(3) (2022). https://dl.acm.org/doi/10.1145/3572751.3572757
9. Lindner, M., Straub, S.: Datentreuhänderschaft Status Quo und Entwicklungsperspektiven. Whitepaper, Begleitforschung Smarte Datenwirtschaft (2023). https://www.iit-berlin.de/wp-content/uploads/2023/02/SDW_Datentreuhand.pdf
10. OECD: Enhancing access to and sharing of data: reconciling risks and benefits for data re-use across societies. In: OECD Publishing (2019). https://www.oecd-ilibrary.org/science-and-technology/enhancing-access-to-and-sharing-of-data_276aaca8-en

[12] https://www.kwh40.de/s3i-x/ (only in german).

[13] https://www.kwh40.de/dtmforst/ (only in german).

[14] https://www.kwh40.de/co2for-it/ (only in german).

11. Reiberg, A., Appelt, D., Kraemer, P.: Datentreuhänder, Datenvermittlungsdienste und GaiaX. Whitepaper, Gaia-X Hub Deutschland (2023). https://gaia-x-hub.de/wp-content/uploads/2024/02/WP-GX-Datentreuhaender.pdf
12. Schinke, L., Hoppen, M., Atanasyan, A., Gong, X., Heinze, F., Stollenwerk, K., Roßmann, J.: Trustful data sharing in the forest-based sector - opportunities and challenges for a data trustee. In: VLDBW 2023: Joint Proceedings of Workshops at the 49th International Conference on Very Large Data Bases (VLDB) (2023). https://doi.org/10.18154/RWTH-2023-08214
13. Vashishth, T.K., Sharma, V., Kumar, B., Sharma, K.K.: Optimization of data-transfer machines and cloud data platforms integration in industrial robotics. In: Machine Vision and Industrial Robotics in Manufacturing (2024). https://www.taylorfrancis.com/books/9781003438137/chapters/10.1201/9781003438137-26
14. von Ditfurth, L., Lienemann, G.: The data governance act:- promoting or restricting data intermediaries?. Compet. Regul. Netw. Ind. **23**(4) (2022). https://journals.sagepub.com/doi/10.1177/17835917221141324
15. Wendelborn, C., Anger, M., Schickhardt, C.: What is data stewardship? towards a comprehensive understanding. J. Biomed. Inform. **140** (2023). https://linkinghub.elsevier.com/retrieve/pii/S1532046423000588
16. Xia, S., Zhu, Z., Zhu, C., Zhao, J., Chard, K. Elmore, A.J., Foster, I., Franklin, M., Krishnan, S., Castro Fernandez, R.: Data station: delegated, trustworthy, and auditable computation to enable data-sharing consortia with a data escrow. Proc. VLDB Endow. **15**(11) (2022). https://dl.acm.org/doi/10.14778/3551793.3551861

Enhancing Setup Group Optimization in Matrix Production Through Quantum Computing-Based Bin Packing Modelling

Yufei Feng, Christopher Sowinski, Simon Schlichte, Param Patel, Tilmann Schwenzow, Christoph Niedermeier, Jörg Franke and Sebastian Reitelshöfer

Abstract

This paper addresses the challenge of optimizing matrix production in Surface Mount Technology (SMT) Manufacturing, a process characterized by its beneficial flexibility and inherent complexity. Traditional computing systems struggle with the exponential complexity increase in bin packing models as dataset sizes grow. To overcome this, we introduce a Quantum Computing (QC)-based bin packing model. Our approach involves the development of a Quadratic Unconstrained Binary Optimization (QUBO) model tailored for QC applications, employing D-Wave's Hybrid Quantum solver for implementation. We validate our model's reliability by comparing its solutions with those obtained from a brute force algorithm and the SAT-solver in ORtools. The result demonstrates its capability to handle a complex setup group optimization task.

Keywords

Setup group • SMT manufacturing • QUBO • Quantum optimization

Y. Feng (✉) · C. Sowinski · S. Schlichte · P. Patel · T. Schwenzow · J. Franke · S. Reitelshöfer
Lehrstuhl für Fertigungsautomatisierung und Produktionssystematik,
Friedrich-Alexander-Universität Erlangen-Nürnberg, Erlangen, Germany
e-mail: Yufei.Feng@faps.fau.de

T. Schwenzow
Siemens AG, Karlsruhe, Germany

C. Niedermeier
Siemens AG, Munich, Germany

M.-C. Wanner et al. (eds.), *Annals of Scientific Society for Assembly, Handling and Industrial Robotics 2024*, https://doi.org/10.1007/978-3-031-91463-8_10

1 Introduction

In current SMT manufacturing, Siemens' matrix production strategy, which provides the use case for this study, establishes several production lines in parallel, separating the production and inspection areas, thereby significantly enhancing production efficiency. The PCB assembly machines within an SMT line contain a fixed number of feeders, which supply materials to be assembled on the PCBs. Because of the restricted quantity of slots of assembly machines, the capacity of an SMT line is constrained by the limited number of feeders. When a material of the PCB is not equipped on the SMT line, and the remaining slots can not hold new feeders, the setup of the SMT line has to be changed. While many distinct PCBs can be produced across all lines, it may increase the number of setup changes. This, in turn, leads to more downtime, consequently diminishing production efficiency.

The primary objective of this study is to minimize the downtime of setup changing. However, quantifying this downtime is a challenge in matrix production. Therefore, reducing downtime as an optimization objective is unrealistic in the current production environment. This study chooses an alternative by reducing the number of setup groups. In this context, PCBs within the same setup group share similar designs and can be produced on a single SMT line without any setup changeovers, thus indirectly reducing setup downtime.

Several surveys ([9, 11, 12, 16]) highlight the powerful global search ability and stability of quantum optimization algorithms in addressing challenges posed by large-scale datasets and multivariate problems. The Quantum Approximate Optimization Algorithm (QAOA) and Quantum Annealing (QA) are among the widely discussed quantum optimization methods. QAOA specifically addresses solving combinatorial optimization problems characterized by a Polynomial Unconstrained Binary Optimization model, and QA with a Quadratic Unconstrained Binary Optimization (QUBO) model. Metaheuristics such as simulated annealing have the disadvantage that a high solution quality is not guaranteed. Quantum computing, on the other hand, has the potential to provide high-quality solutions compared to those classical approaches. Under the setup optimization problem, we aim to utilize the advantages of quantum approaches. Therefore, another objective of this study is to formulate the setup optimization model into QUBO form, which sets the stage for future investigations that allow us to explore the efficiency of quantum approaches to this problem.

This study aims to develop an optimization model that can correctly describe the setup optimization problem, and then ensure that the model can be run theoretically with currently popular quantum algorithms(QAOA, QA). This paper is organized as follows: Sect. 2 provides a binary quadratic linear model to describe the setup optimization problem. Based on the binary quadratic model, a QUBO model is then derived. A QUBO form ensures that quantum algorithms can solve the problem. Section 3 provides a static analysis to evaluate the feasibility of the binary quadratic linear model using SAT and D-Wave quantum solver. Section 4 summarizes our results.

1.1 Related Works

There are different approaches available focusing on a similar setup optimization problem. The explanations would be too extensive to fit this frame, but further information can be found in the references. For example [10, 13] both use heuristic approaches. While [10] utilizes a hybrid of a genetic algorithm and particle swarm optimization, [13] applies a greedy heuristic method. Both approaches have the advantage that they can solve optimization problems with a large amount of data. However, as these approaches didn't provide mathematical optimization models for their respective use cases, they cannot be adopted for a quantum annealing approach that requires a QUBO model.

Additionally, [4, 6] use a binary optimization model and a heuristic algorithm. Still, they employ non-linear constraints which cannot be used when formulating a QUBO as required for our problem description. This prevents the formulation of a QUBO, which is also necessary for our problem description.

Lastly, [3, 8] deal with a similar use case and provide a quadratic model, which can be formulated as QUBO form. Hashiba and Chang [8] tackles a problem that is similar to ours. The authors also consider the single-feeder changeover problem, which introduces too many variables. In practical production, different PCB designs usually use identical components. Therefore, there is no need to assemble additional feeders. This situation is considered in the algorithm developed in [8]. However, they don't provide any linear mathematical formula to describe it. Chauhan et al. [3] described this situation using two inequalities. Chauhan et al. [3] focuses on group count optimization for feeder trolleys. Considering that the number of feeder trolleys is fixed in Siemens production, its objective does not meet our use case. Nevertheless, [3] utilizes two linear inequalities and simultaneously introduces an auxiliary parameter with a very large value. Based on the modeling method in [8] and the constraint formulas in [3], we provide a quadratic linear model that meets the requirements of our use case and then optimize the number of variables and constraints to make the problem easier to solve.

2 Setup Group Optimization Problem

Considering the numerous complicated technical constraints within Siemens' matrix production, this study provides the following foundational assumptions to simplify the optimization model.

Assumption 1 The feeder trolleys of an SMT line can be seen as a whole container.

Assumption 2 All the SMT lines have the same capacity.

Table 1 Variables for the setup group optimization

Variable	Meaning	Type
x_{pg}	If PCB p is in the g-th set-up group	Binary
y_g	If g-th group is used	Binary
x_{mg}	If material m is used in g-th setup group	Binary

Assumption 3 A feeder contains only one kind of material (i.e. electronic components). The amount of this material is ignored.

Assumption 1 and 2 are derived from real production, in which the SMT lines are identical. Assumption 3 ignores the effect of the amount of material in the feeder. This allows the optimization model to consider only types of material for the grouping problem, thus significantly reducing the number of constraints. In practice, workers can attach additional material belts to fill the feeder, so the number of materials on a given feeder can be considered infinite. Assumption 3 is therefore reasonable. The optimization purpose of our use case is to reduce the number of setup group by fitting as many slots (of varying widths) as possible considering the capacity of the SMT lines. Therefore, our optimization problem is equivalent to a one-dimensional bin packing problem.

Given a set of PCBs $p \in \{1, \ldots, P\} = \mathbf{P}$, a set of materials $m \in \{1, \ldots, M\} = \mathbf{M}$, and a set of initial groups $g \in \{1, \ldots, G_{init}\} = \mathbf{G_{init}}$, the decision variables that indicate the location of the PCB are introduced below:

$$x_{pg} = \begin{cases} 1, & \text{if PCB} \, p \text{ is in the } g\text{-th set-up group,} \\ 0, & \text{otherwise,} \end{cases} \tag{1}$$

$$y_g = \begin{cases} 1, & \text{if } g\text{-th group is used,} \\ 0, & \text{otherwise.} \end{cases} \tag{2}$$

An auxiliary variable is provided to denote the materials that are needed for a setup group, as

$$x_{mg} = \begin{cases} 1, & \text{if material } m \text{ is used in } g\text{-th set-up group,} \\ 0, & \text{otherwise.} \end{cases} \tag{3}$$

The variables and parameters used for the optimization problem are summarized in Tables 1 and 2. The parameter B_{mp} can be generated from the bill of materials (BOM) of PCB p.

Table 2 Parameters for the setup group optimization. G_{init} normally equals the total number of PCBs by default. The rest of the parameters can be derived from the industrial software

Parameter	Meaning	Type
P	The number of PCBs	Integer
M	The number of types of materials	Integer
G_{init}	The given number of set-up groups	Integer
s_m	The slot width of the material m	Integer
$C_{\max}$	The capacity of a SMT line	Integer
B_{mp}	If PCB p contains material m	Binary

The objective function of the setup optimization problem is formulated in the following way:

$$\min \sum_{g=1}^{G_{init}} d_g \cdot y_g. \tag{4}$$

As mentioned in [5], the offset between the suboptimal and optimal solutions is just one unit by the bin-packing problem. Therefore, a coefficient d_g is artificially introduced to enlarge the difference between suboptimal and optimal solutions and thus make it easier to find the real optimum. We chose $d_g = 10g$ with $g \in \mathbf{G_{init}}$ as an arithmetic progression to enlarge the gap between suboptimal and optimal solutions appropriately. The constraint set is given as

$$\mathbf{c1}: \sum_{g=1}^{G_{init}} x_{pg} = 1, \forall p \in \mathbf{P}, \tag{5}$$

$$\mathbf{c2}: x_{pg} \le y_g, \forall p \in \mathbf{P}, g \in \mathbf{G_{init}}, \tag{6}$$

$$\mathbf{c3}: \sum_{m=1}^{M} s_m \cdot x_{mg} \le C_{\max}, \forall g \in \mathbf{G_{init}}, \tag{7}$$

$$\mathbf{c4}: x_{pg} + B_{mp} - x_{mg} - 1 \le 0, \forall p \in \mathbf{P}, m \in \mathbf{M}, g \in \mathbf{G_{init}}. \tag{8}$$

The c1 implies that every PCB must be placed only in one setup group. The c2 combines variable x_{pg} and y_g which means when PCB p are categorized to g-th group, g-th group must be used. The c3 restricts the total slot width of the material equipped on one SMT line so that it should not break its capacity. The c4 outlines that each setup group must include all the material of the PCBs belonging to it. It can be derived from the following table.

Analyzed by Marquand-Veitch diagrams, the boolean function of Table 3 is given as

Table 3 Truth table of c4, which expresses a logical relationship that if PCB p belongs to setup group g, then the group g must contain its materials m

x_{pg}	B_{mp}	x_{mg}	Value
0	–	–	1
1	0	–	1
1	1	0	0
1	1	1	1

$$f(x_{pg}, x_{mg}) = \overline{x_{pg}} + \overline{B_{mp}} + x_{mg}, \tag{9}$$

which is equivalent to the inequality

$$1 - x_{pg} - B_{mp} + x_{mg} \geq 0, \tag{10}$$

which is the same as c4.

This model has $(P + M + 1) \cdot G_{init}$ variables and $(P + P \cdot G_{init} + G_{init} + M \cdot P \cdot G_{init})$ constraints. It proposes fewer variables and constraints compared to the model in [1] with $(P + M + P \cdot M) \cdot G_{init}$ variables and $(G_{init} + P \cdot M + 2P \cdot M \cdot G_{init})$ constraints.

A necessary step to solve a quadratic linear programming model using QA is to convert this model into QUBO form and further map it to the energy Hamiltonian, which also applies to QAOA [15]. Therefore, we have to integrate all the constraints as penalty terms into the objective function. The c1 is an equation and thus takes the following form,

$$\mathcal{H}_1 = \sum_{p=1}^{P} \left(\sum_{g=1}^{G_{init}} x_{pg} - 1 \right)^2 . \tag{11}$$

The c2 is an inequality constraint. According to the equivalent penalty strategy in [7], it can be transferred as

$$\mathcal{H}_2 = \sum_{p=1}^{P} \sum_{g=1}^{G_{init}} \left(x_{pg} - x_{pg} \cdot y_g \right) . \tag{12}$$

The c3 is no longer a binary inequality as all of its coefficients s_m may not equal to one. In this case, slack variables can be introduced to make the original inequality into an equation [2].

$$\sum_{m} s_m \cdot x_{mg} + \sum_{l=0}^{s-1} 2^l a_{lg} = C_{\max},$$

where $s = \lceil \log_2 C_{\max} \rceil$ and a_{lg} denotes a binary variables. $\lceil \cdot \rceil$ means round up to the nearest integer number. The binary expansion strategy $(\sum_{l=0}^{s-1} 2^l a_{lg})$ significantly reduces the number of slack variables compared with a strategy that uses a separate slack variable for each capacity value. ($\sum_{l=0}^{C_{\max}-1} a_{lg}$). c3 can be then represented as

$$\mathcal{H}_3 = \sum_{g=1}^{G_{init}} \left(\sum_{m=1}^{M} s_m \cdot x_{mg} + \sum_{l=0}^{s-1} 2^l a_{lg} - C_{\max} \right)^2. \tag{13}$$

As c4 is derived from the boolean function 9. According to the Fourier expansion theory in [14], its penalty term can be written as

$$\mathcal{H}_4 = \sum_{p=1}^{P} \sum_{g=1}^{G_{init}} \sum_{m=1}^{M} B_{mp} \cdot x_{pg} \cdot (1 - x_{mg}). \tag{14}$$

Table 3 illustrates that $x_{pg} = 1$, $B_{mp} = 1$, and $x_{mg} = 0$ is prohibited. It implies that if PCB p belongs to group g, group g does not contain material m present in PCB p. In this case, $B_{mp} \cdot x_{pg} \cdot (1 - x_{mg})$ is equal to 1, and $\mathcal{H}_4$ trends to $P \cdot G_{init} \cdot M$. Otherwise, $B_{mp} \cdot x_{pg} \cdot (1 - x_{mg})$ is always 0, and thus $\mathcal{H}_4$ trends to 0.

The QUBO formulation of this model is finally given as:

$$\mathcal{H} = \sum_{g=1}^{G_{init}} d_g \cdot y_g + \sum_{r=1}^{4} \gamma_r \mathcal{H}_r, \tag{15}$$

where γ_r are the so-called Lagrange factors. The variable set of 15 is denoted as $z_i \in [x_{pg}, y_g, x_{mg}, a_{lg}]$. When mapping this QUBO form into the energy Hamiltonian, all variables in 15 must be replaced by

$$z_i = \frac{\sigma_z^i + I}{2}, \tag{16}$$

where σ_z^i is the Pauli-Z matrix and I is the identity matrix.

3 Statistic Analysis of the Optimization Model

In Sect. 2, we mathematically derived a quadratic linear optimization model and its QUBO form. In this section, we generate several synthetic datasets and then use a brute force algorithm (BF), the SAT-solver (SAT) from ORtools, and the hybrid-quantum solver (HybridQ) from D-Wave to evaluate the feasibility of this model statistically.

The real datasets gathered from the industrial production PCB data have a wide variety of materials. Our study would like to employ a quantum annealer and gate-based quantum computer to solve this problem, respectively, and the limitations of the hardware available today prevent the solution of a larger problem (despite our approach being suitable for larger problems). Considering the hardware limitation today, we produced some synthetic datasets with smaller data volumes. A synthetic dataset is denoted by $\{P, M, C_{\max}\}$, with $G_{init} = P$ in our statistic. We developed a data generator to create the synthetic dataset automatically

Table 4 Statistic analysis of the optimization model. The first column denotes the characteristic of a dataset with $\{P, M, C_{\max}\}$

Dataset	BF		SAT			HybridQ		
	Best	Gr	Var	Best	Gr	Var	Best	Gr
{3,4,7}	1	1	24	1	1	24	69	1
{7,22,32}	6	4	210	1	4	210	38	4
{8,25,38}	6	3	272	1	3	272	25	3
{9,28,42}	12	5	342	1	5	342	7	5
{12,42,65}	16	4	660	1	4	660	2	4
{13,50,15}	3538	5	832	1	5	832	14	5
{14,50,15}	11367	5	910	1	5	910	10	5
{15,50,15}	355	5	990	1	5	990	22	5
{16,50,15}	38	5	1072	1	5	1072	1	5

and randomly. In our data generator, the number of PCBs P and materials M can be set manually. The data generator first generates a randomized material catalog based on M and then generates a corresponding number of PCBs and their BOM lists based on P. Then, the data generator calculates the required slot width to produce each PCB based on its BOM list. The capacity $C_{\max}$ is set to the largest slot width in the PCB collection by default, or the customer can set it himself. We also developed a brute force algorithm to find the global minimum solution. This algorithm iterates through all possible PCB grouping plans and verifies if any grouping exceeds the capacity of the SMT line. Finally, it returns the grouping solution with the smallest number of groups.

Table 4 shows some datasets and their statistic analysis, where 'Best' represents the number of feasible solutions with the smallest group number and lowest energy level. 'Var' indicates the number of variables required to process the dataset, and 'Gr' denotes the minimum number of groups found by the algorithm. Since the brute force algorithm detects and compares all possible solutions, its result can be identified as the globally optimal solution. When comparing the data of Gr. columns, we find that solving the model using the SAT and HybridQ solvers gives the same results as that obtained by the brute force algorithm. It shows that the optimal solution of the use case can be found using the model in Sect. 2, i.e. the model is feasible for the use case under the current assumptions.

When comparing the value of 'Best' in each row of the table, we find that the HybridQ solver may find more optimal solutions than the brute force algorithm sometimes. This occurs because the brute force algorithm only evaluates every possible combination of PCBs to select the ones that use the minimum number of setup groups. The unused setup groups contain no materials by default. However, when using the proposed model (Eqs. 2–8), an unused setup group may have some materials, such as ($x_{pg} = 0, x_{mg} = 1$), which does not violate c4. Theoretically, such a solution does not affect the feasibility of the model because in the use case, we are only concerned with the setup group being used. An unused setup group equipped with some materials ($y_g = 0$; $x_{mg} = 1$) doesn't affect the result. Intrinsically, HybridQ does not produce more solutions than brute force but provides some additional unused SMT lines. In addition, the HybridQ solver does not necessarily always find more optimal solutions than the brute force algorithm due to the limitation of the number of HybridQ samples.

4 Conclusion

In our study, we demonstrated a feasible quadratic linear optimization model (Eqs. 2–8) to describe the setup group optimization problem under a matrix production scenario. This model can be transferred in a QUBO formulation (Eqs. 11–15) that allows the quantum algorithms (QAOA, QA) to solve it. We also validate the model with synthetic datasets and find that the model can be solved by existing SAT and hybrid quantum solvers with a global minimum.

In this study, our synthetic datasets remain insufficient compared to the real-world industrial production volumes. As the proposed model already fulfilled the necessary prerequisites to leverage QAOA and QA algorithms, our focus will be on identifying appropriate Lagrangian factors of Eq. 15, and forward discussing the potential of QA and QAOA in solving this model, particularly their efficiency regarding large-scale datasets.

Acknowledgements We acknowledge support from the German Federal Ministry of Education and Research (BMBF), funding program "Quantum Technologies–from Basic Research to Market", grant 13N15647 and 13NI6092 (LS, WM) and 13N16093 (KW). WM acknowledges support by the High-Tech Agenda Bavaria.

References

1. Bhaskar, G., Narendran, T.T.: Grouping PCBs for set-up reduction: a maximum spanning tree approach. Int. J. Prod. Res. **34**(3), 621–632 (1996)
2. Bozhedarov, A., Usmanov, S., Salakhov, G., Boev, A., Kiktenko, E., Fedorov, A.: Quantum and quantum-inspired optimization for solving the minimum bin packing problem. J. Phys. Conf. Ser. **2701**, 012129 (2024)
3. Chauhan, V.K., Bass, M., Parlikad, A.K., Brintrup, A.: Trolley optimisation: an extension of bin packing to load pcb components (2022)
4. Chen, W.-S., Chyu, C.-C.: A minimum setup strategy for sequencing PCBs with multi-slot feeders. Integr. Manuf. Syst. **14**(3), 255–267 (2003)
5. Falkenauer, E., Delchambre, A.: A genetic algorithm for bin packing and line balancing. In: ICRA, pp. 1186–1192. Citeseer (1992)

6. García-Nájera, A., Brizuela, C.A., Martínez-Pérez, I.M.: An efficient genetic algorithm for setup time minimization in PCB assembly. Int. J. Adv. Manuf. Technol. **77**(5), 973–989 (2015)
7. Glover, F., Kochenberger, G., Du, Y.: A tutorial on formulating and using QUBO models (2018). arXiv:1811.11538
8. Hashiba, S., Chang, T.-C.: PCB assembly setup reduction using group technology. Comput. & Ind. Eng. **21**(1–4), 453–457 (1991)
9. Herman, D., Googin, C., Liu, X., Galda, A., Safro, I., Sun, Y., Pistoia, M., Alexeev, Y.: A survey of quantum computing for finance (2022). arXiv:2201.02773
10. Kuo, R.J., Lin, L.M.: Application of a hybrid of genetic algorithm and particle swarm optimization algorithm for order clustering. Decis. Support Syst. **49**(4), 451–462 (2010)
11. Li, Y., Tian, M., Liu, G., Peng, C., Jiao, L.: Quantum optimization and quantum learning: a survey. IEEE Access **8**, 23568–23593 (2020)
12. Luckow, A., Klepsch, J., Pichlmeier, J.: Quantum computing: towards industry reference problems. Digitale Welt **5**, 38–45 (2021)
13. Narayanaswami, R., Iyengar, V.: Setup reduction in printed circuit board assembly by efficient sequencing. Int. J. Adv. Manuf. Technol. **26**(3), 276–284 (2005)
14. O'Donnell, R.: Analysis of Boolean Functions. Cambridge University Press (2014)
15. Venegas-Andraca, S.E., Cruz-Santos, W., McGeoch, C., Lanzagorta, M.: A cross-disciplinary introduction to quantum annealing-based algorithms. Contemp. Phys. **59**(2), 174–197 (2018)
16. Zwingel, M., May, C., Reitelshöfer, S., Mauerer, W.: Optimization problems in production and planning: approaches and limitations in view of possible quantum superiority. In: Proceedings of the 8th MHI Colloquium. Springer (2024)

Deviation Analysis of a Process for the Validation of the Commissioning of Environmental Sensors in Automated Vehicles

Kevin Jungbluth, Philipp Litzenburger, Lennard Margies, and Rainer Müller

Abstract

The proliferation of driver assistance systems within vehicles has necessitated an augmented deployment of sensors, including radar, LiDAR, and cameras. These sensors are subjected to commissioning post-assembly, commonly referred to as End-of-Line (EoL) processes. The growing number of sensors and long process times require the parallelization of stationary test stands which is uneconomical. In order to enhance the efficiency and cost-effectiveness of EoL procedures, certain commissioning activities, such as wheel alignment or sensor calibration, are being considered for integration into the assembly line. This article commences with an overview of the current state-of-the-art in the domain of inline commissioning. Subsequently, a process to validate the commissioning of LiDAR sensors is presented, which is currently being developed at the Centre for Mechatronics and Automation Technology (ZeMA). A noteworthy challenge involves the preservation of stringent tolerances specifications for the product and process during the commissioning process. It is imperative to ascertain the deviations within individual process steps. Beyond geometric deviations, such as deviations in position or orientation, non-geometric deviations are gaining prominence due to exacting tolerance specifications. These deviations may stem from the data processing

K. Jungbluth (✉) · L. Margies · R. Müller
Center for Mechatronics and Automation Technology gGmbH, Saarbrücken, Germany
e-mail: kevin.jungbluth@zema.de

P. Litzenburger · R. Müller
Saarland University, Saarbrücken, Germany

M.-C. Wanner et al. (eds.), *Annals of Scientific Society for Assembly, Handling and Industrial Robotics 2024*, https://doi.org/10.1007/978-3-031-91463-8_11

software or the sensor systems themselves. To achieve this, the accuracy of the extraction of the targets from the sensor data and the calculation of the sensor orientation is analyzed.

Keywords

Advanced driver assistance systems • Deviation analysis • Automotive commissioning process

1 Introduction

The increasing number of driver assistance systems in modern vehicles requires a large number of different environmental sensors, such as Light Detection and Ranging (LiDAR), Radio Detection and Ranging (RaDAR) or camera systems. These sensors are commissioned in the End-of-Line (EoL) as part of vehicle assembly. Due to the growing number of sensors and the cycle time overrun of the commissioning processes, parallelization is necessary (Fig. 1). This results in increased space requirements, higher investment costs, as equipment must be purchased multiple times and a higher personnel requirement. In addition to the number of sensors and process times, tight tolerances are another challenge during commissioning [1, 2]. Current approaches, such as Otto [3] and Scheppe [4], focus on commissioning the environmental sensors in the flow assembly line to reduce the need for parallelization.

Due to the safety–critical function of driver assistance systems, it is essential to ensure that commissioning is safe and reproducible. In the future, additional processes will be necessary to validate the commissioning of environmental sensors as a last step in the production line, as demonstrated by Margies et al. [6]. Their publication presents the method

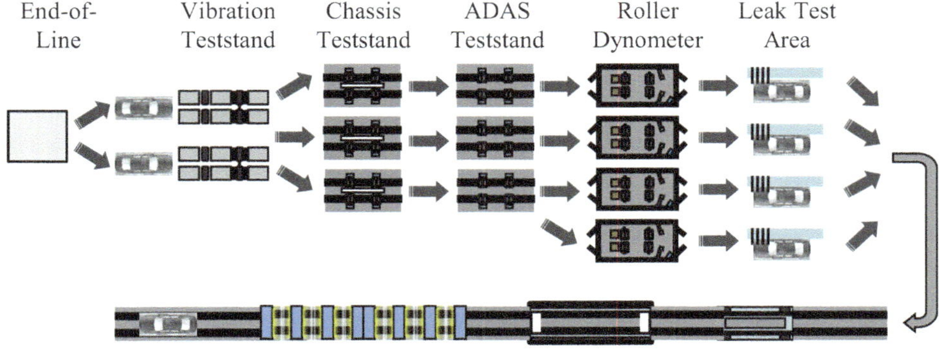

Fig. 1 Structure of an actual EoL [1, 5]

for the validation of the commissioning in greater detail. Given the strict tolerance specifications, the validation process must also be highly accurate. Therefore, the deviations that may occur in the various sub-processes are relevant for designing the validation process and considering its tolerance. The purpose of this article is to analyze deviations that occur during the determination of the target center points from sensor data and the subsequent calculation of the sensor orientation by corresponding software solutions.

The article is structured as follows: an introduction is followed by the current state of inline commissioning of environmental sensors and its validation. Section 3 provides a detailed description of the validation process, which serves as the basis for the experiments presented in this publication. Section 4 outlines the planning, setup, and evaluation of these experiments. The paper concludes with a summary in Section 5.

2 State of the Art

The vehicle assembly process is categorized into five distinct stages: handling, joining, commissioning, auxiliary processes, and special operations. Particularly in the realm of mechatronic products, commissioning, comprising adjustment, parameterization, and function testing, is regarded as an integral facet of assembly. Of paramount significance is the function testing during commissioning, aimed at ensuring adherence to specified specifications. Currently, commissioning takes place at the EoL, where vehicles are assessed for functionality in commissioning test stands. Due to protracted process durations and the rising number of sensors in contemporary automotive design, the parallelization of test stands is undertaken to concurrently evaluate multiple vehicles for functionality (Fig. 1). Contemporary OEM processes, for instance, facilitate chassis adjustments within a duration of 90–180 s [3]. In addition, current environmental sensor validation methods performed on stationary test benches are inefficient. Specifically, existing LiDAR test benches span 4 m, often requiring up to three repetitions before successful sensor calibration is achieved [4].

At ZeMA, research is focused on the vision of "In-Line commissioning" (Fig. 2) aimed at reducing process times and costs by replacing the parallelization of commissioning test stands with a seamless, automated single-line approach. The objective is to automate the determination of the geometric vehicle axis derived from the track and camber values of the vehicle rear axle in continuous operation. This is crucial as it serves as the geometric reference for the commissioning of environmental sensors [3].

The alignment and orientation of the coordinate system of the geometric vehicle axis, captured by ZeMA's wheel adaptation unit (WAU) in continuous operation, primarily serves as a reference for validating the vehicle's environmental sensors. Additionally, another vehicle feature is identified using the WAU, which directly relates to the geometric vehicle axis, facilitating automated capture of the vehicle coordinate system later in the EoL process. Through vehicle position recognition, the previously recorded feature,

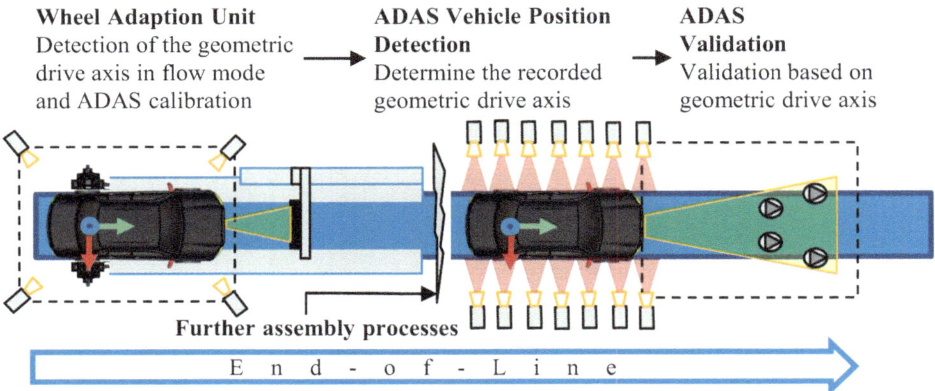

Fig. 2 Vision of in-line commissioning

along with its own feature coordinate system, is detected. The subsequent coordinate transformation yields the recorded vehicle coordinate system or the geometric vehicle axis. Utilizing the vehicle position recognition method, the geometric vehicle axis can be determined entirely automatically at every point in the EoL process. This enables for a validation of the commission of environmental sensors as a last step in the production line. In the following chapter, the necessity for this process, as well as the detailed process steps are described.

3 Validation of the Commissioning of Environmental Sensors

Throughout the entire product development process, various measures are taken to ensure the functionality of the subsystems of an automated vehicle. The aim is to create a fail safe product as early as the product development stage through a standardized procedure and defined development methods [7, 8]. Various simulation methods such as hardware-in-the-loop (HiL) and software-in-the-loop (SiL) approaches are used from prototype development to final product development [6]. Both in the pre-assembly of the individual sensors, in the pre-assembly of the corresponding modules such as the front end and in the main assembly, only standard quality monitoring and product tests are performed. The functionality of the sensors, which form the basis of automated driving functions, is tested exclusively during the vehicle commissioning described above. In these processes a slip of error is possible and manipulation of the sensor system can not be excluded in subsequent processes until the vehicle leaves production [3, 4]. Accordingly, it can not be ensured that automated vehicles leave production in a fail safe condition [6]. Against this background, a process for functional validation was developed at ZeMA as the final

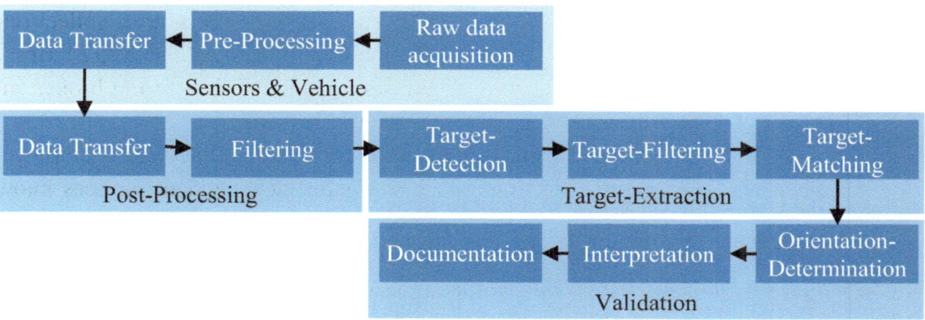

Fig. 3 Main steps of the validation process

step in production [6]. This process uses hybrid multitargets to determine the relative orientation of all sensors in the environment in relation to a defined vehicle reference. The process is divided into several key steps. These are shown in Fig. 3.

As depicted, as a first step the vehicle acquires sensor data of the process scene including the known multi targets. Followed by a preprocessing step, where intrinsic deviations of the sensors are eliminated, the raw data set of every sensor is sent to the process infrastructure via the V2X interface. In the following, the acquired data is filtered by a region of interest to remove disruptive background data. On this filtered dataset a target extraction is performed, comprising the steps of target detection, the filtering of the actual targets from the set of all potential targets in the data and a final matching of the detected targets to the known reference points. Finally, from this matched pointcloud the relative orientation of each sensor is determined, using the well-known algorithm presented by Kabsch [9]. For each sensor type the step of the target extraction is specific to RaDAR, LiDAR and camera. As this paper focusses the examination of LiDAR, these are shown below. The target extraction is performed by a concatenation of different cluster algorithms and RANSAC for detection the circular targets in the point cloud of a LiDAR sensor. From the acquired potential targets, each target is filtered by its properties, namely circularity, size and point density. Hereafter, the matching of the detected targets to the known reference points is performed using a knn-matching algorithm.

4 Experiments to Determine Deviations

4.1 Realisation of the Experiments

An experimental design is conducted to evaluate errors introduced by software-based data processing. The results of the described validation process are subject to deviations resulting from the deviations and uncertainties of the individual process steps. While the uncertainty and deviations of some links of the process chain are known, e.g. the measurement uncertainty of the LiDAR, other links have not yet been considered at all. This includes the influence of possible deviations in the software-based data processing. Specifically, this paper focuses on the algorithm for extracting the target center points from the LiDAR data and the algorithm for calculating the sensor orientation based on the target center points. The aim of the experiments is to determine the distribution and scatter of the errors introduced by these algorithms.

Three experiments are conducted. In **Experiment 01** the algorithm determines the target center points 120 times using the exact same sensor data to demonstrate the scattering of results. **Experiment 02** evaluates the algorithm that calculates the sensor orientation in the system's coordinate system based on the target center points. The test is repeated 120 times using the same target center points. **Experiment 03** examines the combined behavior of both algorithms. The algorithm for calculating the sensor orientation includes a 'best fit' function. The aim is to investigate whether the scatter of the overall results can be reduced by further processing. For this purpose, the algorithm uses the 120 target centers determined in **Experiment 01** to calculate the sensor orientation in the system coordinate system. The rotations are particularly important in **Experiments 02** and **03**, which is why they are given priority in the subsequent evaluation. Section 4.2 describes the planned evaluation of the experiments.

A Blickfeld Qb2 LiDAR sensor, which is attached to a vehicle in the test setup, is available for carrying out the tests. A total of eight multitargets are arranged in the sensor's field of view. The multitargets and their arrangement are shown in Fig. 4. The multitargets are calibrated by a laser inferometer to the above-mentioned system coordinate system.

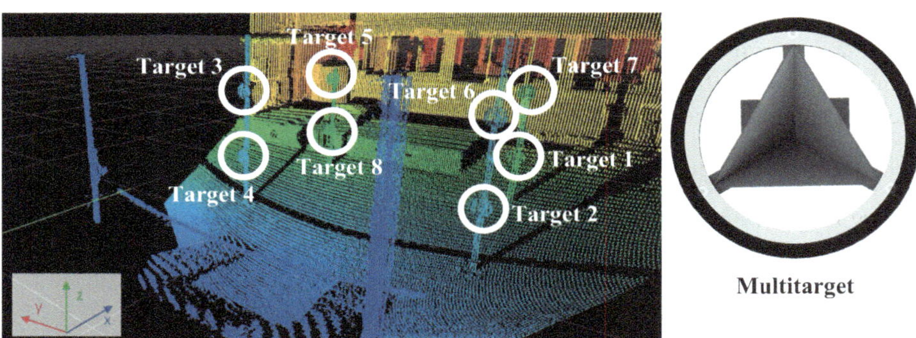

Fig. 4 Experiment environment—view and recording of the LiDAR

4.2 Evaluation Method

To achieve meaningful results, the data is evaluated both analytically and graphically. Initially, it is important to analyze whether the data corresponds to a normally distributed or non-normally distributed population. If the distribution corresponds to a normally distributed population, it can be assumed that there is a low dispersion ratio. In this case, the data is examined using the Shapiro–Wilk test. The test statistic of the Shapiro–Wilk test is defined as: [10]

$$W = \frac{\left(\sum_{i=1}^{n} a_i * x_{(i)}\right)^2}{\sum_{i=1}^{n} (x_i - \overline{x})^2} \quad (1)$$

with n = sample size, $X_{(i)}$ = sorted sample values, x_i = sample values, $\overline{x}$ = average and a_i = coefficients of the covariances ranking.

Hypotheses
Null hypothesis (H_0): Sample is normally distributed.

Alternative hypothesis (H_1): Sample is not normally distributed.

Rejection Range
To reject the null hypothesis, the calculated test statistic W is compared with a critical value (see standard normal distribution table) or the p-value is computed.

Interpretation
If the test statistic W is significantly small (i.e., smaller than the critical value) or if the p-value is less than the significance level α ($\alpha = 0.05$), the null hypothesis is rejected, and it is concluded that the sample does not come from a normally distributed population.

To assess the significance of the average deviations from the average among replications in experimental series conforming to a normally distributed population, a t-test is conducted. Here, the null hypothesis H_0 posits that the average deviation between replications is not significantly different. The test statistic t of the t-test is calculated as follows: [11]

$$t = \frac{\overline{x} - \mu}{\frac{s}{\sqrt{n}}} \quad (2)$$

with s = sample standard deviation, $\overline{x}$ = Average, μ = hypothesized average, n = Sample size.

In addition to the average and median, the standard deviation (STD) σ is of greater importance. The standard deviation indicates how far individual data points typically are from the arithmetic average. It is used to quantify the dispersion or variability of the data around the average.

4.3 Results

Experiment 01

In total, 120 measurements were conducted, so that data were calculated for each spatial axis of the eight targets in the first experiment. According to Shapiro–Wilk, all datasets are to be rejected due to non-normally distributed populations ($p < \alpha$). Exemplarily, the graphs of Target 1 for each axis are depicted (Fig. 5).

Furthermore, a standard deviation of the y-axis of 0.007 m was determined, which corresponds to an angular uncertainty of 0.047° (Fig. 6).

Experiment 02

For experiment 2, approximately identical data were calculated for each rotation axis. Thus, a normal distribution was demonstrated, as well as an undetectable dispersion (Fig. 7).

Experiment 03

In total, 120 measurements were conducted, so that in the third experiment data were calculated for each spatial and rotational axis of the sensor. According to Shapiro–Wilk, all datasets are to be accepted due to normally distributed populations ($p > \alpha$). Exemplarily, the rotational graphs of the sensor for each axis are depicted (Fig. 8).

The normal distribution of the data is evident in the box plots, with the majority of values clustering between the 25th and 75th percentiles (Fig. 9). The standard deviation around the y-axis measures 0.1537°. Furthermore, the average deviation from the average among the replicates in all tests is not significant, as $p > \alpha$ ($p(xRot) = 0.998$, $p(yRot) = 0.998$, $p(zRot) = 0.997$, $p(x) = 0.971$, $p(y) = 0.983$, $p(z) = 0.980$).

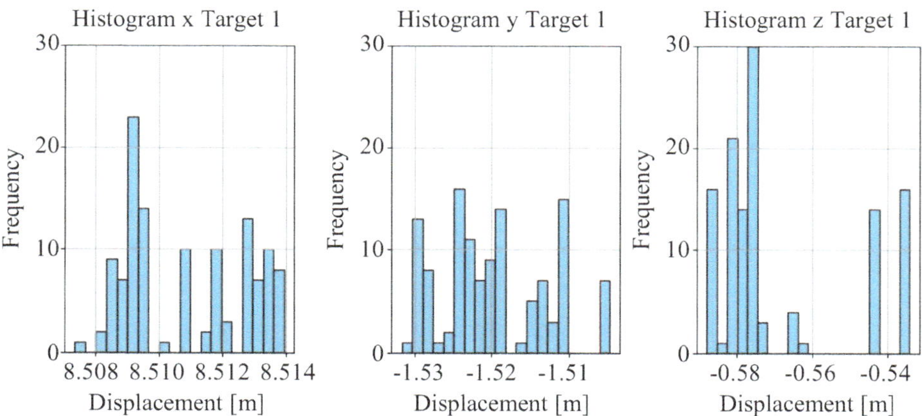

Fig. 5 Histograms of the target 1 (1st experiment)

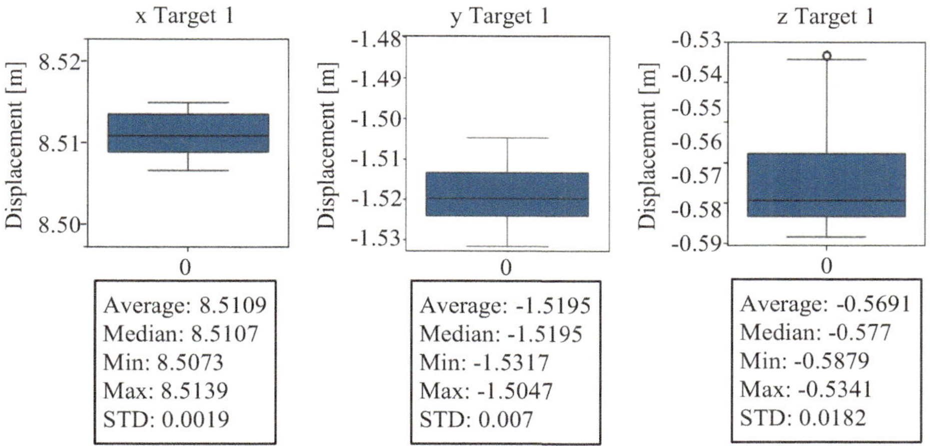

Fig. 6 Boxplots of the target 1 (1st experiment)

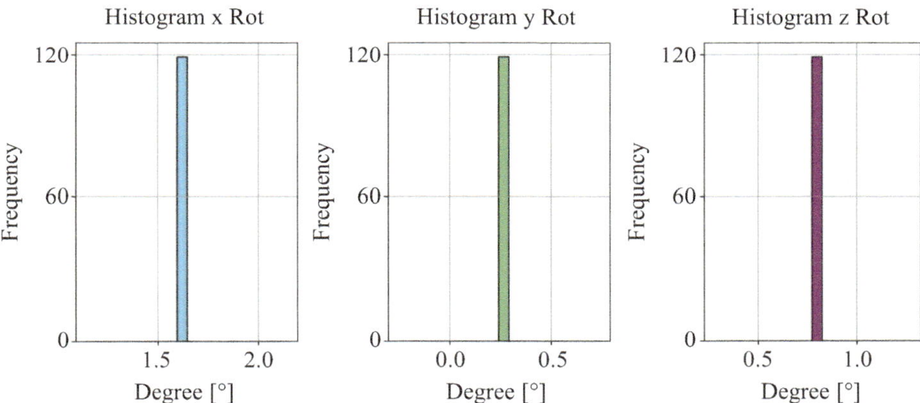

Fig. 7 Histograms of the rotational axis (2nd experiment)

4.4 Discussion of the Results

The experimental results suggest that there is scatter in the algorithmic results of the target center determination. In contrast, the scatter of the algorithm for calculating the sensor orientation is too small to determine. This is primarily due to the standard nature of the orientation computation, which yields consistent results for identical inputs, except for numerical noise. The process of determining target center points involves the use

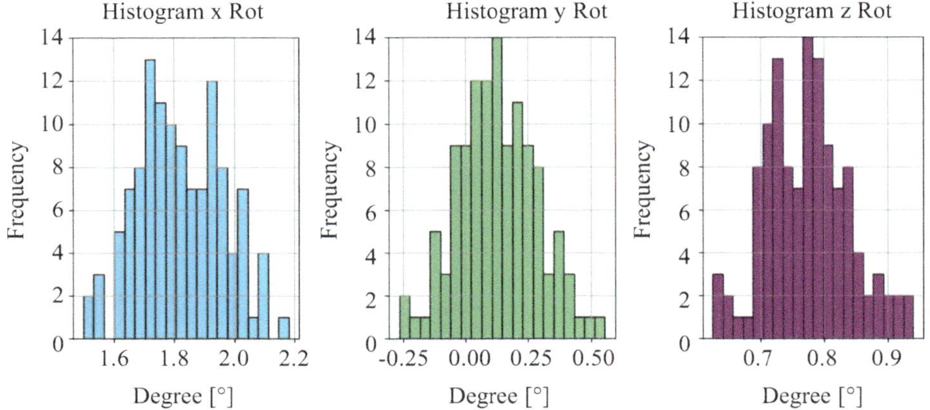

Fig. 8 Histograms of the rotational axis (3rd experiment)

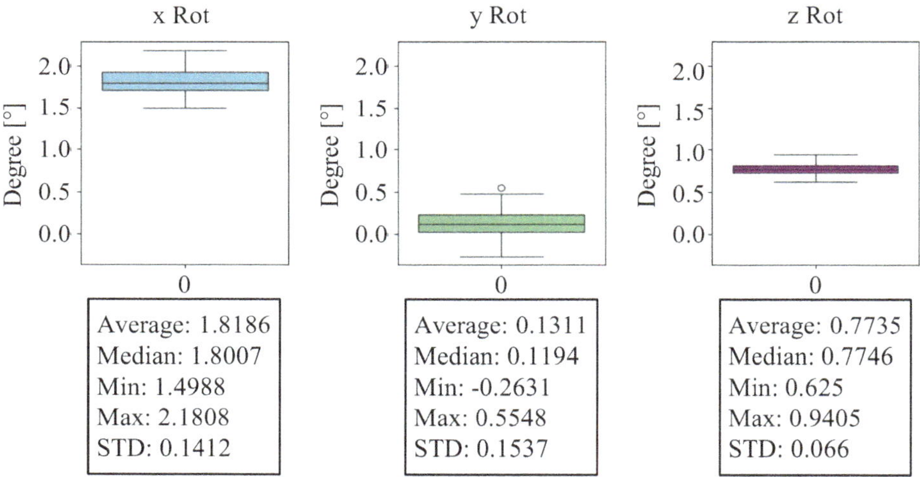

Fig. 9 Boxplots of the rotational axis (3rd experiment)

of various filtering and clustering techniques across executions, resulting in varying outcomes. The center points obtained through this process may exhibit a significant degree of variation.

It is noteworthy that the dispersion of the x-coordinate of the center points across all targets remains consistently lower than that of the y- and z-coordinates. This discrepancy is due to the time of flight (ToF) principle of the sensor and the flat and level nature of the targets, which allows for more precise filtering in the x-direction compared to the y- and z-directions, thereby reducing scatter.

The azimuth is a crucial variable for subsequent driving functions. It corresponds to the rotation around the z-axis, making its scattering particularly important. The third experiment shows that the scattering of the rotation around the z-axis in the overall results after software-side data processing is approx. $4'$. For example, according to Otto, currently used wheel alignment systems have a measurement uncertainty of approx. $0.6'$ when determining the geometric driving axle angle in azimuth [5]. The overall azimuth process tolerance is around $\pm 6'$ [2]. The scatter of the z-axis rotation results indicates that their magnitude is comparable to the metrological and mechanical process components. This demonstrates that in future tolerance analyses of commissioning processes, it is necessary to consider deviations of the software-based data processing, in addition to mechanical and metrological deviations, contrary to previous opinions.

5 Conclusion

In the present study, the measurement algorithm for validating the LiDAR sensor using multitargets was investigated. For this purpose, a static reference image of the sensor was assumed. While the determination of the target center points in Experiment 1 showed dispersion, the calculation of the sensor orientation remained stable. The x-coordinates of the centroids were the most accurate due to the ToF principle of the sensor and the flat structure of the targets. The dispersion of the azimuth rotation (z-axis) was about $4'$, comparable to mechanical tolerances of $\pm 6'$. Future analysis needs to consider the software influences on the process results. The research indicates that software processing of LiDAR data leads to variable results, especially in the determination of target centroids. The scatter of $4'$ in azimuth demonstrates the impact of the software on the overall process and highlights the need to consider it in the future tolerance analysis, contrary to previous opinions.

References

1. Müller, R., Schirmer, L., Otto, M., Scholer, M.: Neue Ansätze für die Inbetriebnahme moderner Fahrzeuge. In: Fachtagung Mechatronik 2019. Paderborn, 27.03.-28.03.2019, pp. 109–114. VDI Verlag, Darmstadt (2019)
2. Gresser, J.S.: Ganzheitliche Absicherung der Inbetriebnahme. Eine Methodik zur Absicherung der Inbetriebnahme von mechatronischen Komponenten und Systemen am Beispiel des autonomen Fahrens. Herzogenrath: Shaker (2018)
3. Otto, M.: Effizienzsteigerung und Absicherung automobiler Inbetriebnahmeprozesse. Diss. Universität des Saarlandes (UdS) Saarbrücken (2021)
4. Scheppe, P.: Inline Inbetriebnahme und Kalibrierung von Fahrerassistenzsystemen. Diss. Universität des Saarlandes (UdS) Saarbrücken (2023)

5. Müller, R., Otto, M.: Challenges for safeguarding the function of autonomous vehicles in the production. Bad Nauheim, 18–19, February 2020. ACI Monatgesysteme, Reihe (2020)
6. Margies, L., Otto, M., Müller, R.: Vernetzte Inbetriebnahme und Funktionsabsicherung hochautomatisierter Fahrzeuge in der Produktion. In: Fachtagung Mechatronik 2022. Darmstadt, 22–24, March 2022, pp. 1–6. VDI Verlag, Aachen (2022)
7. Norm ISO 26262-1:2018: Road Vehicles—Functional Safety. Part 1: Vocabulary (2018, Dezember)
8. Norm ISO 21448:2022: Road Vehicles. Safety of the Intended Functionality (2022, Juni)
9. Kabsch, W.: A solution for the best rotation to relate two sets of vectors. Acta Crystallogr. Sect. A **32**(5), 922–923 (1976)
10. Monter-Pozos, A., González-Estrada, E.: On testing the skew normal distribution by using Shapiro–Wilk test. J. Comput. Appl. Math. **440**, 115649 (2024)
11. Schuster, T., Liesen, A.: Statistik für Wirtschaftswissenschaftler. Ein Lehr- und Übungsbuch für das Bachelor-Studium. Springer, Berlin (2013)

Accelerated Market Entry of Fuel Cells Through Digital Technologies in Stack Assembly

Sarah Zimmer, Fabian Klaus, Lennard Margies, and Rainer Müller

Abstract

Fuel cell systems will play an important role in the achievement of global climate goals, particularly in the building and transportation sectors. However, manufacturers face several challenges in the production of these systems, arising primarily from the complexity of products, processes, and equipment as well as the associated data processing requirements. This paper briefly outlines the challenges in the assembly of fuel cell stacks and introduces a specific design of a Digital Twin that addresses them with the help of tailored data management, data evaluation and information utilization. This is followed by the presentation of two potential application scenarios: the use of the Digital Twin as the basis for an expert system that accelerates scaling from prototype to series assembly and as a tool for end-to-end quality monitoring to continuously optimize the products and processes.

Keywords

Fuel cell assembly • Digital twin • Expert systems • Quality management

S. Zimmer (✉) · L. Margies
Center for Mechatronics and Automation gGmbH (ZeMA), Saarbrücken, Germany
e-mail: sarah.zimmer@zema.de

F. Klaus · R. Müller
Chair of Assembly Systems, Saarland University, Saarbrücken, Germany

M.-C. Wanner et al. (eds.), *Annals of Scientific Society for Assembly, Handling and Industrial Robotics 2024*, https://doi.org/10.1007/978-3-031-91463-8_12

1 Introduction

1.1 Motivation

The significant challenges arising from the global climate targets exert a major impact on industry, politics and the economy. Due to the increasing shift away from fossil fuels, hydrogen is gaining in importance as a crucial element contributing to the success of a sustainable energy transition [1]. In this dynamic context, fuel cells are at the center of this pioneering technology and provide a promising solution to the demands of the sustainable energy generation.

Meeting the rapidly increasing future demand for fuel cells, while ensuring their commercial viability, presents a major challenge. The fuel cell stack is often considered the heart of the fuel cell. Its efficient and scalable production is crucial for the widespread adoption of fuel cell technology. One significant hurdle is transitioning the assembly of fuel cell stacks from prototype to series production. Throughout this transformation, numerous additional challenges arise, particularly in processing large amounts of assembly data coupled with the high complexity of products, processes, and equipment. These data are crucial for identifying issues during the prototype phase and thus ensuring the high quality demanded in series production.

The Digital Twin proves to be a valuable tool to face these challenges. It enables the efficient management, processing, and utilization of relevant data, as demonstrated in this paper. In the context of fuel cell stack assembly, the Digital Twin is of particular importance, as it allows for real-time monitoring and optimization of the complex assembly processes. This ensures that potential issues are identified at an early stage, and consistent quality is maintained from prototype through to mass production.

1.2 Scope of the Paper

This paper initially describes the fuel cell stack and the corresponding assembly processes while highlighting the arising challenges. Subsequently, the Digital Twin is introduced as a pertinent solution for addressing these challenges effectively. After providing a general overview of the technology, a specific framework for a Digital Twin within the "H_2SkaProMo" project, supported by the German Federal Ministry for Economic Affairs and Climate Action (BWMK), is introduced. The structure and rationale for using a Digital Twin to address the challenges are validated through the implementation of two use cases: the use of the Digital Twin as the basis for an expert system that accelerates scaling from prototype to series assembly and as a tool for end-to-end quality monitoring to continuously optimize products and processes.

2 State of the Art

2.1 Fuel Cell Stack Assembly

The fuel cell stack represents a highly complex product, not only in terms of its function but also in its structure. This results in particularly stringent requirements for assembly. These challenges are elaborated on in the publicly funded research project “H_2SkaProMo”, addressing the Proton Exchange Membrane Fuel Cell (PEMFC), which is primarily used in the mobility sector [2]. The results will be incorporated into a technology demonstrator, illustrating scalable assembly in manual, semi-automated, and fully automated degrees. In Fig. 1, the product is depicted based on its schematic and highly simplified structure. The entire stack assembly is shown on the left side of the figure, consisting of the cell stack, the head module, the base module, media supply, and bracing mechanism. The fuel cell stack itself comprises 250 stacked individual bipolar plates and membrane electrode assemblies, as depicted on the right side of the figure.

The assembly process for this product, as summarized in Fig. 2, starts with the pre-assembly of the head and foot module. This initial process consists of various joining operations, including the insertion and bolting of a spring assembly, along with precision measuring and leveling tasks that are conducted manually. After that, the bipolar plates and MEAs are stacked onto the head module alternately, a high-accuracy process repeated 250 times. Once the stacking is complete, the base module is placed on top and the stack is compressed. After compression, the stack is braced using bracing bands. The next step is a testing series, including a leak test via flow and pressure drop assessments, as well as a load test. If necessary, rework is scheduled. Finally, the stack undergoes labeling and packaging.

In previous work, the challenges in the assembly of PEM fuel cell stacks were thoroughly outlined. At the outset, a subdivision of the challenges into general challenges and process-specific topics took place. Many of the identified challenges, especially regarding process-specific issues, could be attributed to the lack of familiarity of the operators and

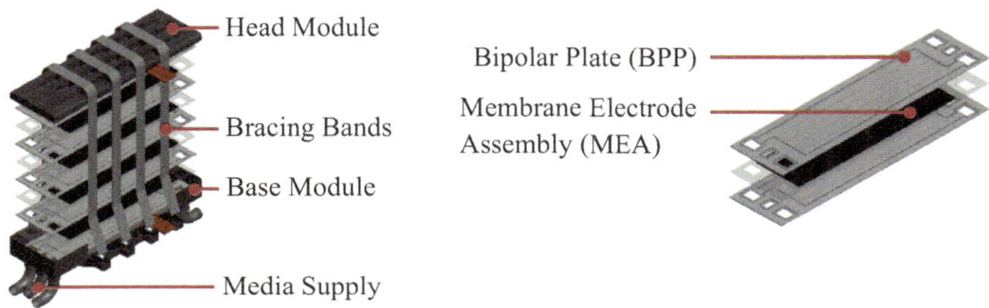

Fig. 1 Simplified structure of a PEM fuel cell stack

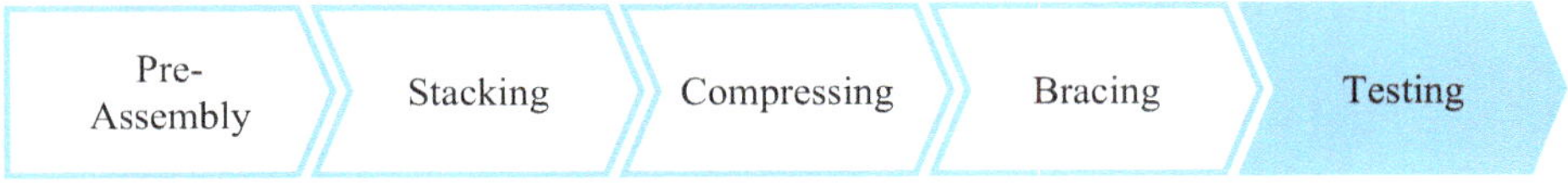

Fig. 2 Fuel cell stack assembly process

engineers involved with the product (lack of experts) or the complexity of the fuel cell and its assembly processes [3].

A brief overview is provided below.

- Pre-Assembly: The pre-assembly includes the assembly of the head and the base module. As previously mentioned, operations in this step are conducted manually. This can be an issue as PEM fuel cell stacks are often relevant as a substitute or complement to established technology, e.g. regarding the combustion engine. In the early stage of the PEMFC, this leads to a lack of necessary product and process knowledge among the employees, thus requiring individual assistance to accelerate the learning process and gain proficiency faster. Being in its early phase, the PEMFC product design is presumably subject to major changes in the future, which will mostly impact pre-assembly due to the number of components being handled. Without assistance, this could pose another significant issue [2].
- Stacking: The fuel cell stack is the core of a fuel cell system. It composed of multiple fuel cells, each consisting of two bipolar plates (BPP) and a MEA. This process requires high repeatability in the range of tenths to a few hundredths of a millimeter with a high-speed stacking frequency (0.5–2 Hz). Deviations in positioning BPP and MEA are critical to quality.
- Compressing: The process of compressing presents specific demands for process control and is a significant factor for product quality and performance [4]. Handling a system of more than 250 individual parts requires high product and process data availability and transparency.
- Testing: A defective cell which is not detected before stacking or compressing will lead to the necessity of disassembling the entire stack after load or leak testing and must be avoided.
- Rework: Currently, there is no general rework process, but various approaches are being assessed for their feasibility.

The majority of the mentioned issues can be addressed by obtaining and utilizing the right data of product, process and equipment at the right time, e.g. regarding worker assistance, real-time alignment of BPP and MEA, calculation of the optimal compression force or prediction of failure probability for individual components. For this, a framework is necessary that allows for storing and merging product, process and equipment data, as well as incorporating logic. The Digital Twin provides a solution for this.

2.2 Digital Twin

"A Digital Twin is a virtual information construct of a physical product [5], a service system [6], a system behavior [6], a process [7] or other existing or planned elements, which contains a complete description of the respective object" [8]. The ability to exchange data bidirectionally between the real and the virtual world is distinctive for the Digital Twin [5]. Furthermore, a Digital Twin is composed of a Digital Model and its corresponding Digital Shadow [7]. The Digital Model comprises the master data (e.g. geometric models or other core data) and thus represents the basis of the Digital Twin [9]. The transfer of data from the real world to the Digital Model is carried out manually. The Digital Shadow, on the other hand, enables the automated data transfer from the real to the virtual world [8].

Digital Twins are used in a wide range of industries and can be useful for a variety of tasks. In terms of assembly, their use can range from planning and operation of the assembly system to long term optimization.

Fundamentally, Digital Twins can be categorized into three main types: The Digital Product Twin, the Digital Production Twin, and the Digital Performance Twin [10, 11]. However, this paper focuses on Digital Performance Twins functioning as a protocol of all relevant data and empowering users to enhance products, processes and production systems throughout their life cycles [10, 12].

3 Design of the Digital Twin

In order to achieve the overarching goal of the "H_2SkaProMo" project, there are certain requirements for the structure and the implementation of the Digital Twin.

According to the definition given in Sect. 2.2, a bidirectional data exchange must be ensured, with the Digital Twin comprising both a Digital Model and its corresponding Digital Shadow. To accurately reflect reality, adaptability, especially modularity and scalability, are other important requirements that not only the assembly system but also the associated Digital Twin must fulfill. Furthermore, fully mapping the reality in the Digital Twin requires the capability to represent products, including assemblies and components, along with processes, resources, and their interrelationships. Another crucial capability of the Digital Twin within the project is its ability not only to store data but also to transform that data into useful information. To implement the specialized application of an expert system within the Digital Twin, a depiction of the components of knowledge input, knowledge storage, and knowledge utilization is required. This entails specific requirements such as a channeling input interface, which saves articulated knowledge in a standardized form, the machine-readability of the storage enabling the system for further processing of knowledge, and the capability of incorporating algorithms to effectively utilize the knowledge. To make the Digital Twin accessible to all stakeholders, the development of a user-friendly graphical interface is necessary. Acknowledging the varying objectives among stakeholders adds a layer of complexity to the design and implementation of the Digital Twin, requiring a flexible and adaptable system to accommodate their individual needs.

Meeting these requirements involves several key components and technologies. Within this scope, the Digital Model incorporates the data model, encapsulating the entirety of all products, processes and resources, along with their interrelationships and attributes, each endowed with corresponding target values. To map the data model a graph database is used. A pivotal advantage supporting the utilization of this type of database is its flexibility and adaptability in data modeling, coupled with its easily comprehensive structure. This not only facilitates efficient representation of complex relationships within the Digital Model but also ensures seamless adaptability to evolving requirements, making it a valuable asset in projects with complex structures.

The bidirectional data exchange between the real and virtual world can be implemented using a client–server architecture. In the scope of the project an MQTT broker is used as the communication protocol. Sensors, such as cameras or pressure sensor, collect real-time data from the assembly line and transmit it to the MQTT Broker. The MQTT Broker facilitates a publish-subscribe architecture, where sensor publish data to the broker, and various applications can subscribe to receive this data in real-time.

In order to implement various functionalities, a microservice architecture is employed. In this context, the application is decomposed into smaller, independent services known as microservices, each designed to fulfill a specific function. The architectural approach enables a flexible development process and enhances scalability within the system. Integration with containerization allows the autonomous deployment of the different functions in isolated containers, promoting operational flexibility and scalability [13].

To realize various applications of the Digital Twin, such as enhancing user interactions or supporting specific functionalities, the development of a graphical user interface (GUI) is essential. This can be technically implemented using frameworks, enabling the creation of user-friendly interfaces in Python. The integration of these GUIs facilitates user access and efficient control over the functions of the Digital Twin, contributing to its versatile application in different contexts.

4 Application Scenarios of the Digital Twin in Fuel Cell Stack Assembly

4.1 Expert System

To quickly generate high-quality output in prototype assembly and thereby accelerate the transition to mass production, the accumulation and utilization of expert knowledge is essential, especially regarding new products. Both can be conceptually validated with an innovative approach using the Digital Twin.

As outlined in the state of the art, a deficit of common knowledge regarding product, process, equipment and their cause-and-effect-relationships in fuel cell stack assembly can still be assumed. Individual process experts may also have varying levels of knowledge. Conventional knowledge building, for example based on historical data, proves challenging as data availability during the prototype phase is initially limited, even though it obviously increases with each assembly operation conducted. However, to ensure high-quality output during prototype assembly, knowledge must be made accessible and utilized. Therefore, in H_2SkaProMo the functionality of an expert system was implemented by using the Digital Twin.

An expert system in artificial intelligence refers to a software system capable of delivering problem-solving solutions in a limited field of expertise that qualitatively match or exceed those of a human expert [14].

The process chain of stack assembly offers many interesting opportunities for the use of the expert system within the Digital Twin. In the following, this is discussed using the example of the compression process, due to its high relevance.

Compressing the stack is force-controlled and path-monitored. Once the target force is reached, compliance with the process tolerances is checked. Thus, reaching the target force ensures the tightness and performance of the stack to the fullest extent, while meeting the displacement interval guarantees the stack's fit within the customer housing and compliance with demanded volumetric performance. Due to the multitude of components per stack, the process outcome depends on many individual factors. However, these influences and their strength are not yet fully considered in the parameterization of the press at an early stage. The possibly uncertain knowledge about individual influence strengths lies with experts, such as the test engineer who conducted pressing tests or the machine operator. This knowledge must be extracted and harnessed.

Knowledge Acquisition and Storage: To extract knowledge about cause-and-effect-relationships between product, process, and equipment attributes, the knowledge engineer initially conducts structured interviews with all individuals involved in the assembly process. Structured expert interviews are employed to ensure comprehensive information gathering. The interviews are conducted in a systematic manner, with the objective of ensuring the collection of all relevant information. Additionally, measures are taken to encourage experts to share their complete knowledge, with particular emphasis on the collaborative nature of the project and the importance of confidentiality.

Before the interview, the knowledge layer of the Digital Twin is called up, containing all stack components, processes, and equipment as nodes. While being talked through the assembly process step by step, the interviewed experts highlight the process outcome attributes relevant to quality. For compression, for example, the attribute "StackHeight_compressed" of the process node "Compress" is highlighted, which is then connected to the product node "FC Stack". The expert noticed that during previous compression trials, especially with older bipolar plates, stack height deviates upwards from the target value after compression, suspecting the cured seal of the bipolar plate as the probable cause. This information is shared with the knowledge engineer. The engineer creates a knowledge node "compression_force-prod_date", which connects "Gasket_BPP" and "Compress" using edges. The relevant attributes "compression_force" and "prod_date" are stored in the knowledge node, along with their assumed relation as shown in Fig. 3. The relation can be qualitative or quantitative. In this case, it is stored as "WHEN prod_date INCREASING THEN compression_force INCREASING". The maximum force is 35.4 kN, the production date up to which the force is to be further increased (today–180 days) and the progression of the increase, which is assumed to be linear.

Knowledge Utilization: In the Process Layer of the Digital Twin, a microservice continuously monitors the Knowledge Layer for newly added knowledge nodes. In the present case, the new node "compression_force-prod_date" is recognized. The relation contained

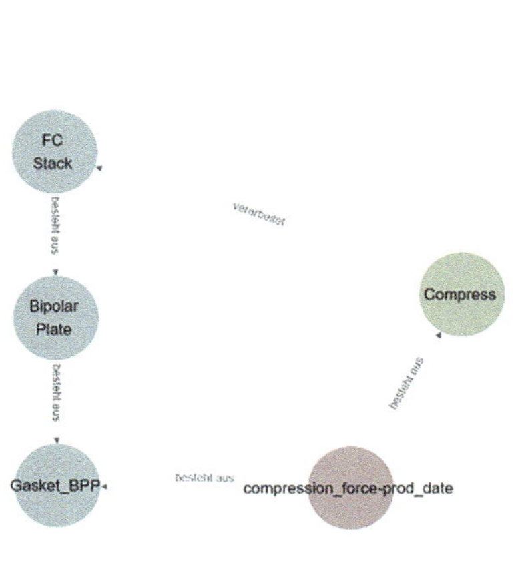

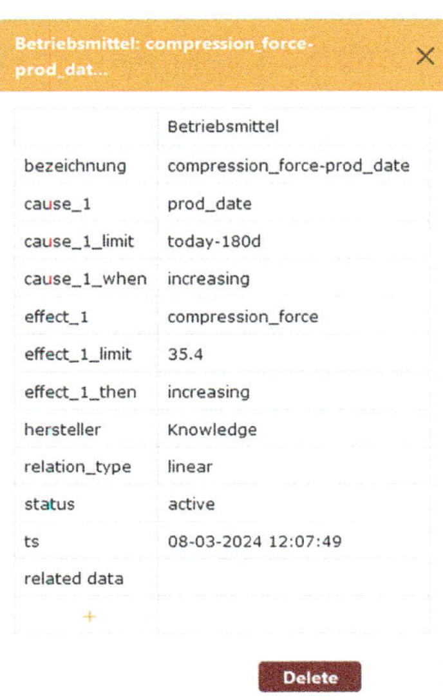

Fig. 3 Knowledge layer

therein, along with the linked components and associated process, is extracted. When a new assembly run is started, the cause attribute "prod_date" is monitored for all bipolar plates. For each plate, an individual pressing force is derived based on the production date, which is then converted into a total compression force. This is then output for the operator as a parameterization recommendation on a display at the workstation of the respective machine, in this case the press.

To determine the validity and actual strength of the relation assumed by the expert, a control instance based on the principle of four eyes must be installed, and feedback based on the process results must be implemented. This is accomplished using data recordings from the Quality Monitoring and feedback from the worker during rework. This allows the process data to be read in real-time after changing the parameterization, and the improvement or deterioration of the target values after following the recommendations can be tracked using another microservice. If a deterioration occurs, the system provides a notification for adjustment or deletion of the established relation based on this information or feedback from rework.

4.2 Quality Monitoring and Rework System

In the context of fuel cell stack assembly, as previously emphasized, quality plays a pivotal role. To ensure quality and enable the efficient reaction to errors, particularly during the early stages of the prototype assembly, the Digital Twin emerges as a suitable tool. Its application for this purpose will be described in the following.

The design of the Digital Twin begins with the setup of the Digital Model. The prescribed quality values, along with their respective attributes, are entered into the Digital Model. The determination of these target values is based on design specifications or, in some instance, on expert knowledge.

Throughout the process, real-world data is preprocessed with an IPC and transmitted via MQTT. The data is continuously associated with the Digital Shadow and mapped against the specifications in the Digital Model. If necessary, data is analyzed into useful information using the microservices. The results are linked to the target values in the Digital Model. In the event of a deviation two possibilities emerge. Upon exceeding the predefined warning threshold, the subsequent process requires meticulous monitoring. Depending on the process, adjusting parameters for subsequent steps may also be beneficial. However, if the intervention threshold is exceeded, the product needs to be removed from the assembly process, as soon as possible, and transported to a rework station.

An example of this process is stacking, where bipolar plates and MEAs are layered on top of each other. They must be precisely positioned relative to each other to ensure the stack is leakproof and maximizing overall performance. In order to achieve this, the centroids of the MEA and bipolar plate are determined. If the centroids deviate from each other, but the deviation is within tolerance, the next process step can be adjusted so that

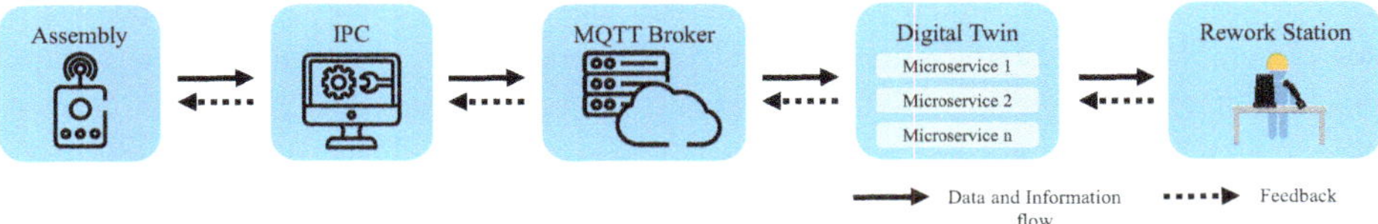

Fig. 4 Data flow and feedback loop in the digital twin framework

the positioning of the subsequent part is altered in the opposite direction. However, if the tolerance limit is exceeded, the process must be halted, and the sub-product needs to be directed to the rework station.

In particular, during the prototype phase of the assembly system the occurrence of errors is inevitable. This makes it all the more important to be prepared and able to react in such situations. The Digital Twin, with its ability to encapsulate predefined scenarios and corresponding workflows, emerges as a valuable tool for anticipating and addressing specific situations or conditions that require rework. By embedding predefined processes for rework within the Digital Twin, potential challenges can be proactively identified and actively managed.

In the context of stacking, the worker receives information about the error through a dedicated GUI created for rework. The GUI displays the specific error linked with the affected part, and the corresponding process flow for addressing the error, which was previously defined, is presented. In the case of the incorrectly positioned MEA or bipolar plate, the process involves unstacking and realigning the components accurately.

The insights gathered during rework are collected to continuously optimize processes, products, equipment, end even the rework workflow itself.

Figure 4 illustrates the continuous flow of data and information from assembly line sensors to the IPC, leveraging the MQTT Broker and microservices for processing within the Digital Twin framework as well as the feedback loop, crucial in the initial stages.

A novel aspect of this approach is that the knowledge gained from rework is used to further optimize the assembly. Common defects can be identified, leading to design improvements that make products easier to disassemble, thereby reducing Mean Time to Rework. In addition, predictive quality algorithms can be implemented to detect defective products earlier or prevent defects from occurring in the first place. The Digital Twin and its microservices facilitate these optimizations by providing real-time data analysis, automated feedback loops and adaptive process adjustments.

5 Conclusion and Outlook

The results of this paper illustrate that the Digital Twin plays a crucial role in facing today's challenges of fuel cell stack assembly. Successful application scenarios were demonstrated by implementing an expert system and a rework system within the Digital Twin. The expert system facilitates rapid, high-quality output in prototype assembly, leveraging expert knowledge effectively by revealing, utilizing and validating assumed quality-relevant relations in the data. Simultaneously, the rework system ensures quality assurance and efficient error correction during the assembly process, thus contributing to accelerate the transition from prototype to mass production and continuously optimize products and processes.

While the implemented systems show promising results, their further development and adaption require thorough testing and evaluation. Further work will focus on assessing the performance and robustness of the expert system and rework system under various conditions. This may include the integration of additional sensors, assessing new algorithms and refining them as well as accommodating different assembly scenarios.

The versatility of the Digital Twin extends beyond the systems presented, demonstrating its potential for broad applications. In essence, the presented design of the Digital Twin stands as a dynamic solution, ready to meet the evolving challenges of fuel cell stack assembly.

Acknowledgements This work is supported by the German Federal Ministry for Economic Affairs and Climate Action (BMWK) with funds provided through the project "H_2SkaProMo" (03EN5014I).

References

1. Bundesministerium für Wirtschaft und Energie (BMWi): Die Nationale Wasserstoffstrategie, Berlin (2023)
2. Führen, D., Graw, M., Kröll, L., Ilsemann, J. et al.: Wertschöpfungskette Brennstoffzelle. NOW GmbH (2022)
3. Klaus, F., Margies, L., Müller, R.: Latest Challenges in the Development of Scalable Assembly Systems for Fuel Cell Stacks, Chemnitz (2023)
4. Kampker, A., Ayvaz, P., Schön, C., Reims, P., Krieger, G.: Produktion von Brennstoffzellensystemen. RWTH Aachen; VDMA, Aachen, Berlin
5. Grieves, M., Vickers, J.: Digital twin: mitigating unpredictable, undesirable emergent behavior in complex systems. In: Kahlen, F.-J., Flumerfelt, S., Alves, A. (eds.) Transdisciplinary Perspectives on Complex Systems, pp. 85–113. Springer International Publishing, Cham (2017)
6. Stark, R., Anderl, R., Thoben, K.-D., Wartzack, S.: WiGeP Positionspapier: "Digitaler Zwilling." ZWF - Zeitschrift für Wirtschaftlichen Fabrikbetrieb **115**, 47–50 (2020)
7. Lindow, K.: Smarte Fabrik 4.0—Digitaler Zwilling. Fraunhofer IPK, Berlin

8. Zimmer, S., Molter, J., Margies, L., Karkowski, M., Müller, R.: Development of a Workflow for the Aggregation and Usage of Data in Digital Twins of Adaptable Assembly Systems, Chemnitz (2023)
9. Zeballos Raczy, J.-P.: Digital Models, Digital Shadows, and Digital Twins Help Transform and Optimize Industrial and Business Operations, pp. 20–21. InTech (2022)
10. Siemens: Digitaler Zwilling. https://www.plm.automation.siemens.com/global/de/our-story/glossary/digital-twin/24465. Accessed 07 April 2024
11. Krack, M.: Der Digitale Zwilling—Kern der Fabrik der Zukunft. topsoft Fachmagazin, pp. 4–5 (2021)
12. Mertens, P., Bodendorf, F., König, W., Schumann, M., Hess, T., Buxmann, P.: Grundzüge der Wirtschaftsinformatik, 12th edn. Springer, Berlin Heidelberg (2017)
13. Manditereza, K.: Event Driven Microservices Arhictecture for IoT Soluitions Using MQTT (2021). Accessed 09 March 2024
14. Styczynski, Z.A., Rudion, K., Naumann, A.: Einführung in Expertensysteme. Grundlagen, Anwendungen und Beispiele aus der elektrischen Energieversorgung. Springer Vieweg, Berlin, Heidelberg (2017)

Towards Process-Driven Industry 4.0 Components: A Concept for Process Modeling, Control and Simulation of Industry 4.0 Components

Anil Riza Bektas and Jürgen Roßmann

Abstract

This paper presents a comprehensive concept for process-driven Industry 4.0 components, centered around a service-oriented architecture and hierarchical process modeling of Digital Twins. With a comprehensive literature review, we outline key requirements for modeling Digital Twins' behavior and propose a novel methodology for modeling hierarchical behaviors using different process modeling languages. An exemplary process description using the Business Process Modeling Notation 2.0 shows the proposed set of modeling elements and techniques for Digital Twins based on the Asset Administration Shell. By emphasizing the interoperability and flexibility of Industry 4.0 systems, our approach enables automated manufacturing processes.

Keywords

Process modeling • Industry 4.0 component • BPMN 2.0 • Colored petri nets • Asset administration shell

1 Introduction

Technical systems are continuously evolving into more complex cyber-physical systems throughout their lifecycle. With the Industry 4.0 (I4.0) initiative, actors of these systems are typically represented by their I4.0 component, composed of the asset and its Digital Twin (DT), the one-to-one representation of the asset, especially in terms of semantics, struc-

A. R. Bektas (✉) · J. Roßmann
Institute for Man-Machine Interaction, RWTH Aachen, Aachen, Germany
e-mail: bektas@mmi.rwth-aachen.de
URL: https://www.mmi.rwth-aachen.de

M.-C. Wanner et al. (eds.), *Annals of Scientific Society for Assembly, Handling and Industrial Robotics 2024*, https://doi.org/10.1007/978-3-031-91463-8_13

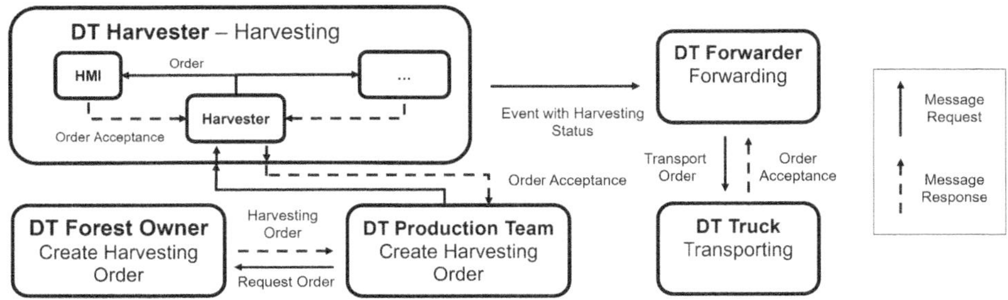

Fig. 1 Illustration of a logging process in the Forestry 4.0 domain

ture, behavior, and interaction. With increased digitization, DTs are intelligently networked through the use of Internet of Things (IoT) technology in various domains, with different requirements for automation, communication, and data privacy. This necessitates not only the management of internal DT processes but also the coordination of cross-organizational processes utilizing various DTs. A typical example is the commissioning and coordination of logging procedures in the Forestry 4.0 domain, shown in Fig. 1. The process describes an exemplary high-level perspective of the logging procedure. While part of the process may be orchestrated and controlled via a centralized actor, other parts, e.g., the communication between the harvester and forwarder shall be executed in a choreographed and thus decentralized manner without the control of a centralized party. Yet, this is only one of the various possible configurations of a logging process. There is the need to handle these different configurations during and beyond the development of each DT without the need to adapt their behavior after the deployment. Instead, each resource shall be dynamically configurable in different scenarios.

Following this introduction, the paper proceeds with the following structure: Sect. 2 gives an overview of process-driven architectures in the field of I4.0. Section 3 presents the core concept of the herein proposed architecture for process modeling, automation, and simulation of I4.0 components. Section 4 summarizes the contribution of this paper.

2 State of the Art

As systems become increasingly digitized and service-oriented, the modeling and orchestration of IoT services in systems of systems is an active area of research [1]. According to Chinosi and Trombetta, process modeling encompasses three primary applications: the textual or graphical description of the process logic, the simulation to understand the interaction dynamics, and the execution of the process itself [5]. The Workflow Management Coalition describes a generic set of workflow primitives as the building blocks of a process model, allowing to map or subordinate various patterns of state-of-the-art process modeling languages [9, 10]. While UML activity diagrams are particularly used in software development [3], business modeling languages like BPMN 2.0 [18], on the other hand, are commonly employed in more abstract processes at the business level for modeling and execution of the process. BPMN 2.0 enables a textual and graphical notation for both scenarios, choreographed and orchestrated processes. In [15], a framework is presented for web-based IoT decision-making in BPMN 2.0. The framework combines BPMN 2.0 and the Decision Model and Notation for orchestrated scenarios using different communication interfaces between the orchestrator and the resources. The choreography of services is addressed in [6, 21]. Corradini et al. present an approach for modeling and executing choreographed IoT processes based on BPMN 2.0 choreography diagrams using blockchain technology and smart contracts [6]. In [21], a protocol is introduced that propagates local changes in service architectures, allowing them to be reflected in the external independent services with the help of negotiations and suggestions. Yet, this does not serve as a solution to dynamically integrate services into different scenarios as changing, testing, and propagating the changed behavior becomes burdensome. Faruk et al. give an overview of remaining challenges of BPMN 2.0 in the field of IoT process modeling and execution [12]. One significant challenge pertains to achieving a trade-off in process modeling, conforming to the standard, while maintaining a detailed description for actual automation. It is further noted that achieving a hard-coded one-to-one mapping between IoT devices and processes may not be satisfactory. Instead, processes should be flexible with dynamically selectable devices.

While business process simulation is naturally applied for simple use cases, such as the investigation of endless loops or deadlocks [5], more recent works reveal more complex applications. In [20], an extension to the BPMN 2.0 standard is utilized for discrete-event simulation in healthcare scenarios. The extension allows for a more complex handling of queues, as opposed to the Business Process Simulation Specification (BPSim). Paolo et al. present a framework using model-to-code transformation by extending the BPSim standard with performance- and reliability-related parameters [2]. The framework in [7] on the other hand, uses model interpretation for simulation and execution in multi-robot-systems. Further, the BPMN 2.0 standard is utilized to increase interoperability. Yet, there is no process model of the low-level time-continuous behavior, which is rather implemented directly and thus simulated.

Despite its user-friendly modeling language, BPMN 2.0 is limited in other scenarios. While Petri Nets can be used to model quasi-continuous aspects of simulations (robot kinematics, physics simulation, etc.), BPMN 2.0 is not suitable for these, due to their restriction to a single-time paradigm [17]. Colored Petri Nets (CPN) introduced by Jensen et al. [14], extend the basic definition of Petri Nets as they make tokens distinguishable by different data types, hence colors. The work in [17] presents a framework to model, execute, and simulate processes in the field of robotics. Here, the Semantic Object-oriented Modeling Language (SOML++) allows modeling object-oriented, hierarchical, and CPN with a direct integration into C++ for simulation or execution engines. Hence, it allows to model more complex behaviors with a quasi-continuous time paradigm. In Context-adaptive Petri Nets, the use of ontology-based descriptions is used on top of CPN to offer more flexible and adaptive systems that can adapt to the application context [23]. Some also consider transformations of BPMN 2.0 or UML to Petri Nets [3, 8], to allow for a user-friendly modeling, while making use of the execution capabilities of Petri Nets. A different approach to the problem is presented in [16], where a combination of CPN at production level and BPMN 2.0 at business level is used for I4.0 applications.

With the I4.0 initiative, the DT of a process gained more attraction in recent research literature. The Association of German Engineers introduces an interaction protocol for bidding procedures as one possible interaction protocol based on the I4.0 language [27]. It describes choreographed interactions between DTs or processes from service offering to execution. A prototypical implementation of the interaction protocol based on state machines is presented as part of a proactive Asset Administration Shell (AAS) in [11]. The capability-based approach involves describing the functionalities of the AAS with the use of semantic machine-interpretable information in ontologies [22], supporting assistance in engineering tasks. However, there is no concept of how to build trust between the I4.0 components during the bidding procedure or capability checking which depicts a risk in enterprise overarching scenarios. The DT of a process typically describes a DT of high-level processes [7, 13, 19]. The work in [19] proposes an architecture to model business processes based on AAS-based DTs and a herein-proposed submodel to describe a REST Interface. Yet, no consideration towards the intended interface description via AAS Capabilities, Operations, or other elements were made. A different approach is presented in [13], where the process itself is modeled AAS-based Process DT, by following the capability-based engineering approach. Yet, no consideration towards other AAS elements, choreographed processes, or complex simulations was made.

Finally, service-oriented architectures are presented in [24–26]. In [25], Theorin et al. present an information system architecture for flexible choreographed processes in I4.0. However, no approach to model these processes using DTs was presented. Steindl and Kastner propose a general microservice-oriented architecture for DTs [24]. Here, service discovery is enabled using a shared knowledge graph instead of current standardized I4.0 services. In [26] a microservice-oriented architecture for choreographed processes in BPMN 2.0 is presented. Here, each microservice is modeled as a process, however, no integration of DTs and IoT was made.

The comprehensive literature review shows that while a lot of research is being done in the field, we identify a set of remaining challenges and gaps. In increasingly digitized processes, interoperability is an indispensable goal for the DT. For this reason, this work explicitly addresses the standardized representation of a DT, the AAS and the associated concepts as propagated by the I4.0 Platform. There is the need to encapsulate building blocks of the process and make them accessible via modern communication technologies. While the AAS already defines Operations (action callable via remote-procedure call), Events (one-directional message propagation of events) and Properties (information about the asset) as part of their interface description, no concept for the modeling of the behavior during the operations is presented. Current literature either considers high-level processes only [7, 11, 13, 19] or does not refer to a digital twin representation or the standard [24–26]. Thus, we attempt to fill this research gap with this paper.

3 Towards Process-Driven Industry 4.0 Components

The herein proposed concept of process-driven I 4.0 components follows the approach of a service-oriented architecture for I 4.0 components. The remaining structure follows the approach from Chinosi et al. [5] to decompose the applications for process modeling into three main goals: modeling, automation, and simulation.

3.1 Process-Driven Behavior

Following the exemplary process from Fig. 1, we can identify that the role of each actor depends on the context. The overall logging process can be considered choreographed, describing two detached processes individually orchestrated. In the first process, the DT Production Team orchestrates external AAS Operations[1] from other DTs. In the second process, the DT Forwarder orchestrates its own and external Operations at the same time. External Operations are encapsulated and called via remote procedure calls which underlines the importance of the input-output interface from an orchestrator point of view. Yet, from the DT Harvester perspective, one hierarchical level below, the harvesting Operation also follows a process-behavior. Here, the DT Harvester orchestrates its own low-level functionalities and Operations as well as external Operations. Low-level functionalities here refer to encapsulated building blocks which are only accessible for the DT itself, as opposed to Operations which are made available via the AAS. Following this basic example we can formulate the following:

- The behavior of all Digital Twins can be described via processes.
- Each Digital Twin acts as an orchestrator of its own low-level functionalities and Operations as well as external Operations.
- Interaction elements such as AAS Operations, Events, and Properties are made available via the AAS for discovery.

[1] For better readibility, AAS Operations will refer to Operations in the following.

3.2 Modeling of Industry 4.0 Components

The model of I4.0 components can be subdivided into two categories, functional and physical. While the functional model describes logical properties via the AAS which are influenced by the high-level behavior processes, the physical properties can be also referred to by the AAS, whereas they are influenced by the low-level physical behavior (Fig. 2). Based on the problem statement and literature review, the paper outlines the following requirements for process-driven modeling of DTs behavior:

1. Process models shall serve as a simple and human-understandable description of the underlying logic of the behavior.
2. All interaction elements of Digital Twins shall be made available for modeling, allowing to create complex interactive processes.
3. Orchestrated and choreographed processes can be modeled.
4. The process model can serve as an executable process model for event-based or time-continuous behavior.

While requirements 1–3 describe rather soft requirements, requirement 4 clearly describes a hard design criteria defining the basis of the herein-proposed concept. The literature highlights certain limitations of BPMN 2.0 in handling complex scenarios [16, 17]. Therefore, our work aligns with and expands upon the approach presented in [16]. This paper proposes CPN, as both the meta-process modeling language and the underlying engine. While various modeling languages are utilized, multiple research papers have demonstrated the feasibility of transforming these process modeling languages into CPNs [3, 8]. In our contribution,

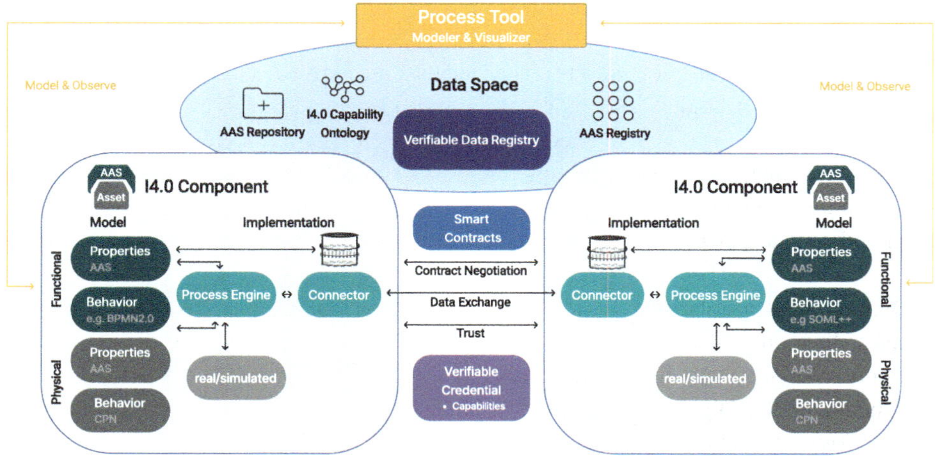

Fig. 2 Overall architecture for process driven Industry 4.0 components

we propose a methodology to model processes using different process modeling languages and subsequently transform them into CPNs. This allows for an user-friendly and flexible modeling in different scenarios.

Following the approach of process-driven behavior, each service and thus process model is composed of low-level functionalities or interaction elements of other DTs, e.g. AAS Events and Operations, leading to hierarchical processes. As of now, there is no concept of how to model these hierarchical I 4.0 processes using all provided interaction elements. A prerequisite for this is the discovery of all AAS-DT interaction elements from the modeling tool. While in [19] already basic discovery services are integrated into the BPMN 2.0 modeler Camunda, we expand on this and [13] and propose to create a one-to-one mapping for each interaction element (AAS Operations, Events, Properties) directly in the modeling language rather than on submodel level. The service discovery should then follow the basic capability-based approach as described in [13] (Fig. 2). A concept on how to gain trust of the capability description is provided in Sect. 3.3.

Figure 3 shows an exemplary representation of the hierarchical process from Fig. 1 using the BPMN 2.0 standard. The choreographed logging process is presented from the DT Production Team perspective, responsible for the high-level modeling and control of the process. Yet, as the harvesting status is not intended to be shared with the DT Production

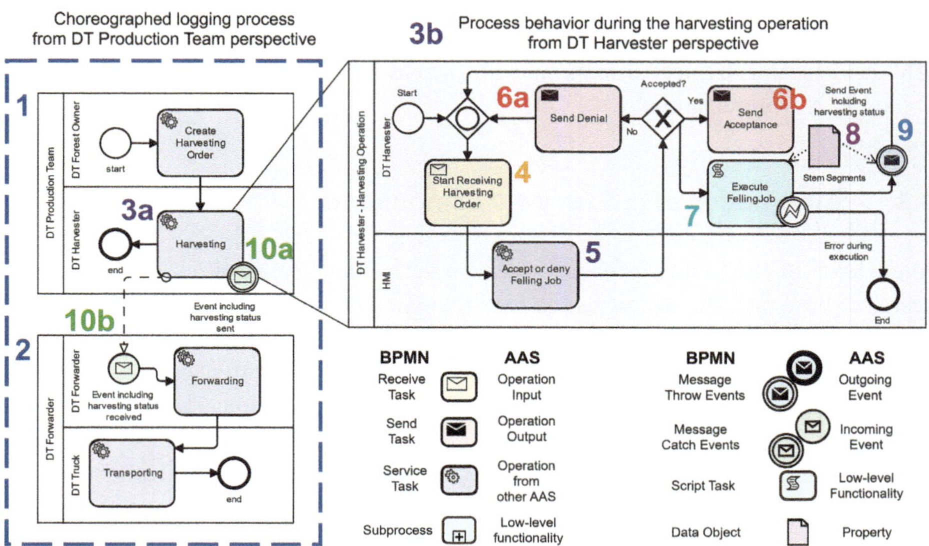

Fig. 3 Process modeling of Digital Twins behavior on different hierarchical levels based on the BPMN 2.0 standard, illustrated for an exemplary logging process in Forestry 4.0. **Left**: Visualization of the high-level choreographed logging process. **Top-right**: Visualization of the lower-level process model of the harvesting operation. **Bottom-right**: Proposed mapping from AAS interaction elements to BPMN 2.0 elements

Team, the process is required to be modeled in a choreographed way, leading to two loosely coupled processes 1 and 2, here modeled as two *Pools*,[2] orchestrating two *Lanes* each. While process 1 is orchestrated by the DT Production Team, process 2 is orchestrated by the DT Forwarder. From the DT Production Team perspective, the *Service Task* in 3a encapsulates the Operation offered by the DT Harvester. This Operation is called via a remote procedure call via any suitable communication protocol. From the DT Harvester perspective, this Operation also follows a process which is orchestrated by the DT Harvester with the same set of available behavioral elements (3b). The input behavior is described in the *Receive Task* element in 4, waiting for the Operation requests from other DTs. Since the DT Harvester orchestrates this process, external Operations of the HMI are called (5) and the corresponding response is handled in the process (6). Depending on the output of 5, an Operation Output with a denial or acceptance is propagated back to the requester (process 1) via a *Send Task*. If the order is accepted, low-level functionalities of the DT Harvester can be called via encapsulated *Script Tasks* or directly modeled using *Supprocesses* (7). As for any hierarchical level, this can be done in BPMN 2.0 or CPN or any other process model with a valid transformation to CPN. A *Data Object Reference* can then refer to AAS Properties of the DT Harvester while referencing any other element (8). The AAS Property Stem Segments is being generated from the low-level functionality and is part of an AAS Outgoing Event, published via a *Message Throw Event* (9). A reference to the same event can be then created via the discovery service. Any *Message Catch Event* allows to catch the AAS Outgoing Event while defining an AAS Incoming Event (10). This extension of [16] facilitates hierarchical process modeling across different languages, converging into a unified CPN representation.

3.3 Automation of Industry 4.0 Components

Automation of the I4.0 components refers to the automatic execution of the previously described behavior. This includes edge-based automation of each DT (e.g. DT Harvester), as well as the automation of high-level business processes (e.g. DT Production Team). Talking about the edge-based automation of DTs, the DT Harvester may offer additional operations following a process-based behavior, while running general operational functionalities. Here the SOML++ and its corresponding engine, can depict hierarchical CPN and allow to run these internal processes with a hybrid time paradigm in parallel [17]. However, for the choreographed logging process, this is not directly applicable. The choreographed process is orchestrated via two individual DTs. Clearly, the individual processes could be propagated

[2] In the following, BPMN 2.0 elements are marked in italics.

in the development phase of the DT allowing them to be part of the implementation, yet this does not serve as a solution to create dynamic processes, e.g. choreographed processes, when no such Interface is provided yet. As of now, there is no solution presented to create these dynamic processes in the field of I4.0. This paper proposes to also make these processes as part of the service negotiation of the I4.0 components [27]. The DT Production Team shares this high-level process with each actor to access their service, here AAS Operation, as part of a digital contract. Yet, the DT Forwarder takes a special role, as it is required to orchestrate the second process. Since the second process is also modeled using the same set of interaction elements, the DT Forwarder can automatically execute and orchestrate this external process in parallel to its own processes, e.g. describing the Operation behavior. Additional conditions may be part of the negotiation leading to conditional-based process termination and thus one form of smart contract (Fig. 2). A prerequisite for this is the establishment of trust between I4.0 components. A suitable solution could be the use of decentralized identifiers and the corresponding verifiable credentials via a Verifiable Data Registry [4], allowing to issue and sign capabilities from third parties which can act as a trustful partner (Fig. 2).

3.4 Simulation of Industry 4.0 Components

The simulation of I4.0 components clearly describes the simulation of the behavior of varying complexity within a virtual testbed. The behavior can be divided into functional and physical behavior. While the simulation of physical behavior typically demands a complex and continuous-time simulation, the functional behavior can be simulated and analyzed using discrete-event simulation. Using CPN as the underlying modeling language of SOML++, both time paradigms are served. Further, during development, the modeling and run-time environment can serve as a virtual testbed, where some building blocks can be simulated while others are already deployed, leading to a Hybrid Twin (Fig. 2).

4 Conclusion

The landscape of I4.0 presents complex challenges as processes with heterogeneous actors of varying complexity get increasingly digitized. Our comprehensive literature review identifies various challenges which were not yet addressed in this field. The main contribution of this research paper is the herein introduced novel methodology for process modeling, automation and simulation of I4.0 components. The proposed concept leverages the use of different process modeling languages and transforming them to CPN to allow for discrete-event or continuous-time behavior. A representative methodology on how to model I4.0 components on different hierarchical levels using the BPMN 2.0 standard was introduced. With the one-to-one mapping to all AAS interaction elements, we aim to address the limitations of existing approaches and facilitate flexible modeling in diverse scenarios such as choreographed processes.

References

1. Achir, M., Abdelli, A., Mokdad, L., Benothman, J.: Service discovery and selection in IoT: a survey and a taxonomy. J. Netw. Comput. Appl. **200**, 103,331 (2022)
2. Bocciarelli, P., D'Ambrogio, A., Cialei, M.M.E.: A MSaaS platform for business process modeling & simulation. In: Annual Modeling and Simulation Conference, pp. 223–234. IEEE, San Diego, CA, USA (2022)
3. Chang, C.H., Lu, C.W., Chu, W.C., Huang, X.H., Xu, D., Hsu, T.C., Lai, Y.B.: An UML behavior diagram based automatic testing approach. In: IEEE 37th Annual Computer Software and Applications Conference Workshops, pp. 511–516. IEEE, Japan (2013)
4. Chen, J., Bektas, A.R., Roßmann, J.: From centralized to decentralized: a DID-based authentication concept in forestry 4.0. In: IEEE International Conference on Omni-layer Intelligent Systems, pp. 1–5. IEEE, Berlin, Germany (2023)
5. Chinosi, M., Trombetta, A.: BPMN: an introduction to the standard. Comput. Stand. Interfaces **34**(1), 124–134 (2012)
6. Corradini, F., Marcelletti, A., Morichetta, A., Polini, A., Re, B., Tiezzi, F.: Blockchain-based execution of BPMN choreographies with multiple instances. Distrib. Ledger Technol.: Res. Pract. 3637555 (2023)
7. Corradini, F., Pettinari, S., Re, B., Rossi, L., Tiezzi, F.: Executable digital process twins: towards the enhancement of process-driven systems. Big Data Cogn. Comput. **7**(3), 139 (2023)
8. Dechsupa, C., Vatanawood, W., Thongtak, A.: Transformation of the BPMN design model into a colored petri net using the partitioning approach. IEEE Access **6**, 38421–38436 (2018)
9. Fortis, A., Fortis, F.: Workflow Patterns in Process Modeling (2009)
10. Grigorova, K.: Process modelling using Petri nets. In: Proceedings of the 4th International Conference Conference on Computer Systems and Technologies E-Learning, pp. 95–100. ACM Press, Rousse, Bulgaria (2003)
11. Grunau, S., Redeker, M., Göllner, D., Wisniewski, L.: The Implementation of proactive asset administration shells: evaluation of possibilities and realization in an order driven production. In: Jasperneite, J., Lohweg, V. (eds.) Kommunikation und Bildverarbeitung in der Automation, vol. 14, pp. 131–144. Springer, Berlin Heidelberg, Berlin, Heidelberg (2022)
12. Hasic, F., Asensio, E.S.: Executing IoT processes in BPMN 2.0: current support and remaining challenges. In: 13th International Conference on Research Challenges in Information Science (RCIS), pp. 1–6. IEEE, Brussels, Belgium (2019)
13. Huang, Y., Dhouib, S., Malenfant, J.: AAS capability-based operation and engineering of flexible production lines. In: 26th IEEE International Conference on(ETFA), pp. 01–04. IEEE, Vasteras, Sweden (2021)
14. Jensen, K., Kristensen, L.M., Wells, L.: Coloured Petri Nets and CPN tools for modelling and validation of concurrent systems. Int. J. Softw. Tools Technol. Trans. **9**(3–4), 213–254 (2007)

15. Kirikkayis, Y., Gallik, F., Reichert, M.: IoTDM4BPMN: an IoT-enhanced decision making framework for BPMN 2.0. In: International Conference on Service Science (ICSS), pp. 88–95. IEEE, Zhuhai, China (2022)
16. Kozma, D., Varga, P., Larrinaga, F.: Data-driven workflow management by utilising BPMN and CPN in IIoT systems with the arrowhead framework. In: 24th IEEE International Conference on Emerging Technologies and Factory Automation, pp. 385–392. IEEE, Zaragoza, Spain (2019)
17. Leonhardt, D.: Prozessmodellierung und -simulation für experimentierbare digitale Zwillinge in der eRobotik. Ph.D. thesis, Rheinisch-Westfälischen Technischen Hochschule Aachen (2019)
18. Object Management Group.: Business Process Model and Notation v. 2.0 (2011)
19. Ochoa, W., Larrinaga, F., Pérez, A.: Architecture for managing AAS-based business processes. Procedia Comput. Sci. **217**, 217–226 (2023)
20. Onggo, B.S.S., Proudlove, N.C., D'Ambrogio, S.A., Calabrese, A., Bisogno, S., Levialdi Ghiron, N.: A BPMN extension to support discrete-event simulation for healthcare applications: an explicit representation of queues, attributes and data-driven decision points. JORS **69**(5), 788–802 (2018)
21. Ortiz, J., Torres, V., Valderas, P.: Microservice compositions based on the choreography of BPMN fragments: facing evolution issues. Computing **105**(2), 375–416 (2023)
22. Plattform Industrie 4.0: Describing Capabilities of Industrie 4.0 Components (2020). https://www.plattform-i40.de/IP/Redaktion/EN/Downloads/Publikation/Capabilities_Industrie40_Components.html
23. Serral, E., De Smedt, J., Snoeck, M., Vanthienen, J.: Context-adaptive Petri nets: supporting adaptation for the execution context. Expert Syst. Appl. **42**(23), 9307–9317 (2015)
24. Steindl, G., Kastner, W.: Semantic microservice framework for digital twins. Appl. Sci. **11**(12), 5633 (2021)
25. Theorin, A., Bengtsson, K., Provost, J., Lieder, M., Johnsson, C., Lundholm, T., Lennartson, B.: An event-driven manufacturing information system architecture for industry 4.0. Int. J. Prod. Res. **55**(5), 1297–1311 (2017)
26. Valderas, P., Torres, V., Pelechano, V.: A microservice composition approach based on the choreography of BPMN fragments. Inf. Softw. Technol. **127**, 106,370 (2020)
27. VDI/VDE-Gesellschaft Mess- und Automatisierungstechnik.: VDI/VDE 2193 Blatt 2: language for industry 4.0 components - interaction protocol for bidding procedures. Guideline 2193 Blatt 2, VDI/VDE (2019)

Horizontal Scaling of Compute-Intensive Real-Time Control Tasks Across Multiple Compute Nodes

Rebekka Neumann, Christian von Arnim, Michael Neubauer, Armin Lechler, and Alexander Verl

Abstract

Industrial control systems are hard real-time systems. The correctness of such a system is determined not only by the result, but also by the time frame in which the result is obtained. Logical controllers receive sensor inputs, onto which they perform logical operations and generate an output to affect the behavior of the system. The end-to-end response time in distributed real-time systems is composed of communication and processing time and is time-constrained. The execution time of a task depends on the node's resources available for execution. Therefore, execution of compute-intensive control tasks with limited edge node resources is challenging. Since meeting deadlines is mandatory, additional resources must be provided. For resource-constrained edge nodes, horizontal scaling must occur at some point and additional nodes must be taken into account to satisfy the requirements. In this paper, we develop a concept for scaling real-time control tasks horizontally across multiple nodes. For this, we apply the approach of software pipelining to multiple nodes. To ensure deterministic communication, a real-time TSN network is used.

R. Neumann (✉) · C. von Arnim · M. Neubauer · A. Lechler · A. Verl
Institute for the Control Engineering of Machine Tools and Manufacturing Units, University of Stuttgart, Stuttgart, Germany
e-mail: rebekka.neumann@isw.uni-stuttgart.de

C. von Arnim
e-mail: christian.von-arnim@isw.uni-stuttgart.de

M. Neubauer
e-mail: michael.neubauer@isw.uni-stuttgart.de

A. Lechler
e-mail: armin.lechler@isw.uni-stuttgart.de

A. Verl
e-mail: alexander.verl@isw.uni-stuttgart.de

M.-C. Wanner et al. (eds.), *Annals of Scientific Society for Assembly, Handling and Industrial Robotics 2024*, https://doi.org/10.1007/978-3-031-91463-8_14

Keywords

Real-time tasks • Horizontal scaling • Distributed system • Pipelining

1 Introduction

Industrial control systems are hard real-time systems. Therefore, the correctness of the system is equally dependent on the result and the time at which it is obtained [1]. In control technology, logical operations are performed on sensor input values to control actuators. This process is generally carried out cyclically [2]. End-to-end latency between sensing and actuation is constrained by the cycle time in which actuation must be performed. Each cyclic real-time task T_i is characterized at least by its execution time Δe_i, period P, and deadline d_i. In addition, the attributes readiness time r_i, starting time s_i, and completion time c_i can be specified.

$$T_i(r_i, \Delta e_i, s_i, c_i, d_i, P)$$

The execution time of the task depends on available resources, such as CPU time and cores and memory. For a functional system, tasks must be completed by executing before its deadline. In dynamic environments with constantly changing conditions, control applications must be adapted to the resulting requirements [3]. If the workload increases, the tasks may exceed its deadline, if the available resources are not scaled accordingly. On the other hand, if the workload decreases, resources should be de-allocated to avoid over-provisioning and with that increased operational costs [4].

In this paper, we develop a concept for multiple node pipelining to execute compute-intensive real-time control tasks, which processed on a single node would exceed the required cycle time.

The remainder of this paper is structured as follows. In Sect. 2 the context is given, and the considered use case is described. In Sect. 3 related work concerning approaches to exploit parallelism at different levels to accelerate execution to counteract exceeding of cycle times is presented. The conceptual design, realization, and verification are presented in Sect. 4. Finally, in Sect. 5, the contents of this work are summarized and an outlook on future work is given.

2 Context

Traditional manufacturing systems are rigid monolithic configurations of hardware and software components combined with communication technologies to perform predefined functionality and, therefore, are limited in adaptability [5]. The software-defined manufacturing (SDM) paradigm aims to overcome these limitations and enable the adaptability of production systems by determining the functionality of the system purely through software [6]. To achieve changeability in production, the underlying control system must adapt to changing requirements. Therefore, the control software must be dynamically updated to adapt the functional scope accordingly. When updating applications, the workload may increase, thus also increasing the execution time of the updated application. To avoid exceeding required cycle times and missing mandatory deadlines due to the increased workload of real-time control applications, scaling must occur.

2.1 Scalability of Distributed Systems

Scalability is the ability of systems to adapt to workload changes while ensuring adequate performance. Scaling mechanisms are differentiated into vertical and horizontal methods. Vertical scaling of an application involves adding resources to the compute node or virtual machine (VM) and is therefore limited by the resources of the physical machine [7]. In contrast, horizontal scaling is not limited by the resources of a single node. Horizontal scaling is realized by adding or removing instances of the application and computing these on additional nodes or VMs. Therefore, horizontal scaling is only applicable to applications that can be decomposed or replicated and parallelized [8]. Evaluation metrics have been developed to measure and compare the scalability of applications from one scale k_1 to another scale k_2 in a distributed system consisting of a number k of processors. Metrics present in the literature are Speedup S, Efficiency E, and Scalability $\psi(k_1, k_2)$ [9]. Speedup S measures the factor by which the execution time T is accelerated by scaling adjustment compared to one processor $k = 1$. Efficiency E measures the work rate per processor.

$$S(k) = T(1)/T(k) \tag{1}$$

$$E(k) = S(k)/k \tag{2}$$

2.2 Use Case

Real-time communication technology is an enabler of distributed processing of compute-intensive real-time tasks in distributed systems since processing power is not limited by a single resource-constrained edge controller. An example of a compute-intensive workload in automation is the determination of the alignment of parts on a conveyor belt based on camera image processing.

In the considered case, a robot is to sort manufactured parts according to their quality. The parts are manufactured and subsequently placed on a conveyor belt that leads to the robot. The result of the image analysis must be communicated to the robot control in time for the robot to perform actuation. To act in time, the robot must receive the image analysis results every millisecond, but the workload is too high for the task to be completed in the controller cycle time. Due to the distance between camera and the robot the deadline of the task is greater than the frequency the images are provided to the processing edge node.

3 Related Work

To meet the quality of service (QoS) requirements of applications, such as response or execution times, different approaches exist in the literature, which are presented in the following. Particular emphasis must be placed on compliance with real-time requirements to enable scaling of control applications.

3.1 Parallelism

The computation time of an application can be reduced through parallelization at different levels: Instructional-level parallelism (ILP), thread-level parallelism (TLP), and data-level parallelism (DLP) [10]. In this work we focus on ILP and TLP. In DLP data-independent fragments of code are identified by the compiler to be parallelized. Therefore, the potential DLP in typical cyclic industrial control tasks is limited. In one thread, logical and arithmetic operations are performed on sensor input values sequentially.

The instruction pipeline parallelizes at the instruction level [11]. Commands in the program code are decomposed into instruction sets which are then processed parallel. Instruction processing requires a sequence of steps, such as instruction fetching (IF), decoding (ID), execution (EX), and write-back (WB). On reduced instruction set processors (RISC), these steps are executed by different functional units, each of which can execute concurrently. The principle is shown in Fig. 1.

Instruction pipelining enables execution time reduction. The speedup S_α resulting from parallel instruction processing with ILP is shown in (3).

$$S_\alpha = \frac{n * (\alpha * \tau)}{(\alpha + n - 1) * \tau} = \frac{n * \alpha}{\alpha + n - 1} \tag{3}$$

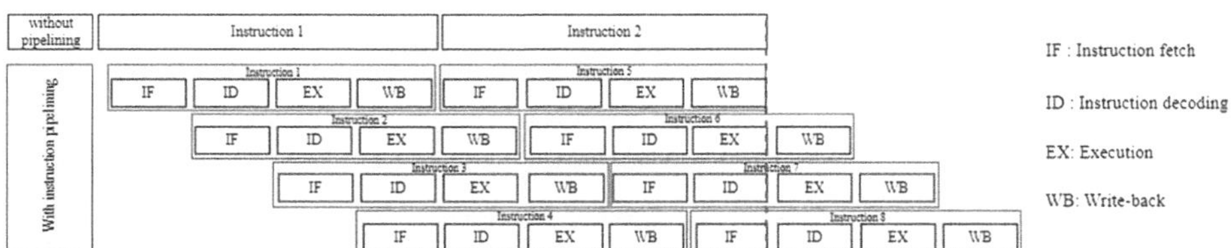

Fig. 1 4-stage pipelining for command on a RISC architecture

$$lim_{x \to \infty} S_\alpha = \lim_{x \to \infty} \frac{n * \alpha}{\alpha + n - 1} = \alpha \tag{4}$$

The productivity of the pipeline can be increased by adding more instructions or steps (deepening or widening of the pipeline). The boundary of the speedup theoretically possible with ILP is linear, which is shown in (4). In reality, it is limited, e.g. by the lack of parallelism in programs, the bandwidth available for instruction fetching, random register access, and the amount of virtual registers needed to be accessed per cycle. In order to overcome the limits of ILP, higher-level parallelism, such as thread-level parallelism, may be applied. Applications are logically structured into threads. Threads are processes with own instructions and data, whereas a thread can be a part of a complex program or an independent program. Karn and Elfadel [12] present an approach for dynamic core auto-scaling at thread level to reduce cycle time. If core utilization is higher than the threshold level, threads are migrated to other cores with less utilization. In [13] Saidi et al. accelerate simulations by parallel execution on several cores while enabling the execution of the sub-simulators at different rates. For this, the authors realized synchronization of time and data exchange. A maximum speedup of $S(4) = 2.91$ was achieved on four cores. Canedo et al. [14] adapt software-pipelining on programmable logic controllers (PLCs) to control loops in industrial control systems to reduce scan cycle time. The three sequential steps, sensing, executing, and actuating of IEC61131-1 control applications are computed partially parallel on multiple cores.

According to Amdahl's law, the performance of parallel execution of applications is limited by bottlenecks, formed by the non-parallel parts of the application. Vertical scaling capability is bounded by the resources of the physical host and is therefore not suitable for resource-constrained edge nodes. Thus, additional computing power in edge environments can only be achieved by horizontal scaling, i.e., by allocating resources of additional edge nodes.

3.2 Multi-Node Scaling

The predominant use case for horizontal scaling in distributed environments is the scaling of web applications. VMs or containers running the applications are replicated to process incoming requests simultaneously to satisfy the desired response times [15]. Unlike web applications, where exceeding the desired response time results in decreased user satisfaction and degradation of the QoS level, control applications have hard real-time requirements that must be satisfied for the overall functioning of the system [16]. Today's cloud scaling frameworks are not suitable for scaling real-time edge applications [17], since important real-time metrics are not monitored and evaluated, monitoring intervals are insufficient, and execution security guarantees are not implemented.

Approaches for multi-node execution for delay-sensitive edge applications are a subject in research today. Wang et al. [18] present a framework for proactive online instance purchase for horizontally scaling delay-sensitive edge applications. According to the predicted workload of the application, VM instances are purchased proactively to avoid time consuming cold starts of additional nodes. The main focus of this work is cost minimization for proactive horizontal cloud resource scaling. The communication and management overhead due to distribution was not considered. Although the start-up latency of the VMs could be reduced, the approach is not sufficient for hard real-time applications. Kronauer et al. [19] analyze the end-to-end latency of multi-node Robot Operating System (ROS) 2 and compare it with single node usage with the result that latency is significantly higher for the distributed applications. The authors pointed out that the latency differences were mainly due to different types of communication and energy saving features of the operating system of the nodes. While inter-process communication via shared memory was used in the single-node samples, data was transferred via non-deterministic publish-subscribe communication over the network in the multi-node environment. Since the authors exclusively used out-of-the-box components of ROS 2 without adaptation, as well as computing nodes without real-time operating systems but with the energy-saving settings activated, the considered tests must be improved with optimized real-time systems and communication.

With horizontal scaling of real-time control applications, synchronization and additional communication overhead, as well as network traffic, are added to the system. Communication speed is limited by bandwidth and network utilization. Multi-node scaling overcomes the limits of vertical scaling.

4 Concept

In this chapter we present a concept for horizontal scaling of real-time control tasks across multiple nodes in a real-time network, connected to a controller (e.g. for a robot). The approach of ILP pipelining shall be applied onto the considered use case. To enable pipelining of the control tasks without modification, parallelism is applied at the application level. Due to the fact that the deadline of the task is greater than the period, since the result must not be available at the controller in the next cycle, pipelining with multiple nodes is reasonable. In our case, the deadline d_i is twice the period P, as shown in (5), allowing higher computation time on each of two nodes by exploitation of functional-level parallelism. Therefore, the tasks shall be executed on two nodes (180° out of phase/with time delay), applying the approach of software pipelining at application level.

$$T_i(\Delta e_i, d_i, P) = T_i(\Delta e_i, 2P, P) \tag{5}$$

In the considered use case, the processing task, if processed by a single edge node, exceeds the cycle time of the robot controller. For the considered robot control to operate, it must receive a result of the processing task every millisecond to act in time. The processing time of the image analysis task exceeds the cycle time of the controller and shall therefore be pipelined on multiple nodes. The data to be processed is sent every millisecond to the edge nodes alternately. Subsequently, the task is executed by the corresponding edge node and the response is sent to the controller.

4.1 Experimential Realization

The experiential setup is depicted in Fig. 2a and corresponds to an industrial network consisting of three computers, representing the robot controller with the camera and the two edge nodes (Edges A and B) for executing the image processing tasks. All three computers run a Linux Debian operating system whereby real-time capability is achieved by kernel preemption implemented by the PREEMPT_RT patch. Time synchronization of the network participants is performed using IEEE 802.1AS-2020. All PCs perform their respective tasks synchronised. In order to realize the transmission in real time, the Time Aware Shaper (TAS) of the IEEE 802.1Qbv standard is used, to ensure the network is exclusively available for the applications for certain time slots. The image data set is sent to Edges A and B alternately. Both nodes communicate the response to the same robot controller. Thus, the controller receives a result every millisecond, see Fig. 2b. As fine-tuning of the schedule is outside the focus of this paper, the time slots are selected to be larger than necessary for the communication of the data so that delays within the switches do not have to be taken into account separately when forwarding.

Each payload of the UDP frames contains an unique identifier (Id) which is identical in request and response, enabling mapping of request and corresponding response. Time stamps when the messages are sent and received by the network card are included in the payload, which is extracted by the controller for verification of the end-to-end latency. Four time stamps are included in every request-response message sequence, see Fig. 3.

The send-receive statistics are analysed to determine the end-to-end latency, the communication time, and the calculation time. The first time stamp SCHED_SEND_TIME corresponds to the scheduled sending time from the controller to the edge nodes. RCV_TIME determines the time the message is received by the network card of the edge node.

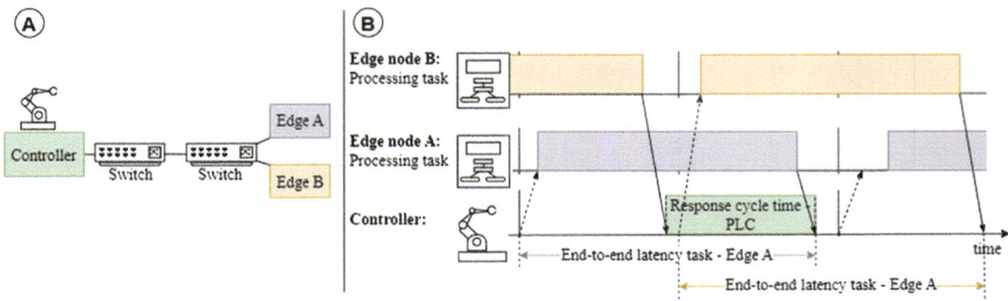

Fig. 2 Executing task across two nodes. **a** Experimental network setup. **b** Principle for pipelining control tasks on two edge nodes

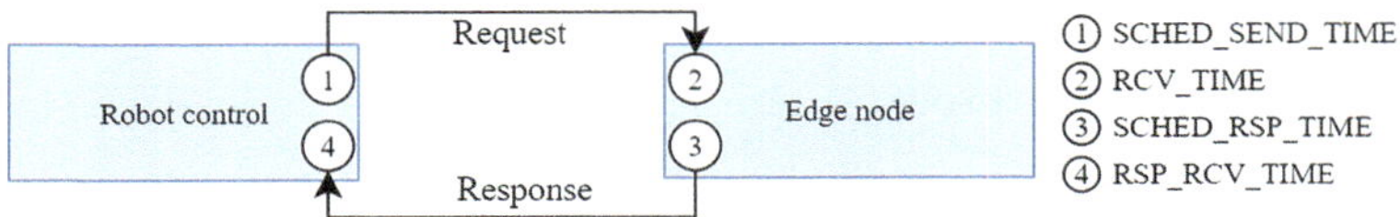

Fig. 3 Time stamps evaluated and corresponding point in transmission process

SCHED_RSP_TIME is the scheduled time for the response sent to the robot controller. RSP_RCV_TIME equals the time stamp of the network card on which the response is received by the controller. The timestamps are evaluated below.

4.2 Verification

To verify the concept presented, the end-to-end latencies of request-response sequences were evaluated. For this, the time stamps in the payload were assigned to points in time of the transmission process, enabling the determination of the end-to-end latency, the cycle time in which results are received by the controller, as well as which percentage of the end-to-end response time consists of transmission time. The end-to-end latency is the time differences between when the response is received by the controller and the request with data is sent to the edge node.

Several time slots of 10 s were evaluated for this purpose. Sensor data is processed every millisecond, resulting in 10000 requests and corresponding responses. No packages were lost. No deadlines were missed. The controller received an evaluation response every cycle with jitter of 60 µs. The end-to-end latency of data processing, including transmission and computation, was always less than the deadline $d_i = 2$ ms. This can be seen in Fig. 4.

4.3 Evaluation

To evaluate the scalability of the system the metrics presented in Sect. 2.1 are used. For this, the speed up and efficiency of the pipelining approach are theoretically and experimentally evaluated. Theoretically, the maximum speedup S_{max} possible is determined by the response time interval the controller receives a response of the task prior T_1 and after T_2 scaling onto

(A)

minimum end-to-end latency	average end-to-end latency	maximum end-to-end latency
1.806 ms	1.843 ms	1.874 ms

(B)

minimum time differences between two responses	average time differences between two responses	maximum time differences between two responses
0.951 ms	0.999 ms	1.049 ms

Fig. 4 **a** End -to-end latency of pipelined tasks. **b** Time differences of responses received by the controller

two nodes. Since the deadline is extended to twice the value, compared to before scaling, the theoretical maximum speedup for the approach with n nodes results in $S_{max} = T_1/T_n$. In the considered case, $S_{max} = 2$. It is assumed that the additional communication overhead caused by multicast from the second node is assumed to be negligible. For the experimental tests, the actual speedup, calculated with the average cycle times resulted in $S_{real} = 1,84$ due to the fact that the maximum end-to-end latencies were not fully utilized. The end-to-end latency of a single processing task was assumed to be equal with one or two edge nodes and is therefore chosen as T_1. The time differences between the responses received by the controller correspond to the controller cycle time, which was achieved with the two-node pipelining approach being T_2.

The multi-node piplining approach for few nodes increases the maximum possible calculation time, since the communication overhead for message transmission is constant. In our experiment the transmission latency resulted in $\Delta t_{trans} = 0,1149$ ms. The maximum amount of computation time tc_{max} available, with two-node pipelining, results from the maximum end-to-end latency and the transmission time. Since only every second response must be provided to the controller by the corresponding edge node, the maximum end-to-end latency for every task results in 2 ms. Therefore, the maximum computation time for the tasks in our experiment is $tc_{p_{max}} = 1,8851$ ms in contrast to $tc_{max} = 0,8851$ ms before scaling. When comparing the ratio $tc_{p_{max}}/tc_{max} = 2,1298$ to the ratio of the task periods $p_1/p_2 = 2$ of the tasks, resulting in more than linear increasing of maximum possible computation time.

4.4 Experiment Discussion and Summary

The experiential results in this chapter show that the pipelining approach at the functional level enables reducing the scan-cycle time at which the responses are received by the controller while allowing longer computation time for each task without modification of the tasks. With the approach, the compute-intensive cyclic real-time task can are processed 180° out of phase on two edge nodes. Therefore, the controller receives a result every cycle even though the processing time of the task on a single node would exceed the cycle time of the controller. Due to constant transmission times and longer deadlines, the maximum possible computation time with this approach is increased further than the quotient of deadline to cycle time.

The limits of this approach are that the concept is only applicable to systems where the response of the processed data must be available several cycles after the request. The number of nodes for pipelining is limited by the time between request and response, the cycle time of the controller and the network bandwidth, limiting the number of requests possible per time instance. Prerequisites for applying the concept to cyclic real-time control tasks are network synchronization, real-time communication and RTOS on nodes to time message sending and processing. In our work, we assumed that the scheduled sending times are similar to the time the frame is actually being sent by the network card. The time stamps of when the message is received by the edges and the controller correspond to the actual receipt times.

Common metrics to evaluate the scalability of distributed systems are not sufficient to evaluate the scalability of real-time systems because important requirements are not assessed and are not included in the evaluation. The Speedup corresponds to the throughput of the system, aiming to maximize the throughput of the system without considering, e.g., deadline-compliance, compliance with the time frame a result must be available, as well as the jitter. No value is placed on adherence to the time frame in which a result is provided. In a cyclic real-time systems results must not be available too early in order to avoid processing wrong values.

5 Conclusion and Outlook

With this work we present and verify an approach to exploit functional parallelism for computation intensive tasks with bounded execution time. For this, we applied the concept of software-pipelining onto real-time tasks in distributed systems, using different nodes as execution units to provide additional resources to the application without being limited by the resources of a single edge node.

Future work could include dynamic horizontal scaling, the development of suitable metrics to evaluate the scalability of distributed real-time systems, and extending the levels of parallelism of the tasks executed.

Acknowledgements The presented work was party funded by the Ministry of Science, Research and the Arts of the Federal State of BadenWurttemberg within the 'Innovation Campus Future Mobility', and by the joint research project SDM4FZI, supported by the BMWK as part of the "Future Investments in the Automotive Industry" funding program, which is gratefully acknowledged.

References

1. Stankovic, J.: Real-time computing. Comput. Sci. Eng. (1992). https://doi.org/10.1007/1-4020-0613-6
2. Liu, C.L., Layland, J.W.: Scheduling algorithms for multiprogramming in a hard-real-time environment. J. Autom. Comput. Mach. **20**(1), 46–61 (1973)
3. Oechsle, S., Walker, M., Fischer, M., Frick, F., Lechler, A., Verl, A.: Real-time capable architecture for software-defined manufacturing. In: Kiefl, N., Wulle, F., Ackermann, C., Holder, D. (eds.) Advances in Automotive Production Technology–Towards Software-Defined Manufacturing and Resilient Supply Chains. ARENA2036, pp. 3–13. Springer International Publishing, Cham (2023). https://doi.org/10.1007/978-3-031-27933-1

4. Singh, P., Kaur, A., Gupta, P., Gill, S.S., Jyoti, K.: RHAS: robust hybrid auto-scaling for web applications in cloud computing. Clust. Comput. **24**(2), 717–737 (2021). https://doi.org/10.1007/s10586-020-03148-5
5. Walker, M., Klingel, L., Oechsle, S., Neubauer, M., Lechler, A., Verl, A.: Safeguarded continuous deployment of control containers through real-time simulation. In: 2023 IEEE 28th International Conference on Emerging Technologies and Factory Automation (ETFA), pp. 1–8. IEEE (2023). https://doi.org/10.1109/ETFA54631.2023.10275659
6. Lechler, A., Riedel, O., Coupek, D.: Virtual representation of physical objects for software defined manufacturing. In: DEStech Transactions on Engineering and Technology Research (ICPR) (2018). https://doi.org/10.12783/dtetr/icpr2017/17652
7. Millnert, V., Eker, J.: Holoscale: horizontal and vertical scaling of cloud resources. In: 2020 IEEE/ACM 13th International Conference on Utility and Cloud Computing (UCC), pp. 196–205. IEEE (2020). https://doi.org/10.1109/UCC48980.2020.00038
8. Al-Dhuraibi, Y., Paraiso, F., Djarallah, N., Merle, P.: Autonomic vertical elasticity of docker containers with elasticdocker. In: 2017 IEEE 10th International Conference on Cloud Computing (CLOUD), pp. 472–479. IEEE (2017). https://doi.org/10.1109/CLOUD.2017.67
9. Jogalekar, P., Woodside, M.: Evaluating the scalability of distributed systems. IEEE Trans. Parallel Distrib. Syst. **11**(6), 589–603 (2000). https://doi.org/10.1109/71.862209
10. Sirhan, N.N., Serhan, S.I.: Multi-core processors : Concepts and implementations. Int. J. Comput. Sci. Inf. Technol. **10**(1), 01–10 (2020). https://doi.org/10.5121/ijcsit.2018.10101
11. Hepola, K., Multanen, J., Jääskeläinen, P.: Dual-is: Instruction set modality for efficient instruction level parallelism. In: Schulz, M., Trinitis, C., Papadopoulou, N., Pionteck, T. (eds.) Architecture of Computing Systems. Lecture Notes in Computer Science, vol. 13642, pp. 17–32. Springer International Publishing, Cham (2022). https://doi.org/10.1007/978-3-031-21867-5
12. Karn, R.R., Elfadel, I.A.M.: Autoscaling of cores in multicore processors using power and thermal workload signatures. In: 2016 IEEE 59th International Midwest Symposium on Circuits and Systems (MWSCAS), pp. 1–4. IEEE (2016). https://doi.org/10.1109/MWSCAS.2016.7870120
13. Saidi, S.E., Pernet, N., Sorel, Y.: A method for parallel scheduling of multi-rate co-simulation on multi-core platforms. Oil & Gas Science and Technology–Revue d'IFP Energies nouvelles **74**, 49 (2019). https://doi.org/10.2516/ogst/2019009
14. Canedo, A., Ludwig, H., Al Faruque, M.A.: High communication throughput and low scan cycle time with multi/many-core programmable logic controllers. IEEE Embed. Syst. Lett. **6**(2), 21–24 (2014). https://doi.org/10.1109/LES.2014.2299731
15. Qu, C., Calheiros, R.N., Buyya, R.: Auto-scaling web applications in clouds. ACM Comput. Surv. **51**(4), 1–33 (2019). https://doi.org/10.1145/3148149
16. Walker, M., Fischer, M., Lechler, A., Verl, A.: Evaluation of isolation and communication mechanisms for real-time containers. In: 2023 IEEE 32nd International Symposium on Industrial Electronics (ISIE), pp. 1–8. IEEE (2023). https://doi.org/10.1109/ISIE51358.2023.10228012
17. Ju, L., Singh, P., Toor, S.: Proactive autoscaling for edge computing systems with kubernetes. In: Bittencourt, L.F., Sill, A. (eds.) Proceedings of the 14th IEEE/ACM International Conference on Utility and Cloud Computing Companion, pp. 1–8. ACM, New York, NY, USA (2021). https://doi.org/10.1145/3492323.3495588
18. Wang, W., Liu, L., Yan, Z.: Proactive auto-scaling for delay-sensitive iot applications over edge clouds. IEEE Internet Things J. 1 (2023). https://doi.org/10.1109/JIOT.2023.3324546
19. Kronauer, T., Pohlmann, J., Matthe, M., Smejkal, T., Fettweis, G.: Latency analysis of ros2 multinode systems. In: 2021 IEEE International Conference on Multisensor Fusion and Integration for Intelligent Systems (MFI), pp. 1–7. IEEE (2021). https://doi.org/10.1109/MFI52462.2021.9591166

Simulating Material Deposit for Fiber Sprayed Composites Using Beta Paint Distribution

Fabian Viessmann, Lukas Wagner, Georg Puchas, Stefan Schafföner and Dominik Henrich

Abstract

Fiber reinforced composites are often manufactured using cost-effective fiber spraying processes. As quality fluctuates during manual manufacturing, a robot-based automation is advantageous. The automation could be greatly improved by a prior simulation of the material deposit to minimize overspray and to ensure a homogeneous material distribution. Even so, currently, specific simulation models for fiber spraying processes are unavailable. Thus, this paper investigates in which respect existing spraying process models are able to predict the material deposit, as they do not include variables such as fibers and compressed air. As a use case, we consider a thermal spraying model for the manufacturing of short fiber reinforced oxide fiber composites. We evaluate the model by simulating the deposit on a planar workpiece. The evaluation reveals that this model can only be applied to a limited extent. For higher accuracy, the model must consider further fiber spraying characteristics.

Keywords

Simulation • Material deposit • Fiber spraying process • Fiber sprayed composites • Industrial robotics • Beta paint distribution

F. Viessmann (✉) · D. Henrich
Chair for Robotics and Embedded Systems, University of Bayreuth, Bayreuth, Germany
e-mail: fabian.viessmann@uni-bayreuth.de

L. Wagner · G. Puchas · S. Schafföner
Chair of Ceramic Materials Engineering (CME), University of Bayreuth, Bayreuth, Germany

M.-C. Wanner et al. (eds.), *Annals of Scientific Society for Assembly, Handling and Industrial Robotics 2024*, https://doi.org/10.1007/978-3-031-91463-8_15

1 Introduction

Fiber reinforced composites are lightweight materials that offer valuable properties, such as a high strength-to-weight ratio, fatigue resistance, and design flexibility [1]. Accordingly, these materials are becoming increasingly important and are used today in numerous fields, such as energy, automotive, space and aeronautics or civil engineering [2]. Often they are manufactured by a *fiber spraying process* (FSP) since an FSP is flexible and cost-effective [3]. Widely established examples are glass fiber composites [4] and recently also ceramic matrix composites, such as *oxide fiber composites* (OFC) [5]. However, these processes have traditionally been carried out manually, leading to a larger variability of the composite quality. As a result, robotic automation has been the focus of research in recent years to improve production costs, throughput, accuracy and reproducibility [3].

In order to obtain ceramic matrix composites with damage tolerant fracture behavior, the coating thickness during spraying must be as homogeneous as possible. Furthermore, to reduce material waste, the amount of material that does not reach the surface of the mold or does not provide any functionality must be minimized (*overspray*). Therefore, the robot must be programmed to fulfill these two criteria. For example, the fulfillment can be achieved by an expensive trial-and-error strategy. However, a simulation of the material deposit resulting from a robot trajectory before the manufacturing can greatly improve the cost-efficiency.

In this paper, we examine the extent to which the material deposit can be predicted in a robot-based FSP using existing models for spraying processes in general. We consider our automation approach for the manufacturing of short fiber reinforced OFC [5] as a use case, in which the programming is based on the intuitive playback programming paradigm [6]. For our automation approach, we therefore apply an existing simulation model used for thermal spraying [7]. As our approach utilizes pressure regulators, we extend the model to include compressed air influence. The goal is to investigate the applicability of the model in terms of the accuracy of the simulated deposit. For an FSP, the deposit mainly consists of fibers which are not taken into account by prior models explicitly. Therefore, Sect. 2 gives an overview of the related work regarding the automation of FSPs and simulation models for spraying processes. In Sect. 3, we explain the concrete model and its application to our FSP, taking compressed air into account. In Sect. 4, we evaluate the model in terms of the accuracy of the simulated thickness. At the end, Sect. 5 summarizes and concludes our paper.

2 Related Work

Fiber spraying systems are widely commercially available [8–10]. However, most of them have to be operated manually like a spray paint gun, which exposes the operator to air pollution. Furthermore, the result is not reproducible and very error-prone. A robot-based automation would overcome these disadvantages.

To the best of our knowledge, there are only two robot-based FSPs. One approach uses off-line programming, which combines graphical programming and virtual reality. The system has a two-axis turntable for greater flexibility [11]. However, this programming framework is not suitable for small batch sizes since it requires experts in robot programming. In our automation approach for manufacturing short fiber reinforced OFC, non-experts program the robot using the intuitive playback programming paradigm [5]. Therefore, the programming consists of two phases [6]. First, the operator guides the robot kinesthetically while the robot configurations are recorded at a constant frequency. Subsequently, the robot program can be played back and edited intuitively [12]. Our approach also uses a two-axis turntable for better flexibility and can be extended to other kinds of FSPs, like glass fiber spraying [5].

Both approaches do not simulate the material deposit before the manufacturing, so that there is no simulation model for FSPs available. Therefore, we consider other general spraying process models. These models are divided into microscopic and macroscopic models [7]. Microscopic models deal with the simulation at the particle level, while macroscopic models abstract from individual particles and consider mass flow. Since it is sufficient for us to consider the mass flow, and since this is possible in a short computing time, we will only deal with this type of model. An overview of robot spray painting models is given by Chen et al. [13].

For simplified situations with a constant spraying distance, orthogonal spraying, and a planar workpiece, modeling the material distribution is sufficient. Thus, the first models describe this distribution by summing several Gaussian functions [14, 15]. Less complex is the beta distribution model, which allows a clear choice of parameters depending on the spray gun [16, 17] and which has been extended to 2D [18]. In order to predict the material deposit on curved surfaces, a deposition model plane is utilized which also considers different spraying distances and angles [14, 19, 20]. Previously, a comparable model was employed that predicted thermal spraying processes well [7].

In summary, there are many models available for material deposit simulations. However, it is still an open question whether these can be applied to an FSP. They also do not consider the influence of fibers and the compressed air. To fill this gap, we are investigating the applicability of the thermal spraying model [7] to our FSP and extend it in terms of the compressed air influence.

3 Methodology

We aim to predict the material deposit on the workpiece as part of our FSP for OFC [5] by applying an existing simulation model. In such an FSP, continuous oxide fiber bundles (roving) are chopped to uniform length by a cutting unit and then ejected out of the spray gun into a slurry spray. The fiber bundles are infiltrated with the slurry during the flight. Immediately afterwards, the fiber slurry mixture reaches the surface of a mold [21]. The resulting thickness results primarily from the fibers and secondarily from the slurry.

The thickness depends on the *robot trajectory* defined as a sequence $\mathcal{C} = (q_0, q_1, \ldots, q_n)$ of joint configurations $q_i \in \mathbb{R}^d$ with d joints of the robot at each timestep i. A joint configuration q_i is transformed into the *tool center point* (TCP) $p_i \in \mathbb{R}^3$ in workspace by using the forward kinematics of the robot. It is assumed that the TCP lies directly in front of the spray gun nozzle and represents the apex of an *elliptic spray cone*. The spray cone has an axis a_i that points into the interior of the cone. Furthermore, to control the material flow and the spray pattern, we use five pressure regulators [6]. The *round jet* and *flat jet* influence the elliptic spray pattern, while the *slurry value* defines the slurry volume flow. The *pilot air* serves for the de-/activation of the material flow ejected out of the spray gun, and the *cutter value* influences the compressed air that controls the cutting unit speed. For the simulation, we only consider the influence of different round and flat jet values. The slurry value is assumed to be constant, as this does not influence the fiber deposit. The pilot air is assumed to be activated, while the cutter value for a uniform fiber length is also assumed to be greater than zero and constant. Thus, we define the round jet value as p_r and the flat jet value as p_f, which are constant for all timesteps.

To simulate the material deposit, a CAD model of the workpiece must be available. As for the thermal spraying model [7], we use a triangular mesh. This allows us to store the material deposit at each vertex. So our workpiece $\mathcal{W} = (V, F)$ is a set of vertices $V = \{v_0, v_1, \ldots, v_m\}$ with $v_j \in \mathbb{R}^3$ and faces $F \subset V^3$. Each vertex v_j has an associated normal n_j and all faces must have a consistent anti-clockwise orientation. Since we store the material deposit at the vertices v_j, we assume that the nodes v_j are uniformly distributed and close to each other, which is a requirement for a precise simulation.

3.1 Footprint Concept

For our FSP, we apply a simulation model used for thermal spraying [7], which utilizes a so-called *footprint*. A footprint describes the material distribution in the spray cone and can later be used to calculate the final material deposit on the workpiece. The footprint is determined by spraying orthogonally with a fixed position and orientation of the spray gun onto a flat surface for a given duration, distance to the workpiece, and pressure values. The material distribution of this footprint is determined by sampling the thicknesses, which, in our case, mainly consists of fibers compared to thermal spraying [7].

Generally, the footprint is modeled using a deposition model plane, called *footprint plane*, which is orthogonal to the axis a_i of the spray cone and shifted from the TCP by an offset λ in the direction of a_i. The shift by λ can be chosen arbitrarily, as this only scales the distribution in the footprint plane. The plane has its own coordinate system, where the intersection of a_i and the plane defines the origin. The x- and y-axis correspond to the half-axes of the elliptical spray pattern and depend on the spray gun. In the following, we call them $x_F, y_F \in \mathbb{R}^3$ defined in workspace. The footprint plane is shown in Fig. 1.

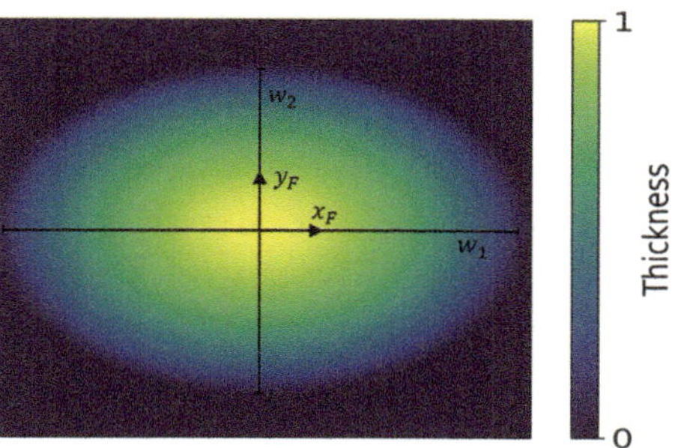

Fig. 1 An exemplary footprint plane with normalized deposit is shown here. The plane has its own coordinate system, whereby the origin is the intersection between the spray cone axis a_i and the plane. The two axes x_F and y_F are also the half-axes of the elliptical spray pattern. Within this plane, the spray cone has a diameter of w_1 along x_F and w_2 along y_F

While the thermal spraying model [7] uses a sum of Gaussian distributions to describe the material distribution, we use a *bivariate beta paint distribution* [18]. The beta paint distribution is sufficient for us, as we assume that our distribution is unimodal and not rotated by any covariances. Furthermore, we modify it slightly so that the material deposit outside the spray cone is zero and that two diameters describe the spray cone size. The formula is given in Eq. 1.

$$T_{C,\beta_1,\beta_2,w_1,w_2}(x, y) = \begin{cases} C \cdot (1 - \frac{4x^2}{w_1^2})^{\beta_1 - 1} \cdot (1 - \frac{4y^2}{w_2^2 \cdot (1 - \frac{4x^2}{w_1^2})})^{\beta_2 - 1}, & \frac{4x^2}{w_1^2} + \frac{4y^2}{w_2^2} < 1 \\ 0, & \frac{4x^2}{w_1^2} + \frac{4y^2}{w_2^2} \geq 1 \end{cases} \quad (1)$$

Therefore, the thickness $T_{C,\beta_1,\beta_2,w_1,w_2}(x, y)$ at point $(x, y)^T$ in footprint coordinates depends on the following parameters: maximum thickness C, the shaping parameters β_1 and β_2, and the diameters w_1 and w_2 of the spray pattern within the footprint plane in x_F-/y_F-direction. In the following, the distribution is abbreviated as $T(x, y)$ if an explicit specification of the parameters is not necessary.

At this point, we extend the footprint concept to model the influence of p_r and p_f. Therefore, we determine our so-called *default footprint* with a fixed round and flat jet. This footprint has the parameters $\hat{C}, \hat{\beta}_1, \hat{\beta}_2, \hat{w}_1, \hat{w}_2$. When the round and flat jet change, we scale the default footprint to obtain a new distribution with the parameters $C, \beta_1, \beta_2, w_1, w_2$. Thus, we calculate new diameters by functions $w_1 = f(p_r, p_f)$ and $w_2 = g(p_r, p_f)$. These functions are defined for our spray gun in Sect. 3.4. To maintain the shape of the distribution, we set $\beta_1 = \hat{\beta}_1$ and $\beta_2 = \hat{\beta}_2$.

However, as the sum of the material deposit remains constant for all possible footprints, we must calculate a new maximum thickness C. Therefore, we choose C so that the sum of the material deposit is always equal to the sum of the material deposit of the default footprint, as the amount of mass flow does not change. Let I_d be the integral value of our default footprint and I_s the integral value of our scaled footprint. We determine the parameter C as given in Eq. 2.

$$C = \frac{I_d}{I_s} \cdot \hat{C} \tag{2}$$

3.2 Simulation Model

Using the footprint plane, we calculate the material deposits at each vertex $v_j \in \mathcal{V}$. This calculation is heavily based on the thermal spraying model [7]. Therefore, we have to sum up the individual material deposits resulting from each robot configuration $q_i \in \mathcal{C}$ for a surface point $v_j \in \mathcal{V}$. Given a robot configuration q_i with corresponding TCP p_i, the spray direction is calculated by $d_{i,j} = v_j - p_i$. The material deposit at point v_j is determined by projecting point v_j onto the footprint plane along the spray direction $-d_{i,j}$ with corresponding coordinates $(\pi_x(p_i, d_{i,j}), \pi_y(p_i, d_{i,j}))^T$. We calculate these as given in Eq. 3.

$$\begin{aligned} \pi_x(p_i, d_{i,j}) &= \left(p_i + \frac{\lambda}{d_{i,j}^T a_i} \cdot d_{i,j}\right)^T \cdot x_F \\ \pi_y(p_i, d_{i,j}) &= \left(p_i + \frac{\lambda}{d_{i,j}^T a_i} \cdot d_{i,j}\right)^T \cdot y_F \end{aligned} \tag{3}$$

These coordinates are used as input for the beta paint distribution (Eq. 1). Overall, this projection generates skewed material distributions on the workpiece, even though the beta paint distribution is not skewed.

After this step, the material deposit is scaled by $a(n_j, d_{i,j})$, given in Eq. 4.

$$a(n_j, d_{i,j}) = \begin{cases} \frac{(-d_{i,j})^T \cdot n_j}{||d_{i,j}||_2 \cdot ||n_j||_2}, & (-d_{i,j})^T \cdot n_j > 0 \\ 0, & (-d_{i,j})^T \cdot n_j \leq 0 \end{cases} \tag{4}$$

As the sprayed area increases with increasing spray angle, less material needs to be applied per point as the mass flow remains the same. If the spray angle is equal to or greater than 90°, the material deposit is zero. In addition, we scale the material deposit with the spray duration $\hat{t}_i = \frac{t_{i-1}+t_i}{2}$ and we also consider the distance from the TCP p_i to the point v_j. The final simulation model for determining a material deposit z_j at v_j is given by Eq. 5. The model is visualized in Fig. 2.

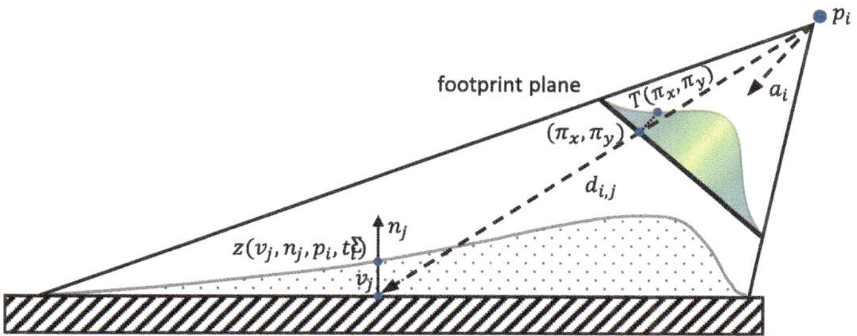

Fig. 2 The material deposit at v_j is determined by projecting v_j onto the footprint plane by using $\pi_x(p_i, d_{i,j})$ and $\pi_y(p_i, d_{i,j})$ (abbreviated as π_x and π_y). These coordinates serve as input to $T(\pi_x, \pi_y)$ to determine the unscaled deposit which is then scaled by further factors to obtain the final deposit $z(v_j, n_j, p_i, \hat{t}_i)$

$$\begin{aligned} z_j &= \sum_{i=0}^{n} z(v_j, n_j, p_i, \hat{t}_i) \\ &= \sum_{i=0}^{n} \left(\frac{a(n_j, v_j - p_i)}{||v_j - p_i||_2^2} \cdot \hat{t}_i \cdot T(\pi_x(p_i, v_j - p_i), \pi_y(p_i, v_j - p_i)) \right) \end{aligned} \tag{5}$$

3.3 Determination of Our Footprint

To determine our default footprint, we sprayed orthogonally with a fixed position and orientation of the spray gun onto a flat surface for two seconds, at a distance of 300mm, and with $p_r = 5$bar and $p_f = 0$bar. The slurry value was set to 1bar, the pilot air was activated, and the cutter value was set to 3.5bar. Afterwards, we sampled the spray pattern consisting of fibers and slurry with a resolution of 100mm^2. The K450 electronic external measuring gauge (Kroeplin Längenmesstechnik, Germany) was used for this measurement. The device has an error of 0.05 mm. The process was repeated three times, and the mean was considered. Figure 3a depicts the final averaged spray pattern.

To determine the bivariate beta paint distribution (Eq. 1) for our FSP, we fitted the simulation model (Eq. 5) to the averaged spray pattern. We used the Levenberg-Marquardt algorithm, minimizing the mean squared error (MSE). For the footprint plane we choose $\lambda = 50$ mm. Figure 3b shows the resulting spray pattern when simulated through the fitted footprint. The residuals are depicted in Fig. 3c. Overall, the MSE is 0.017 mm with a standard deviation of 0.107 mm. The parameters of the fitted bivariate beta paint distribution (our default footprint) are listed in the following: $\hat{C} = 1.293\,614\,37 \times 10^5$ mm, $\hat{\beta}_1 = 7.84$, $\hat{\beta}_2 = 7.98$, $\hat{w}_1 = 26.67$ mm, $\hat{w}_2 = 23.33$ mm.

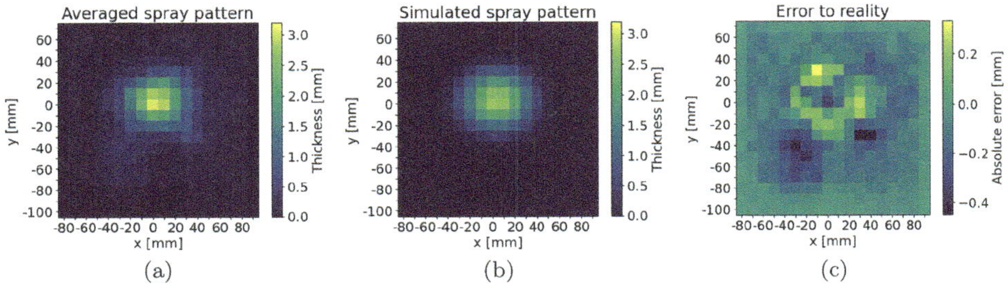

Fig. 3 We fitted the model (Eq. 5) into our averaged spray pattern (Fig. 3a). The fitted footprint generates the simulated spray pattern (Fig. 3b). The residuals are depicted with a differently scaled color bar for a better visualization (Fig. 3c)

3.4 Determination of the Round and Flat Jet Influence

As mentioned, the round jet value p_r and flat jet value p_f influence our default footprint. Therefore, we describe w_1 and w_2 by functions: $w_1 = f(p_r, p_f)$ and $w_2 = g(p_r, p_f)$. To determine these two functions for our FSP, we carried out several experiments with all possible combinations of $p_r \in \{4, 5, 6\}$ bar and $p_f \in \{0, 0.25, 0.5, 0.75\}$ bar. A further change is not useful because the overspray becomes larger, leading to slurry back splash on the robot. This could block the cutting unit of fiber rovings. All other parameters are the same as in Sect. 3.3.

We measured the diameters of all spray patterns and calculated the opening angles of the spray cones in both the x_F- and y_F-direction. We used these angles to determine w_1 and w_2 within the footprint plane. Overall, it resulted in a sampling of the two functions in a rectangular grid, which is why we use bilinear interpolation to determine further values. The results are shown in Fig. 4.

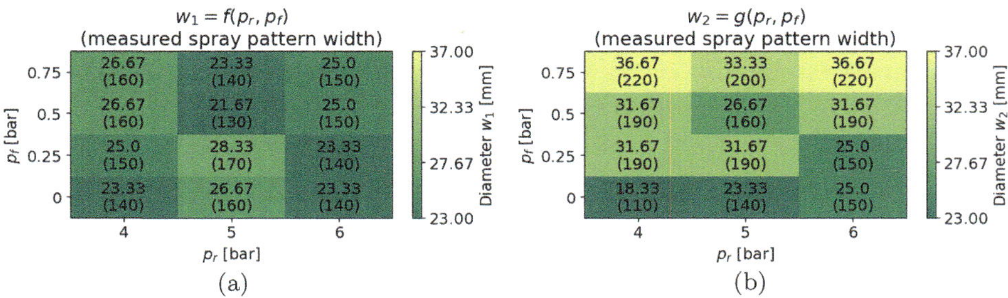

Fig. 4 We carried out several footprint experiments (Sect. 3.3), but with variation of the round/flat jet to measure the influence on the diameters $w_1 = f(p_r, p_f)$ (Fig. 4a) and $w_2 = g(p_r, p_f)$ (Fig. 4b). These are shown in each cell, with the measured spray pattern size at a spraying distance of 300 mm in brackets

In Fig. 4a, it is noticeable that w_1 remains relatively constant regardless of p_r and p_f. The fluctuations in the values were probably due to noise. On the one hand this is due to the fact that we only performed each experiment once for a first trend correlation. On the other hand, measuring the diameter is very difficult as it is not possible to identify the boundaries of the spray pattern clearly. In Fig. 4b, the effect of p_r on the diameter w_2 is similar. However, there is a clear tendency for the diameter w_2 to increase as p_f increases.

4 Evaluation and Discussion

We evaluated the model with our real FSP based on a six degree of freedom robot [5] and a meander trajectory size of 300x300 mm^2 to spray a 500x500 mm^2 plate. The trajectory was parallel to the plate at a distance of approximately 430 mm and was traveled at an average speed of 0.16 m/s. The angle between the spray cone axis and the plate was 65°. The round/flat jet was $p_r = 5$ bar and $p_f = 0.25$ bar. The slurry value was set to 1 bar, the pilot air to activated and the cutter value to 3.5 bar. We scanned the deposit using the Shining 3D EinScan Pro HD 3D scanner (Shining 3D Tech Co., Ltd, China) with a scan accuracy of up to 0.045 mm. Afterwards, we determined the real thickness by comparison with a reference scan. The relevant sections of the plate are shown in Fig. 5a for the real deposit and in Fig. 5b for the difference between the simulated to the real deposit. In both figures, the motion started at top left and ended at bottom right, with vertical paths. The robot was positioned at the bottom center.

The maximum thickness of the plate (Fig. 5a) is 3.56 mm, with an average of 1.6 mm. There are accumulation points at the lower and upper ends of the meander path as the robot's speed decreases and more material is applied. The difference of the simulation to reality (Fig. 5b) is on average −1.24 mm and at maximum −3.43 mm at the accumulation points. This means that too little material is simulated consistently. Therefore, we scaled the simulated thickness to minimize the difference (Fig. 5c). For this, scaling the simulated

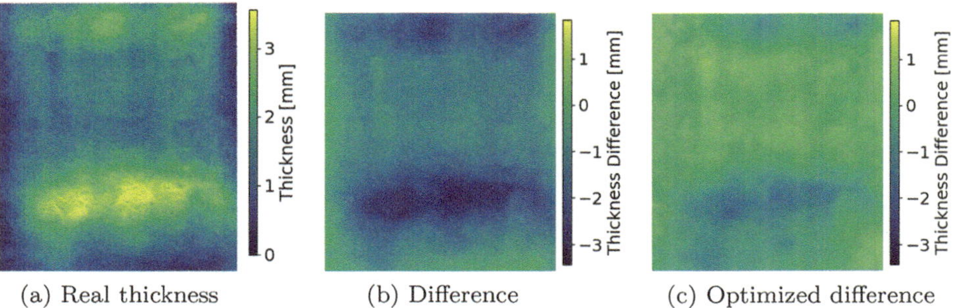

(a) Real thickness (b) Difference (c) Optimized difference

Fig. 5 The difference of the (scaled) simulated thickness (Fig. 5b, c) to the real thickness (Fig. 5a) resulting from a meander trajectory is shown here

thickness with a factor of 4.9 is optimal. The maximum deviation is −0.71 mm on average and 1.84 mm at most.

Thus, scaling the deposit results in an error reduction. In order to map this increased mass flow more realistically, the footprint calibration should be carried out with a moving robot. The good results of the calibration (Sect. 3.3) can only be transferred to stationary sprayings. However, as this scaling also introduced an error, FSPs are not entirely represented by this model. For example, we observed that the deposit did not scale linearly with the spray duration, and the pattern diameters w_1 and w_2 increased with the spray duration.

5 Conclusion

In this paper, we contributed the application of a footprint based simulation model to our fiber spraying process (FSP) for short fiber reinforced oxide fiber composites. We used the beta paint distribution for our footprint, which yields good calibration results. The evaluation shows that for a moving robot and a planar workpiece, the application of the model is limited, as too little deposit is simulated. As a result, stationary spray gun calibration is insufficient when spraying is performed during motion. In addition, the model must consider further characteristics of FSPs, like the spray duration's impact on the material deposit and on the spray cone size. Further experiments are needed to evaluate the effect of the compressed air.

Acknowledgements This work was funded by the Deutsche Forschungsgemeinschaft (DFG, German Research Foundation) - 518255159 (FlexFiber).

References

1. Prashanth, S., et al.: Fiber reinforced composites - a review. J. Mater. Sci. Eng. **06**(03), 341 (2017)
2. Erden, S., Ho, K.: Fiber Technology for Fiber-Reinforced Composites. Woodhead Publishing, Fiber Technology for Fiber-Reinforced Composites (2017)
3. Winkelbauer, J., et al.: Short fiber spraying process of all-oxide ceramic matrix composites: A parameter study. Int. J. Appl. Ceram. Technol. **20**(2), 754–767 (2022a)
4. Witten, E., Mathes, V.: Der europäische Markt für Faserverstärkte Kunststoffe/Composites 2023. AVK e.V, Marktentwicklungen, Trends, Herausforderungen und Ausblicke (2023)
5. Schmidt, E., et al.: Robot-Based Fiber Spray Process for Small Batch Production. Annals of Scientific Society for Assembly, Handling and Industrial Robotics (2020)
6. Schmidt, E., Henrich, D.: Playback Robot Programming Framework for Fiber Spraying Processes. Annals of Scientific Society for Assembly, Handling and Industrial Robotics (2021)
7. Hegels D (2017) Optimierung thermischer Verhältnisse bei der Bahnplanung für das thermische Spritzen mit Industrierobotern. Dissertation, Universität Dortmund

8. Faserspritzanlagen Wolfangel. https://wolfangel.com/faserspritzanlagen/
9. Faserspritzanlagen BÜFA Composite Systems. https://buefatec.de/product_info.php?products_id=956
10. Graco FRP Faserspritzanlage. https://www.pultex.de/de/Graco_FRP_Faserspritzanlage/
11. Robotergesteuerte Faserspritzanlage. https://www.wki.fraunhofer.de/de/fachbereiche/hofzet/profil/technische-ausstattung/Faserspritzanlage.html
12. Schmidt, E., Henrich, D.: Intuitive Optimization of Kinesthetic Programmed Trajectories for Fiber Spraying. Presented at the (2022)
13. Chen, Y., et al.: Paint thickness simulation for painting robot trajectory planning: a review. Ind. Robot: An Int. J. **44**(5), 629–63 (2017)
14. Conner, D.C., et al.: Paint deposition modeling for trajectory planning on automotive surfaces. IEEE Trans. Autom. Sci. Eng. **2**(4), 381–392 (2005)
15. Zhou B, et al.: Off-line programming system of industrial robot for spraying manufacturing optimization. In: 33rd Chinese Control Conference (CCC) (2014)
16. Balkan, T., Arikan, M.A.S.: Modeling of paint flow rate flux for circular paint sprays by using experimental paint thickness distribution. Mech. Res. Commun. **26**(5), 609–617 (1999)
17. Arikan, M.A.S., Balkan, T.: Modeling of paint flow rate flux for elliptical paint sprays by using experimental paint thickness distributions. Ind. Robot: An Int. J. **33**(1), 60–66 (2005)
18. Zhang, Y., et al.: New model for air spray gun of robotic spray-painting. Chinese J. Mech. Eng. **42**(11), 226–233 (2006)
19. Xia, W., et al.: Paint deposition pattern modeling and estimation for robotic air spray painting on free-form surface using the curvature circle method. Ind. Robot: An Int. J. **37**(2), 202–213 (2010)
20. Andulkar, M.V., Chiddarwar, S.S.: Incremental approach for trajectory generation of spray painting robot. Ind. Robot: Int. J. **42**(3), 228–241 (2015)
21. Winkelbauer, J., et al.: Short fiber-reinforced oxide fiber composites. Int. J. Appl. Ceram. Technol. **19**, 1136–1147 (2022)

Robotics in Production and Service

Topology Optimization of Large Robot Structural Components

Andre Siegrist, Alexander Jentsch and Jan Sender

Abstract

Robots must have a high degree of rigidity in order to achieve high absolute accuracy. This depends on both the drive trains and the structural components itself. The rigidity is pose- and orientation-dependent for both serial kinematics and hybrid structures, meaning that different load cases must be considered when developing structural components. In addition, hybrid structures have more complicated boundary conditions due to larger interference contours, which do not occur in this form with serial kinematics. In order to counter this complex design space overlap of the components involved in the development of new robot kinematics, a systematic approach using topology optimization was established. The topology optimization generates a load-optimized material distribution within boundary conditions. By superimposing separate load case solutions, a load-optimized component design can be generated efficiently. In the present case, the component design of a welded steel robot component already in use with is compared with the manual design of a cast aluminium structure and a topology-optimized cast aluminum structure. With the topology optimization, the component mass was significantly reduced and the rigidity greatly increased, so that an economical use has been proved.

A. Siegrist (✉) · A. Jentsch
Fraunhofer-Institut für Großstrukturen in der Produktionstechnik, Rostock, Germany
e-mail: andre.siegrist@igp.fraunhofer.de
URL: https://www.igp.fraunhofer.de

J. Sender
Lehrstuhl für Produktionsorganisation und Logistik, Universität Rostock, Rostock, Germany

M.-C. Wanner et al. (eds.), *Annals of Scientific Society for Assembly, Handling and Industrial Robotics 2024*, https://doi.org/10.1007/978-3-031-91463-8_16

1 Introduction

The development of structural components is an essential part of the prototype development of highly rigid large robot systems with a payload of 2,500 kg at the Fraunhofer IGP in cooperation with the University of Rostock. The stiffness in the entire workspace is dependent on position and orientation and can lead to chatter marks and poor surface qualities in the context of the intended application of the robot system, mechanical post-processing [10, 12]. The rear use of coupling gears for robot actuation on the second and third main axes creates overlapping working spaces for the coupling elements, resulting in strong dependencies in the shaping of the individual components. Adjustments to the structural components required in the advanced design process therefore have an enormous influence on the surrounding structure, which can only be countered with great effort, which is why an efficient development process to save resources is all the more important. A systematic design process for the integration of numerical optimization was used for the design changes made late in the development process with regard to the manufacturing process and material selection.

Numerical optimization can be performed in a variety of ways using different approaches. A study of publication content over the last 20 years has shown that FE-based methods account for the largest proportion of publications in the field of shape optimization, followed by Iso Geometric Analysis-based (IGA) methods [16]. Topology optimization (TO) can be assigned to both areas depending on the design [6]. FE-based topology optimization uses finely resolved solids and the applied boundary conditions to produce a load-compliant contour, which results from the boundary conditions and the target functions in the form of e.g. stress, strain or weight restrictions. In [17] the authors optimized parts of a vehicle wheel suspension with the aid of TO and achieved better dynamic properties by saving 41 % in weight. In the field of optimization of robot structures, [11, 15] dealt with the optimization of individual structural components of serial kinematics and were able to show that topology optimization is a feasable solution. The authors in [15] used the stress and deformation information of the individual components and at the tool center point (TCP) as an indicator for increased absolute accuracy of a robot system with solid structural components. In [18], the authors achieved higher stiffnesses for a serial robot system by splitting the optimization into a structural component optimization and a parametric system optimization based on a stiffness model of the robot and a metamodel of the structural components. Furthermore, in [14] the authors showed the general potential of topology optimization with respect to robots with degrees of freedom ranging from 1 to 25, where the achieved mass reductions ranged from 5% to 80%.

2 Methodology

For the application of the development process (Fig. 2) for new structural components based on an existing prototype (Fig. 1), boundary conditions are necessary in a first phase, which are transferred from the first prototype. Thereafter, numerical methods are applied in the phases "2. concept development" and "3. design implementation" in the form of multi-body simulation, topology optimization and FE simulation [1, 2, 13].

The kinematics and the respective installation spaces of the individual robot components have to be defined in the concept development phase. A multibody simulation (MBS) is carried out for the kinematics development. After the MBS, installation spaces must be developed based on the CAD model of the whole robot, which form the basis for optimization through topology optimization. The robot consists of spindle-driven linkage mechanisms for the main axes 2 and 3 (Fig. 3) and also a modular wrist. In the course of the further development of the system, the stiffness of the coupling gears proved to be essential for the overall stiffness of the system. At first conservative methods for the development of the mechanical structure were used to achieve a reference design based on the previous prototype (Fig. 1), which then had to be revised due to changes in the boundary conditions. Due to the moving interference contours in the form of structural components and the components of the spindle drives, the design space of the individual structural components was determined using a MBS and topology optimization was then used to revise the structural components in order to ensure optimum space utilization of the compact design space.

The design space defines the space that is available for a component. For static assemblies, the installation space is simple to define. In robot development, this is much more complex. The problem is that components occupy several positions in different configurations in relation to each other. The procedure developed for optimizing large robots also includes the point of design space definition. As shown in Fig. 1, the large robot is equipped with

Fig. 1 Heavy-duty robot prototype of the Fraunhofer IGP with rear coupling gears for actuating the second and third main axes

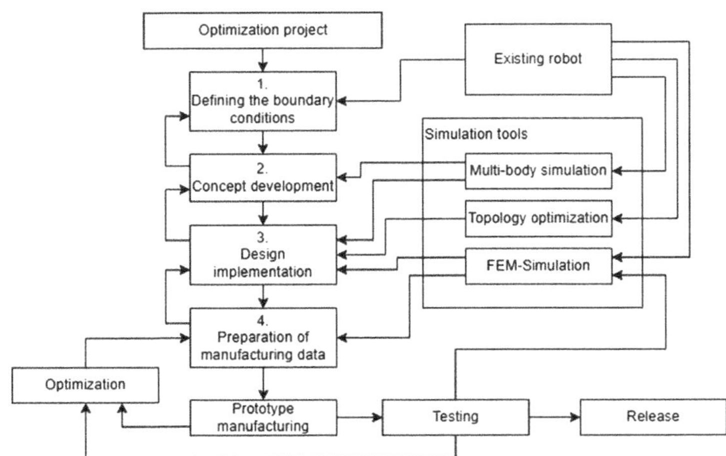

Fig. 2 Part of the systematic approach to the material-efficient optimization of robotic structures based on the development process of lightweight design projects [13]

a slewing ring, tower, tower jib, three-point link 1, three-point link 2, two-point link 1, two-point link 2, linear axis 1 (spindle drive axis 2), linear axis 2 (spindle drive axis 3). The components tower, three-point link 1 and 2, two-point link 1 and 2 as well as the liner axles are located in the same design spaces in the different poses and therefore have overlapping spaces. In order to be able to identify these overlapping spaces, the design components and their corresponding design spaces are analyzed in a multi-body analysis in the different poses and enveloping surfaces are extracted. The enveloping surfaces represent the movement space of the respective components. Subsequently, the enveloping surfaces are examined for overlaps with each other. The example three-point link 2 is shown on the right-hand side of Fig. 3. Due to the position- and orientation-dependent stiffness of the system, reference poses were defined in order to evaluate the overall stiffness of the robot system over the course of development. These are shown in Fig. 4. These poses resulted from the previous prototype in which ship propellers were marked with a tool mounted on the TCP. The propellers are marked on both the pressure and suction sides in a single clamping operation, resulting in the main poses shown for the robot system. In addition to the load on the system with an outstretched hand, the case of the angled robot hand was also taken into account in order to consider torsional influences due to the later free orientability of a payload. For a stiffness evaluation, both the payload of 2,340 kg and additional external loads on the TCP of 8.5 kN in x-, y- and z-direction were taken into account. The first two natural frequencies and the deformations on the TCP were evaluated for comparison between different design revisions.

During the development process, the structural optimization ensures an optimal mechanical shape of the parts in relation to their requirements and boundary conditions. The overall

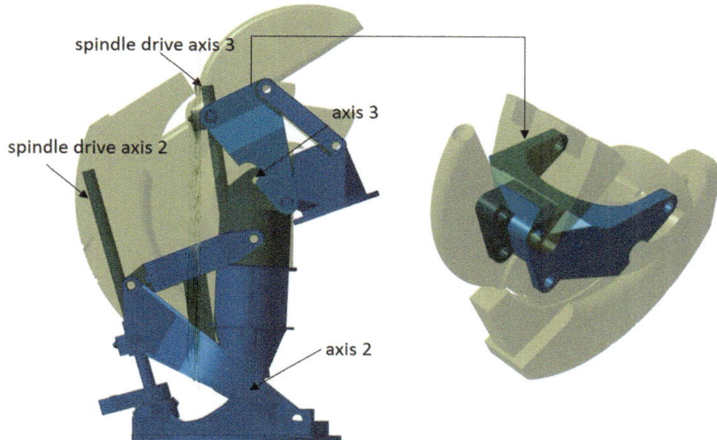

Fig. 3 Intermediate development status of the robot structure without the arm and hand axis structure (blue) including the swivel areas of the structural components when moving over the full axes' angle range (beige), right: three-point linkage 2 (blue) within the interference contour areas of surrounding structural components (beige)

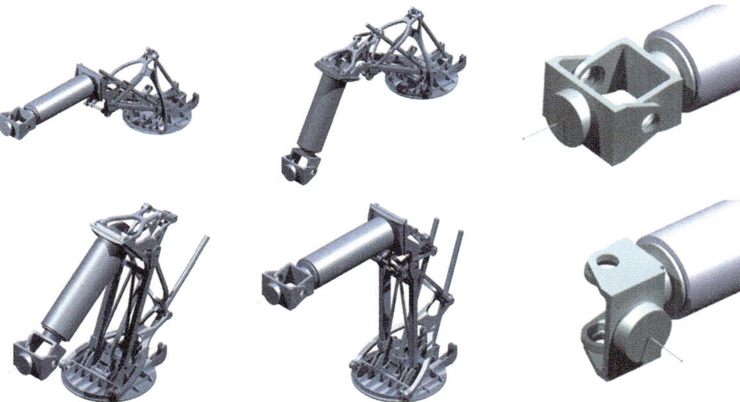

Fig. 4 Model of the first prototype whose 4 poses were used as reference poses for the structural design (top left, top center, bottom left, bottom center) to evaluate the stiffness of the system under static load; top right and bottom right: Illustration of the hand axis orientation used to evaluate the torsional stiffness under static load conditions

stiffness of the robot has to be broken down to the single part stiffness. The necessary stiffness of three-point linkage 2 is derived from the reference robot. With the design space and the load case definition, the three parameters form the basis for optimization. There are different optimization methods in structural optimization. The simplest method describes dimensioning; this achieves optimization by varying the component parameters. Shape optimization varies the contour of the component. The method used in the present case, topology optimization, considers the entire design space and attempts to achieve an optimal distribution of the material in the design space [8, 13].

There are a number of other methods within topology optimization. The SIMP method uses the density of the elements as a design parameter and links this to the young's modulus. In order to obtain a clear interpretation of the material distribution, the density and the young's modulus are normalized to their initial values and a penalty value is used. The penalty value p controls the presence of elements with lower density [7].

The basic relationship between initial young's modulus E_0, young's modulus E, initial density ρ_0, density ρ and the penalty factor is described in Eq. 1 [3].

$$E_e = E_0 * (\frac{\rho}{\rho_0})^p \tag{1}$$

The design space is defined at the initial stage of every topology optimization. Afterwards, the optimization loop starts. The first step is the FE analysis followed by a sensitivity analysis (Fig. 5). A low-pass filter is used to prevent the formation of a checkerboard pattern. If the criteria are met, the optimization loop ends [3–5].

Three load cases are defined as relevant for three-point link 2.

The three load cases are considered separately in the topology optimization and then combined as an overall solution. The three partial solutions can be seen on the left-hand side of Fig. 6. The superposition of these three individual solutions leads to the overall solution of the optimization (Fig. 6). The generated geometry is interpreted manually by the designer. The manufacturing restrictions of the casting process must be considered in the design.

To avoid stress peaks due to unfavorable force dissipation, the focus in the area of highly loaded components is on load paths that are as linear as possible. If this is not possible or

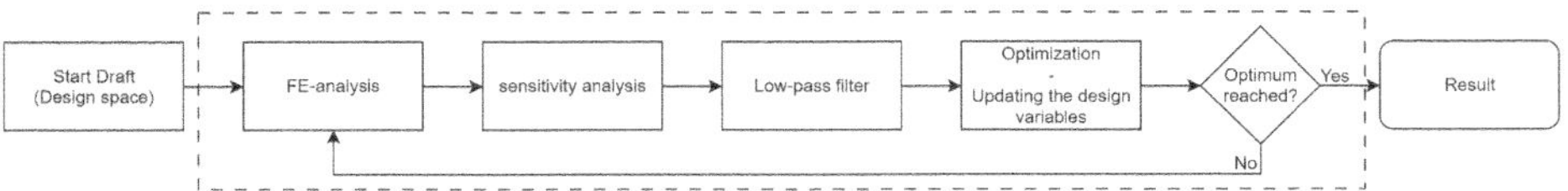

Fig. 5 Sequence of the topology optimization of the SIMP approach [3]

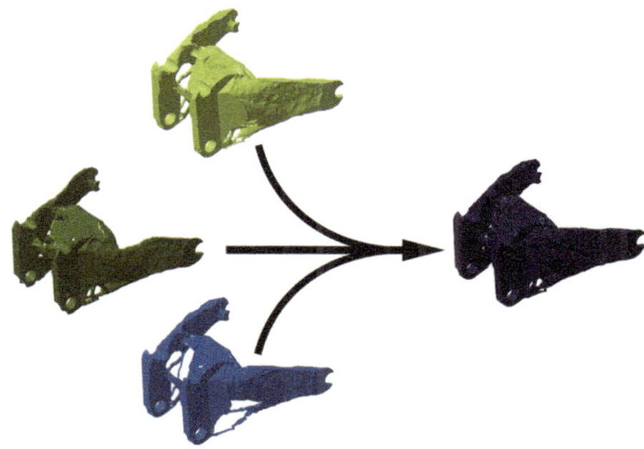

Fig. 6 Superposition of the partial solutions (left) as a basis for the design (right)

if the direction of force causes moments in the component, tangent-continuous curves are used to avoid stress peaks [13].

Results

The interpretation of the optimized topology results in the part on the right-hand side of Fig. 7. In Fig. 7 there are also shown on the left-hand-side the three-point-linkage 2 of the reference robot and the manual designed linkage 2. Considering the displacement shown in Fig. 7, it can be noted that the optimized three-point-linkage 2 has a lower displacement as the manual designed and the reference linkage. To check the optimization results, in Table 1 the welded design (reference robot) is compared in normalized weight and displacement with the manually designed and optimized designed linkage. An increase in mass was deliberately accepted in order to significantly improve the component stiffness of the

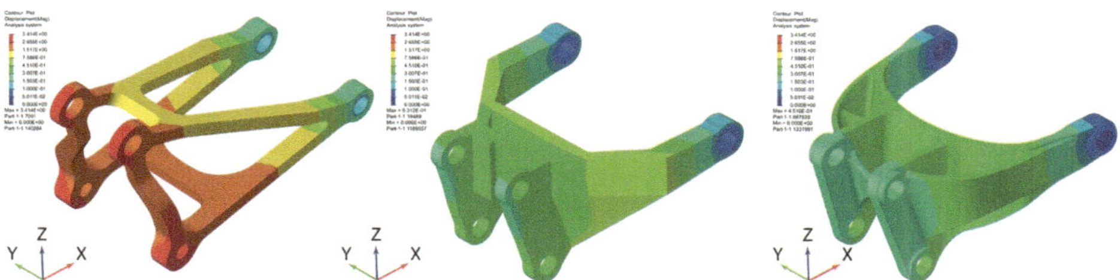

Fig. 7 Deformation illustration with the same scale of iteration steps in structural component development, left: Welded steel structure from an initial development, center: manually optimized cast aluminum structure without taking the manufacturing process into account, right: topology-optimized and derived cast aluminum structure [9]

Table 1 Comparison of normalized component weights and deformations

	Welded design (St)	Cast design (Al, manual designed)	Cast design (Al, optimized designed)
Normalized weight	1	2,44	2,16
Normalized displacement	1	0,17	0,13

reference component. The results show that component optimization carried out by an experienced designer can already achieve significantly improved results in terms of component stiffness (see Table 1). However, the use of a topology optimization leads to an additional increase in stiffness and significantly lighter components. In the present case, a mass saving of 11.5% and a gain in stiffness of 23.5% compared to the manually optimized design was achieved. The results show, numerical methods, like the topology optimization, can make an enormous contribution to maximizing component stiffness while minimizing material usage. Compared to the welded steel construction, the component deformations could be reduced by up to 87 % with almost double the weight. Compared to the manually produced cast construction, the topology-optimized construction saved around 11.5% in weight and increased rigidity by around 23.5%.

References

1. Andeen, G.B. (ed.): Robot design handbook. McGraw Hill, New York, NY (1988)
2. Auat Cheein, F.A., Cabrera, P.P., Fantoni, G. (eds.) Rapid roboting: Recent advances on 3D printers and robotics, Intelligent Systems, Control and Automation: Science and Engineering, vol. 82. Springer, Cham (2022)
3. Bendsøe, M.P.: Topology Optimization: Theory, Methods, and Applications. Engineering Online Library, 2nd edn. Springer, Berlin, Heidelberg (2004)
4. Fiebig, S.: Form- und Topologieoptimierung mittels Evolutionärer Algorithmen und heuristischer Strategien. Logos Verlag Berlin GmbH, Berlin (2016)
5. Franke, T.: Fertigungsgerechte Bauteilgestaltung in der Topologieoptimierung auf Grundlage einer integrierten Gießsimulation. Dissertation, Technische Universität Braunschweig and Logos Verlag Berlin, Berlin (2018)
6. Gao, J., Xiao, M., Zhang, Y., Gao, L.: A comprehensive review of isogeometric topology optimization: Methods, applications and prospects. Chin. J. Mech. Eng. **33**(87) (2020)
7. Harzheim, L., Graf, G.: Topshape: an attempt to create design proposals including manufacturing constraints. Int. J. Veh. Des. **28**(4), 389 (2002)
8. Henning, F., Moeller, E.: Handbuch Leichtbau: Methoden, Werkstoffe, Fertigung. Hanser, München, 2., überarbeitete und erweiterte auflage edition (2020)

9. Jentsch, A., Dryba, S., Klötzer, C., Siegrist, A., Vinçon, A. :Methodische Strukturentwicklung eines großroboters/research project flexgrind. wt Werkstattstechnik online, **111**(09), 628–632 (2021)
10. Josler, L.N., Möllensiep, D., Kuhlenkötter, B.: Einfluss der Frässpindel-Orientierung auf die Bearbeitungsgenauigkeit beim fräsen mit industrierobotern. Ind. 4.0 Manag. **2022**(5), 53–56 (2022)
11. Kouritem, M.I., Abouhef, S.A., Nahas, N., Hassan, M.: A multi-objective optimization design of industrial robot arms. Alexandria Eng. J. **61**, 12847 – 12867 (2022)
12. Produktionstechnik Hannover informiert. Fräsen und bohren: Ein roboter lässt die späne fliegen - produktionstechnik hannover informiert (2018)
13. Schumacher, A.: Optimierung mechanischer Strukturen: Grundlagen und industrielle Anwendungen. Springer Berlin Heidelberg, Berlin, Heidelberg, 3., aktualisierte auflage 2020 edition (2020)
14. Srinivas, G.L., Javed, A.: Synthesis and performance evaluation of manipulator-link using improved weighted density matrix approach with topology optimization method. Eng. Sci. Technol. Int. J. **24**, 1239–1252 (2021)
15. Shanmagusandur, G., Sivaramakrishnan, R., Meganathan, S., Balasubramani, S.: Structural optimization of an five degrees of freedom (t-3r-t) robot manipulator using finite element analysis. Mater. Today: Proc. **16**(2), 1325–1332 (2019)
16. Upadhyay, B.D., Sonigra, S.S., Daxini, S.D.: Numerical analysis perspective in structural shape optimization: a review post 2000. Adv. Eng. Softw. **155** (2021)
17. Viqaruddin, M., Ramana Reddy, D.: Structural optimization of control arm for weight reduction and improved performance. Mater. Today: Proc. 9230–9236 (2017)
18. Wang, X., Zhang, D., Zhao, C., Zhang, P., Zhang, Y., Cai, Y.: Optimal design of lightweight serial robots by integrating topology optimization and parametric system optimization. Mech. Mach. Theory **132**, 48–65 (2019)

Integrative Multi-robot System for Advanced Manufacturing OPTIMA: Combining Additive and Subtractive Processes

Jan Schachtsiek, Patrick Adler, and Bernd Kuhlenkötter

Abstract

This paper introduces an integrative multi-robot system for advanced manufacturing, presenting an innovative approach that addresses contemporary challenges in the manufacturing landscape. By seamlessly integrating additive and subtractive manufacturing processes, namely wire-based laser deposition welding and robotic milling, within a highly automated cell, the system offers a novel solution with the potential to advance industrial production. Delving into the system's architecture, the paper explores the intricate details, emphasizing key characteristics such as functional integration, joint programming, and digital twin. Specific functionalities are elucidated to provide a clearer understanding of the system's capabilities. Research perspectives outlined in the paper shed light on potential advancements in simultaneous additive and subtractive operations, providing insights into the system's capabilities. The process is exemplified through a detailed case study of a steel beam connection piece, illustrating the integration of LMD-w and milling. This example highlights the practical application and addresses challenges such as precise parameter generation, adaptive control strategies, and load- and pose-dependent stiffness compensation. The comprehensive analysis of the OPTIMA system underscores its versatility in aerospace applications, the automotive industry, and customized production. This paper serves as a detailed guide to the integrative multi-robot system, offering in-depth insights into its design, functionalities, and specific contributions to the field of advanced manufacturing. Exploring both research perspectives and practical applications contributes to a holistic understanding of the system's significance in shaping the future of industrial production.

J. Schachtsiek (✉) · P. Adler · B. Kuhlenkötter
Ruhr-University Bochum, Bochum, Germany
e-mail: schachtsiek@lps.ruhr-uni-bochum.de

M.-C. Wanner et al. (eds.), *Annals of Scientific Society for Assembly, Handling and Industrial Robotics 2024*, https://doi.org/10.1007/978-3-031-91463-8_17

Keywords

Multi-robot system • Additive manufacturing • Subtractive manufacturing • Robot-milling • Production systems

1 Introduction

In response to the evolving demands of modern manufacturing, this paper introduces an integrative multi-robot system for advanced manufacturing—an innovative approach poised to redefine the landscape of industrial production. This innovative system seamlessly intertwines wire-based laser deposition welding (LMD-w) for additive manufacturing and robotic milling for subtractive operations within a singular, highly automated cell, which is named "Optimized production through integrated multi-robot additive and subtractive processes" (OPTIMA) [1]. Addressing contemporary manufacturing challenges is centered on efficiency, complexity, and automation. The integration of laser deposition welding in conjunction with machining processes has already established itself as a promising technology. This has been demonstrated successfully by combining similar methodologies [2].

The OPTIMA system aspires to expedite production processes in response to conventional manufacturing constraints, leading to significant cost savings through faster manufacturing. The simultaneous execution of additive and subtractive processes aims to eliminate sequential bottlenecks, fostering agile and responsive production and further enhancing the system's efficiency and economic benefits.

Transitioning to the additive manufacturing aspect, LMD-w is a sophisticated process that employs a laser heat source to melt and fuse a continuous wire feed onto a substrate, forming a metallurgical bond [3]. This technology offers notable advantages, such as precise control over the deposition rate and the ability to deposit materials with varying compositions [4]. The wire feedstock, typically in the form of a metallic wire, is fed into the laser-generated melt pool, where it undergoes rapid solidification [5]. This method allows for the exact layering of material, facilitating the creation of complex geometries and enhancing the efficiency of material usage [6]. The versatility of wire-based laser cladding makes it a compelling choice for applications ranging from surface modifications and repairs to the additive manufacturing of intricate components in various industries [7].

However, one significant drawback of wire-based laser cladding is its tendency to produce suboptimal surface quality. The process may result in surface irregularities and roughness, impacting the final finish of the manufactured components. Hence, additional machining steps, such as milling, are often necessary to refine the surface quality and meet stringent dimensional requirements in post-processing [8].

While wire-based laser cladding offers improvements in additive manufacturing, its true potential emerges when integrated with advanced automation. This constructive interaction emphasizes the pivotal role of robotic systems in modern manufacturing. The

OPTIMA system exemplifies this integration, highlighting the fusion of innovative manufacturing technologies and software-defined robotic automation facilitated by industrial robots from ABB. The degrees of freedom (DoF) granted to these robotic entities hold promise in precisely navigating complex geometries [9]. Including a rotary-tilt positioner, further augments these DoF, establishing a flexible manufacturing environment poised to address the intricacies of contemporary design challenges. In pursuing enhanced automation and efficiency in manufacturing processes, there is a compelling drive to achieve significant cost and time savings. The OPTIMA system emerges as a transformative solution, not merely automating processes but introducing the concept of simultaneous additive and subtractive operations. Simultaneous processing leads to time and, therefore, cost savings. The system's capacity to mitigate idle times and optimize resource utilization aligns with the imperative for increased automation.

The development of this integrative robotic cell is primarily motivated by the recognition that traditional manufacturing system constraints can be overcome through flexibility and adaptability. The need to innovate stems from increasingly complex design requirements, which demand manufacturing solutions capable of efficiently handling intricate geometries and producing components. This innovative approach not only fulfills the automation demand but also underscores the significance of technological advancement and adaptability in addressing the evolving challenges of contemporary manufacturing.

2 System Architecture

This section delves into the system's core elements, emphasizing the seamless integration of LMD-w and robotic milling while also highlighting the extensive DoF afforded by the industrial robots.

Integrating two pivotal manufacturing processes, LMD-w and robotic milling, is central to the system's architecture. The industrial robots (*ABB IRB 6700-150/3.20*) are strategically configured to specialize in either additive or subtractive operations. This specialization avoids redundancy, with one robot dedicated to the laser welding process and the other to milling.

The robot dedicated to laser welding maneuvers a laser welding head, allowing for intricate layer-by-layer material deposition. As previously elucidated, wire-based laser cladding is an additive manufacturing process employing a laser heat source to fuse continuous wire onto a substrate precisely [5].

The *Precitec COAX Printer laser deposition head* (6 kW power input) is crucial in the examined manufacturing cell. The integrated Crossjet serves a dual role, supplying protective gas and safeguarding the optics. Near the wire exit point, there is also an exit nozzle through which shielding gas, such as helium, argon, or nitrogen, can flow (see Fig. 1). The quotient pyrometer at the wire output ensures precise temperature measurements between

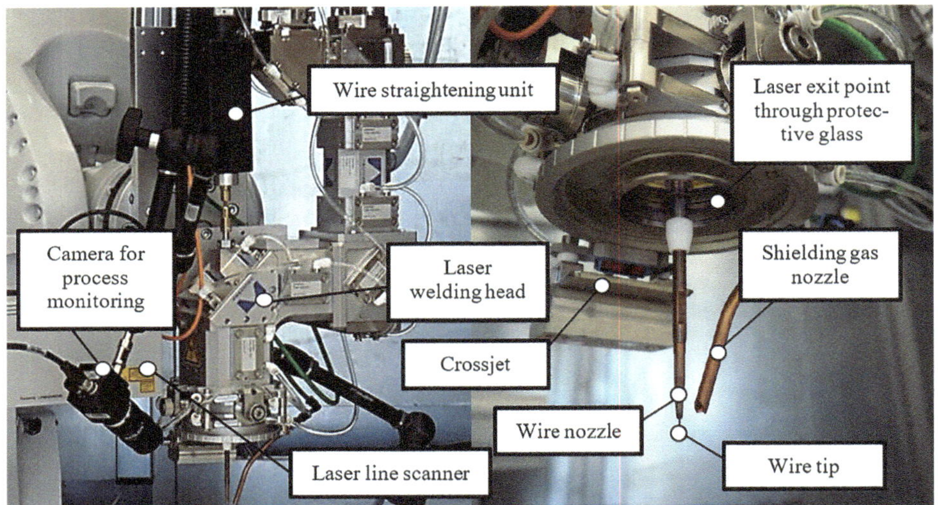

Fig. 1 Components of the laser welding head

600 and 2,300 °C. Additionally, the deposition head is equipped with a laser line scanner for component recognition, facilitating automated comparisons between intended and actual structures. This functionality also allows for the identification and subsequent additive repair of damaged components. Moreover, a process monitoring camera is mounted to ensure continuous surveillance during the welding process.

Within the capabilities of the manufacturing cell, the second process is the subtractive procedure, executed by the *HITECO PX-2 15/12 24 63E NC* milling spindle. This milling spindle is affixed to the second industrial robot, complementing the LMD-w executed by the first robot. The milling spindle integrates Minimal Quantity Lubrication (MQL) and a Force-Torque Sensor (FTS) to achieve a high-quality milling outcome. MQL, a precision lubrication method, minimizes friction, prolongs the machining tool's lifespan, and enhances the overall efficiency of the milling process [10]. Simultaneously, the FTS ensures monitoring of forces and torques exerted during the machining operation.

Both robots, mounted on a shared linear axis within the cell, provide exceptional DoF. This configuration allows the system to navigate intricate geometries, opening new possibilities for creative design solutions and versatile production capabilities. Additionally, incorporating the shared linear axis enables each robot to move linearly by approximately five meters within the enclosed workspace. The increased workspace, spanning up to 2 m × 3 m, further underscores the system's ability to process large-scale components with precision.

In addition to its physical manifestation, the OPTIMA system is complemented by a digital counterpart, facilitating simulation and process validation. This digital counterpart mimics the manufacturing cell virtually, allowing for extensive testing and optimization

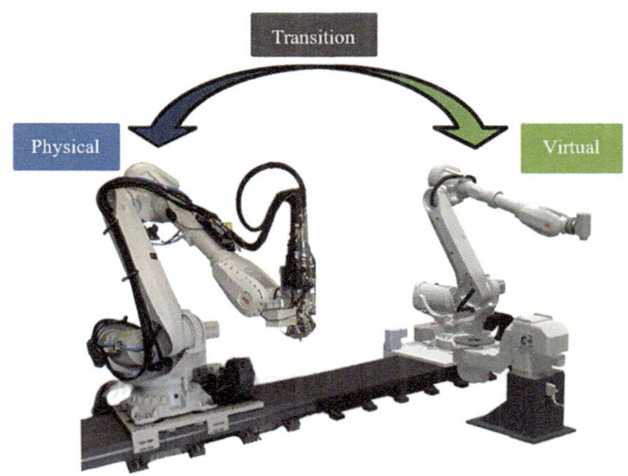

Fig. 2 Synthesized representation of the robotic cell: uniting physical and virtual realms

of production processes. It plays a crucial role in improving efficiency, minimizing downtime, and ensuring optimal production outcomes. Figure 2 provides a depiction of the interior of the robotic cell, capturing both its physical and virtual environment.

Ensuring the safety of operators and environmental mitigation, the entire system is enclosed within a protective structure. This enclosure not only safeguards personnel from potential hazards associated with the 6 kW laser, particularly in relation to laser radiation but also supports the loading of sizable components through strategically placed L-doors. The dimensions of the protective enclosure, approximately 8 m $\times$ 4 m $\times$ 4 m, are designed to balance the need for safety and the requirement for ample space for robot maneuverability.

3 Research Perspectives

The OPTIMA system introduces a change in thinking in manufacturing, raising intriguing research questions at the intersection of additive and subtractive processes. This section delves into potential research avenues, leveraging the features of the system's architecture while combining virtual and physical representation.

3.1 Detailed Methods for Robot-Based Machining: Process Parameters and Adaptive Control

The OPTIMA system highlights the need for more detailed methods in robot-based machining, especially concerning the generation of process-specific parameters and adaptive control strategies. Key aspects include the precise definition of machining parameters such as feed rates, jerk, and acceleration to ensure process stability and efficiency. Moreover, there is a significant opportunity to advance adaptive process control strategies that adjust parameters in real-time based on dynamic machining conditions. Indispensable methods for successful robot-based milling, such as pose-dependent controller parameterization, are crucial in this context [11]. These methods enable the adaptation of control parameters to the specific pose of the robot, enhancing precision and mitigating errors associated with variable robot positions. Additionally, the integration of mechatronic models "in the loop" is essential to determine the chatter limits, providing a dynamic framework that can predict and manage chatter during the milling process. This integration is vital for maintaining surface quality and tool life by avoiding destructive vibrations.

Furthermore, the development of correction fields for the (passive) compensation of load- and pose-dependent stiffness addresses the inherent flexibility issues of robotic arms [12]. By compensating for these variations, achieving higher accuracy and repeatability in machining tasks is possible. These correction fields are calculated based on real-time data, allowing the system to dynamically adjust and compensate for any deviations. These innovations are essential for achieving high precision and stability in robot-based machining processes. Research in this domain aims to establish comprehensive models and real-time control mechanisms to optimize machining performance and address dynamic process variations.

The primary objective of these efforts is to address the skilled labor shortage by developing a process that requires no human intervention, either in the machining process or in parameter selection. The aim is to create a fully autonomous system that improves efficiency and precision in manufacturing, thereby contributing to the advancement of industrial production technologies.

3.2 Optimal Process Parameters for Simultaneous Operations

Investigating the optimal parameters for concurrent additive and subtractive operations within an integrated robotic system presents a significant research endeavor. Achieving a balance between these processes requires a thorough exploration of factors such as laser power, feed rates, and toolpath strategies. Simultaneously executing these divergent processes introduces complexities and challenges. For instance, the production of machining chips during milling may disrupt the laser deposition process, potentially affecting the quality of the deposited material. Moreover, vibrations generated by the milling process can impact the stability and precision of laser welding operations. A comprehensive analysis of the interaction between additive and subtractive processes is essential to address these challenges. Minimizing interference and optimizing overall system performance are the primary goals. Understanding the interdependence of these processes is crucial to enhance the efficiency of the integrated manufacturing cell.

This research dimension is pivotal for improving the efficacy of concurrent operations and advancing the fundamental comprehension of the intricate dynamics between additive and subtractive manufacturing within integrated robotic systems.

3.3 Implementation of Topology-Optimized Structures

Pursuing topology-optimized components within the OPTIMA system introduces distinctive challenges that deviate from conventional manufacturing paradigms. Traditional manufacturing methods often struggle or entirely fail to produce components with complex, topology-optimized geometry due to inherent limitations in toolpath generation and material removal technologies.

In the context of the presented system, the challenge lies not only in the additive and subtractive processes individually but in their combined efforts to realize intricate, topology-optimized structures. Topology optimization often results in unconventional geometries characterized by intricate internal features, undercuts, and varying material densities. The challenges encompass the programming intricacies of toolpaths and the coordination of material deposition and removal to align with the desired topology. This innovative manufacturing approach's considerations for heat dissipation, material properties, and structural integrity become paramount.

4 Potentials and Applications: A Comprehensive Analysis

Chapter Four delves into a detailed examination of practical applications and potentials inherent in the OPTIMA system, following our exploration of potential research avenues in Chapter Three. This chapter provides an overview of the possibilities and real-world implications of the described technology.

In the domain of aerospace applications, the OPTIMA system stands out as an exemplary solution, uniquely suited to meet the rigorous demands of the sector. Its capacity to achieve lightweight structures through topology optimization is pivotal for aerospace components where weight considerations are paramount. Furthermore, the system's ability to navigate complex and intricate geometries, independent of conventional constraints, enhances its attractiveness for aerospace applications. This characteristic aligns with the need for diverse and intricately designed components within the aerospace domain, providing a compelling solution that seamlessly integrates advanced manufacturing capabilities with the specific requirements of space exploration.

Additionally, the system's integration of LMD-w and robotic milling offers substantial advantages to the automotive industry. This integration provides unprecedented DoF in the production process, enabling the fabrication of intricate and topology-optimized structures. The system's flexibility is significant in addressing the automotive landscape's

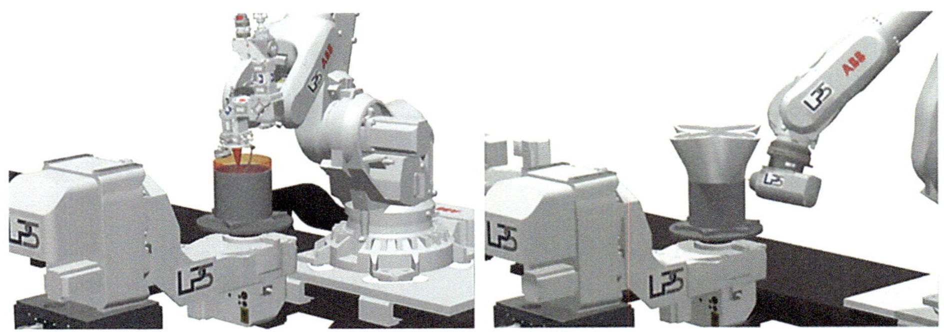

Fig. 3 Visualization of the manufacturing process for a complex steel beam connection piece: on the left-hand side, the LMD-w-process displays the gradual buildup of the component with the indicated welding path. On the right-hand side, the milling process is illustrated

evolving design challenges. Moreover, the integration of real-time process monitoring within the system ensures precision and quality control throughout the manufacturing process. This becomes crucial in industries such as automotive, where stringent safety and performance standards dictate the quality of manufactured components.

Transitioning to a more detailed exploration of the capabilities of the OPTIMA system, the focus narrows to a specific component that exemplifies the system's manufacturing prowess. Illustrated in Fig. 3, a complex steel beam connection piece serves as a prime example of the type of components that could be seamlessly manufactured within the cell. Unlike traditional manufacturing methods, where such components may require multiple steps and assembly processes, the OPTIMA system enables the fabrication of this intricate structure in a single piece, eliminating the need for vulnerable weld seams and reducing the risk of structural weaknesses.

A systematic approach to the steel beam connection piece demonstrates the integration of LMD-w and milling processes. The process begins with the LMD-w process, where the robotic system constructs the initial structure according to a CAD model. This phase builds up the geometric foundation and complex features necessary for the final component. Once the additive process is completed, the component transitions to the milling process. The milling process serves several specific functions depending on the requirements of the component. For the described example, milling is utilized primarily for tolerance adjustment of critical features such as connection interfaces to ensure the component meets the exact geometric and dimensional specifications necessary for its intended application.

This approach is particularly advantageous for manufacturing the steel beam connection piece compared to other traditional methods. Techniques such as machining from solid stock or casting are less suitable for this type of component. Machining from solid stock involves substantial material removal, leading to significant material waste and increased energy consumption. Casting, on the other hand, requires the creation of molds, which is not only time-consuming and costly but also less flexible for producing complex geometries and custom designs. In contrast, the combined LMD-w and milling process is resource-efficient and minimizes material waste. Additive manufacturing precisely deposits material only where needed, reducing excess material usage. This method also allows for energy to be used more efficiently, as it is applied directly to the creation of the structure rather than being expended in removing unnecessary material. Consequently, the approach not only conserves resources but also aligns with sustainable manufacturing practices by optimizing material and energy usage.

The milling process is directed by a digital twin model, which enables real-time adjustments to milling paths, tool speeds, and feed rates based on sensor feedback to achieve the desired surface finish and dimensional accuracy. In order to achieve high tolerances through the milling process, it is necessary to determine and adaptively control the milling parameters, for example, to reduce chatter marks and other deficits (see Sect. 3.1). By carrying out a natural frequency analysis of the robot in advance, for example, natural frequencies can be determined [12]. If the milling process is then controlled in such a way that these frequencies are not excited, the milling result is noticeably improved. In the future, this process is intended to be fully automated in a closed-loop system, eliminating the need for experienced specialists. This advanced approach will leverage real-time feedback and adaptive control strategies to continuously optimize the manufacturing process, ensuring consistent quality and precision without human intervention.

The OPTIMA system's control architecture is central to this integration, managing the coordination of LMD-w and milling processes. The control system schedules operations and adjusts process parameters to ensure a seamless transition between LMD-w and milling. Real-time sensor data maintains dynamic alignment and precision throughout the manufacturing process. This approach demonstrates how the OPTIMA system effectively combines LMD-w and milling into a cohesive manufacturing strategy, addressing advanced manufacturing challenges and offering a versatile solution for high-precision component production across various industrial applications.

5 Conclusion

In conclusion, this paper has delved into the intricacies of the OPTIMA system, elucidating its architecture, capabilities, and potential applications. The seamless integration of LMD-w and robotic milling within a highly automated cell, without human interaction, presents a promising solution to contemporary challenges in industrial production. The examination of simultaneous additive and subtractive operations through a systematic approach applied to a steel beam connection piece demonstrates the system's practical application and addresses challenges such as precise parameter generation, adaptive control strategies, and compensation for load- and pose-dependent stiffness. These insights highlight the system's versatility and its contributions to advancing manufacturing processes. The comprehensive analysis of potentials and applications further underscores the system's adaptability in the aerospace, automotive, and consumer product industries.

Looking forward, several avenues for future research emerge. Firstly, exploring advanced control strategies to enhance the coordination between additive and subtractive processes could optimize overall system performance. Investigating materials beyond metals, such as composites, could expand the system's applicability to a broader range of industries. Additionally, the impact of the OPTIMA system on sustainability and environmental considerations warrants further investigation. Assessing the system's energy efficiency and potential for reducing material waste could contribute to more sustainable manufacturing practices.

Further research could also delve into refining process parameters, leveraging artificial intelligence like machine learning algorithms for real-time optimization, and addressing any limitations that may arise during the system's implementation.

In conclusion, the OPTIMA system significantly advances current manufacturing capabilities and catalyzes future innovations. The outlined research perspectives and potential applications pave the way for a more in-depth exploration of this transformative technology in the years to come.

Acknowledgements Parts of this Project were supported by the Federal Ministry for Economic Affairs and Climate Action (BMWK) on the basis of a decision by the German Bundestag under grant number KK5055223 and by the German Federal Ministry of Education and Research (BMBF) and the Ministry for Culture and Science of the State of North Rhine-Westphalia (MKW) under grant number NW1081004. The acquisition of the robotic cell was financially supported by the German Research Foundation (DFG) under grant reference number INST 213/1027-1 FUGB.

References

1. Glogowski, P., Thiele, M., Kuhlenkötter, B., Ostendorf, A.: Roboterbasiertes Laserauftragschweißen und Fräsen—Additive und subtraktive Fertigung. In: DVS Congress 2023 Manuskript (2023)
2. Du, W., Bai, Q., Zhang, B.: A novel method for additive/subtractive hybrid manufacturing of metallic parts. Procedia Manuf. **5**, 1018–1030 (2016)
3. Qingqing, L., Chen, J., Wang, X., Yang, L., Jiang, K., Yang, S., Liu, Y.: Process, microstructure and microhardness of GH3039 superalloy processed by laser metal wire deposition. J. Alloys Compd. **877**, 160330 (2021)
4. Brandl, E., Michailov, V., Viehweger, B., Leyens, C.: Deposition of Ti–6Al–4V using laser and wire, part II: hardness and dimensions of single beads. Surf. Coat. Technol. **206**(6), 1130–1141 (2011)
5. Bernauer, C., Zapata, A., Zaeh, M.: Toward defect-free components in laser metal deposition with coaxial wire feeding through closed-loop control of the melt pool temperature. J. Laser Appl. (2022)

6. Shaikh, M.O., Chen, C.-C., Chiang, H.-C.: Additive manufacturing using fine wire-based laser metal deposition. Rapid Prototyp. J. **29**(3), 473–483 (2020)
7. Capello, E., Colombo, D., Previtali, B.: Repairing of sintered tools using laser cladding by wire. J. Mater. Process. Technol. **164–165**, 990–1000 (2005)
8. Peng, X., Kong, L., Fuh, J.Y.H., Wang, H.: A review of post-processing technologies in additive manufacturing. J. Manuf. Mater. Process. **5**, 38 (2021)
9. Urhal, P., Weightman, A., Diver, C., Bartolo, P.: Robot assisted additive manufacturing: a review. Robot. Comput.-Integr. Manuf. **59**, 335–345 (2019)
10. Nor Hamran, N.N., Ghani, J.A., Ramli, R., Che Haron, C.H.: A review on recent development of minimum quantity lubrication for sustainable machining. J. Clean. Prod. **268**, 122165 (2020)
11. Xiong, G., Ding, Y., Zhu, L.: Stiffness-based pose optimization of an industrial robot for five-axis milling. Robot. Comput.-Integr. Manuf. **55**, 19–28 (2019)
12. Glogowski, P., Rieger, M., Sun, J., Kuhlenkötter, B.: Natural frequency analysis in the workspace of a six-axis industrial robot using design of experiments. In: WEP Congress, Applied Mechanics and Materials, pp. 345–352 (2016)

Robot-Based End Shield and Insulation Sleeve Assembly in the Context of Flat Wire Stator Technology

Alexander Vogel, Andreas Morello, Philipp Mathea, Jörg Franke, and Alexander Kühl

Abstract

In the manufacture of electric motors, the stator is the most expensive component of the electric motor, accounting for 35% of the total costs. For this reason, the stator is very interesting in terms of automation and offers various possibilities for increasing production efficiency. In addition to some process steps that are already highly automated, others are still largely manual. These mainly include work steps in the area of insulation assembly. In particular, the fitting of insulation sleeves and the assembly of the end shield should be mentioned here. The main challenges of robotic processing is the flexible behavior of the wires, handling of their wire ends and the detection of those due to its undetermined location. This aspect occurs not only with flat wires, but also with classic round wires or litz wires. Collaborative kinematics paired with an instrumented end effector system are required to raise the level of automation in these work steps. Both are used in the research presented in this paper. The results achieved show the possibilities of a collaborative system to also automate the steps of end shield assembly and wire end insolation which is a further step towards a fully automated stator production.

The results, published in this publication are based on a research project named E|Real—Flexible, roboterbasierte Automatisierung der Fertigung von E-Antrieben für die Luftfahrt, FKZ 20O1941B funded by the German Federal Ministry for Economic Affairs and climate action (BMWK) and is managed by the German Aerospace Center (DLR).

A. Vogel (✉) · A. Morello · P. Mathea · J. Franke · A. Kühl
Institute for Factory Automation and Production Systems (FAPS), Friedrich-Alexander-Universität Erlangen-Nürnberg (FAU), Erlangen, Germany
e-mail: alexander.vogel@faps.fau.de

M.-C. Wanner et al. (eds.), *Annals of Scientific Society for Assembly, Handling and Industrial Robotics 2024*, https://doi.org/10.1007/978-3-031-91463-8_18

Keywords

Robotics • Stator production • Electric drive production

1 Introduction

1.1 Scientific Situation

Due to the high proportion of labor costs, manufacturers in high-wage countries are reaching their limits. Industrial or collaborative robots, promising flexible automation for small to medium quantities could be part of an improvement. Even if robot technology continues to advance, practical applications in electric motor production that go beyond simple handling tasks are still limited in their complexity. Possible application scenarios have already been focused in previous articles from institute FAPS like [1, 2], e.g. by Mayr et al. [3], Kühl et al. [4, 5] or Hultman and Leijon [6, 7]. Another topic, which has been investigated in Vogel et al. [8] is the detection of those wires.

1.2 Research Motivation

One common topic in the context of stator manufacturing is to increase the copper-filling-factor. This can be achieved by using rectangular wired called flat wires instead of classic round wires. This technology is introduced by the E|Drive-Center at the Institute for Factory Automation and Production Systems (FAPS) from Friedrich-Alexander-Universität Erlangen Nürnberg (FAU). Using the more complex flat wire technology also implies to find new solution in subsequent steps of the stator production like the automated end shield assembly and the following insulation of the wire ends.

In this publication, a concept of an automated end shield and insolation sleeve assembly, including the detection of the flat wire ends, is presented. The paper is structured in three main parts. At first, the developed assembly cell used in this automation process is described. Afterwards, the automated end shield assembly is detailed and finally, the detection and insulation of the wire ends are examined. Therefore, two detection methods are presented and their results compared regarding their usability.

2 Laboratory Setup

The laboratory setup is shown in Fig. 1 where two collaborative robots, serving as leader and follower kinematics, are mounted facing each other on a working table with a centered stator-clamping device. Both robots are equipped with a Robotiq 2F-85 gripper. During the end shield assembly process, one cobot has the normal 2-fingers as jaws while the

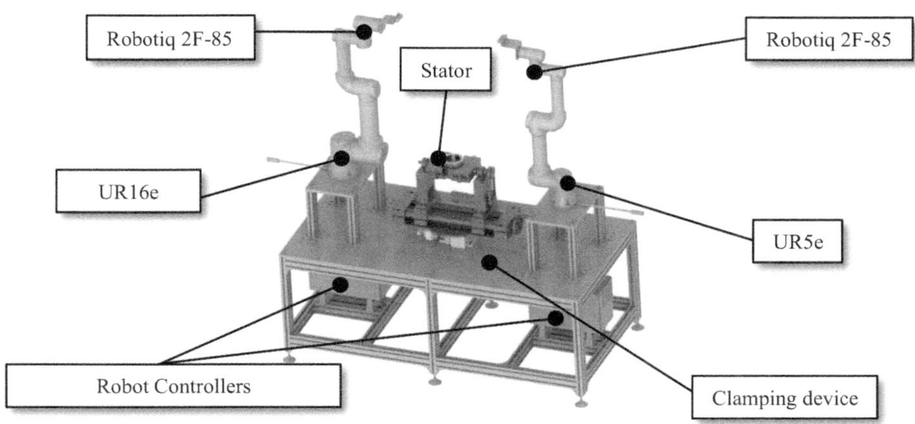

Fig. 1 Laboratory setup

other one has its own joining tool. During the insulation process, only one cobot is used with the same Robotiq gripper but with own gripper jaws.

For the detection of the wire end two approaches are examined. To visually detect the wire ends the MIRAI vision system from Micropsi Industries is used in the intended application scenario. The system consists of a 2D color camera with an attached incident light, the HEXE-E QC force-torque sensor from OnRobot and an additional control unit. The camera is attached directly to the robot arm, which means that the image capture position is flexible. To tactilely detect the wire ends, the force-torque sensor is needed. Although the UR16e has an integrated force-torque sensor in all joins already, however it is not precise enough to detect the limp wire ends.

3 Automated Joining of the End Shield

3.1 Hardware

In order to handle all wires with on kinematic and to be able to join the end shield with the other, a proper tool design for both handling tasks is required. The focus of this publication is located onto the wire handling.

The design of the tool, shown above in Fig. 2 is based on several ideas, listed below:

- The tool must be able to be inserted through the air gap of the stator
- The tool needs top by cylindric therefore
- The tool must be able to grip or fix all wire ends simultaneous
- The tool needs clamping device for each wire end.

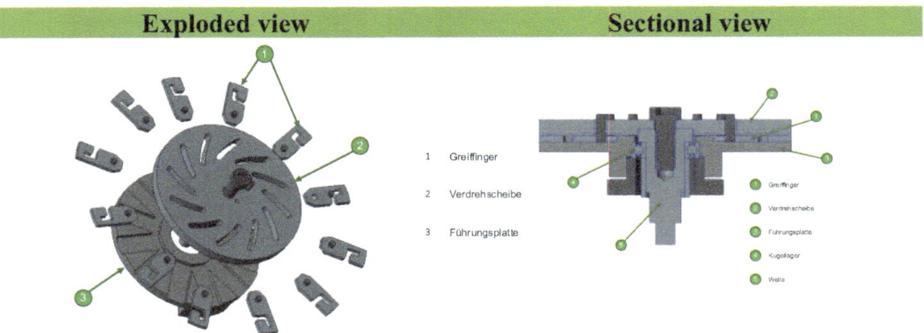

Fig. 2 Detailed assembly of the joining tool

A rotating disk is the central component of the tool. The gripper fingers (1) are located in the square cut-outs of the guide plate (3). The cylindrical pins of the gripper fingers (1) engage in the grooves of the rotating disk (2). The interaction of all three elements converts a rotary movement of the rotating disk (2) into a linear movement of the gripper fingers (1), allowing them to be moved out of and back into the tool.

Figure 3 shows the used angular positions of the tool. In position one, the 0° position, the tool is retracted and can be guided through the stator's air gap. In position two, the tool has its gripper fingers fully extended by 25° and can grip the wire ends. In position three, at 17° of rotation, the gripper's fingers are closed enough to fix all wire ends. The sectional view in Fig. 2 shows how the rotating disk (2) with a shaft (5) is connected to the guide plate in free rotation via a ball bearing (4). A load capacity and service life calculation for the bearing is not necessary due to the low weight and the load to be assumed. The shaft (5) forms the interface to the drive assembly.

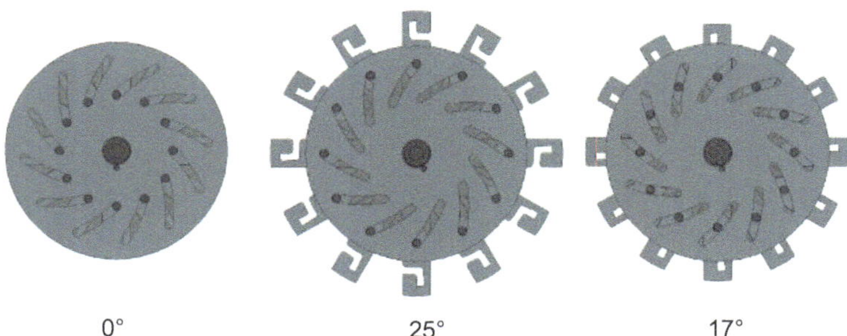

Fig. 3 Different angular positions during processing

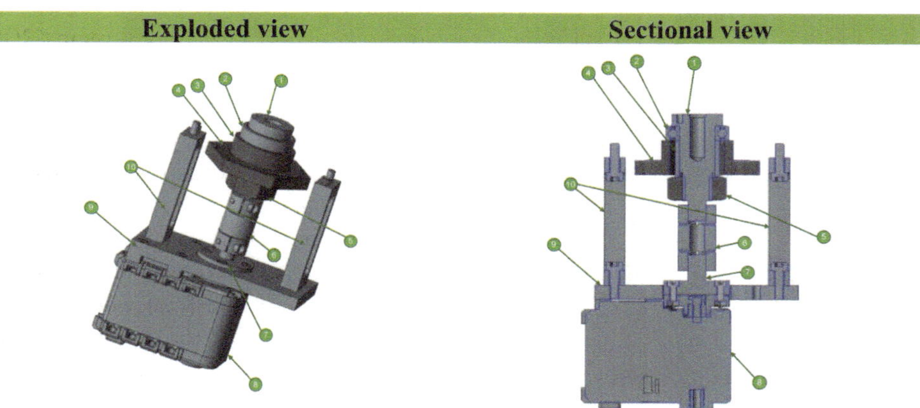

Fig. 4 Assembly group electric drive

The drive assembly contains the actuators of the tool, the servomotor and is the link between the twisting tool and the connection to the cobot. The individual components of the drive assembly can be seen in Fig. 4.

The assembly is connected to the twisting tool via the shaft (1). The ball bearing (2) is fixed to the outer ring via the nut (4) in the guide plate. The inner ring of the ball bearing (2) is attached to the shaft with the nut (5) via a spacer ring (3), which is concealed in this view. The torque is transmitted from the servomotor (8) to the twisting tool via a torsional rigid coupling (6). The coupling (6) serves as a compensating element for angular and tolerance errors in the production of the individual components. The servo adapter (7) is the link between the servomotor (8) and the coupling (6). The mounting plate (9) serves as a connection point for the servomotor (8) and, together with the two retaining struts (10), provides torque support for the motor.

3.2 Joining Process

In the initial position, the stator is placed in the clamping tool, which is rotated into a horizontal position. This allows to work on both sides of the stator simultaneously during the further assembly process. For visual reasons, however, the assembly process is shown vertically in the illustration. As previously mentioned, the developed tool to handle all wire ends is mounted on top of one cobot but only the tool movement is shown below. The end shield joining is done by the other cobot. Here, only the end effector of it is shown due to visual aspects, too.

In Fig. 5 the whole end shield assembly process is depicted. Initially, the twisting tool is inserted into the stator from the backhand side until the tools is just above the stator, see steps a and b. Then, the servomotor rotates the turntable to an angle of 25°, so that the gripper fingers of the tool are fully extended, see step c. The wires are inserted into the tool when the cobot rotates the entire tool, shown in step d. By rotating the turntable back to an angle of 17°, the wires are fixed in the tools gripper fingers, see Fig. 5e. Afterwards, the cobot with the tool moves upwards in the axial direction of the stator until just before the wire ends are reached. Now, since all wire ends are fixed and its position is known, the second cobot joins the end shield exactly above the wire ends, which is sketched in step f. To fulfill the joining process, both cobots now move axial to guide the wired into their holes. When that movement is done, the tool guiding cobot rotates the tool back to a full opened, 25° position, mentioned already in Fig. 3, to fully release all wires. After step g, the whole tool is tuned again which causes the wires to emerge from the tool. Then, the tool's turntable is rotated to the 0° home position and the gripper fingers enter the tool, see Fig. 5h. Finally, the rotating tool is moved axially out of the stator and the end shield is placed the remaining distance onto the stator to be firmly connected to the stator in the next process step.

By following that process chain, it is possible to fulfill the assembly of the end shield without using any vision system.

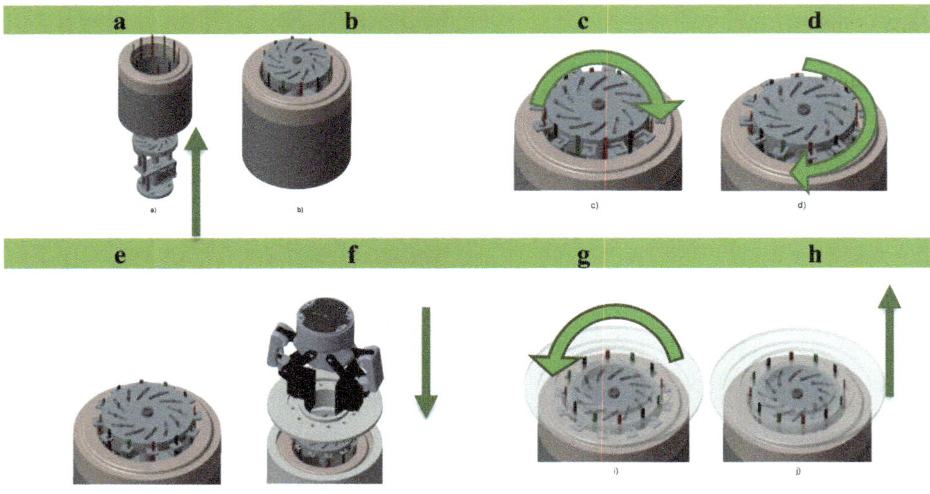

Fig. 5 End shield assembly steps

3.3 Result Discussion

It became apparent, that turning the turntable was only possible with a great deal of force. An initial analysis showed that the dowel pins in the gripper fingers were not positioned perfectly plumb. After adjusting the dowel pins, it was possible to turn the tool without any problems. However, it was not possible to turn back or close the tool. In order to find the cause, the tool was dismantled and all moving components were checked for their partial function. This revealed several complications. When closing the tool, the gripper fingers were pressed against the side wall of the guide disk by the rotating disk. On the one hand, this leads to high friction between the gripper fingers and the wall, on the other hand, the plastic gets caught in the sharp-edged outer geometry of the guide disk. Rounding off the sharp edge led to an initial improvement in the sliding properties of the guide. The plastic gripper finger was then smoothed further on the contact surfaces. It was not possible to improve the surface quality of the guide disk. The use of lubricating oil led to a further improvement in the sliding properties.

4 Insulation Sleeve Assembly

4.1 State of the Art in Manual Sleeve Assembly

After successful winding, an employee with one hand grips the end of the wire to be insulated. In the next step, the second hand is used to push the insulation sleeve onto the end of the wire. In the third process step, one hand begins to push the insulation sleeve onto the wire, while the other hand continues to hold the wire in place. While pushing on the insulation sleeve, it is necessary to reposition the hand holding the wire over the insulation sleeve. This gripping is shown in the fourth process step of Fig. 6. Finally, the insulation sleeve can be pushed all the way down in its final position which is depicted in the fifth step (Fig. 6).

When looking at the process sequence, it becomes clear how complex manual insulation sleeve assembly is, requiring two hands to be used throughout. Threading the insulation sleeve onto the end of the wire is particularly difficult for people, as this process step requires a high degree of precision. Although only a single wire has to be fitted with

Fig. 6 Manually performed process steps during insulation sleeve assembly [9]

an insulating sleeve when installing the insulating sleeve on flat wires, the complexity is still high, because the circular sleeve has to be deformed in a defined manner in order to fit the flat wire, which usually has a width-to-height ratio of ~3:1.

As shown by Mahr et al. [9], a high level of automation can be achieved with a gripper that is precisely adapted to the application. However, if the wire position deviates too much from the expected position, the use of sensors to detect the wire ends is unavoidable.

4.2 Hardware

Based on a demonstrator, shown in Fig. 7, the whole process will be investigated. This is a demonstrator that does not consist of a fully wound laminated core, but only of the original end shield with guides for the flat wires. The demonstrator was used because only the end shield with the protruding flat wires is important for the application. A total of twelve flat wires with a length of 110 mm are arranged protruding in the end shield of the stator. The flat wires have a rectangular geometry with nominal dimensions of 2.50 mm in width and 1.00 mm in depth. Although the wire is coated, it instead has a real width of 2.60 mm and a depth of 1.20 mm.

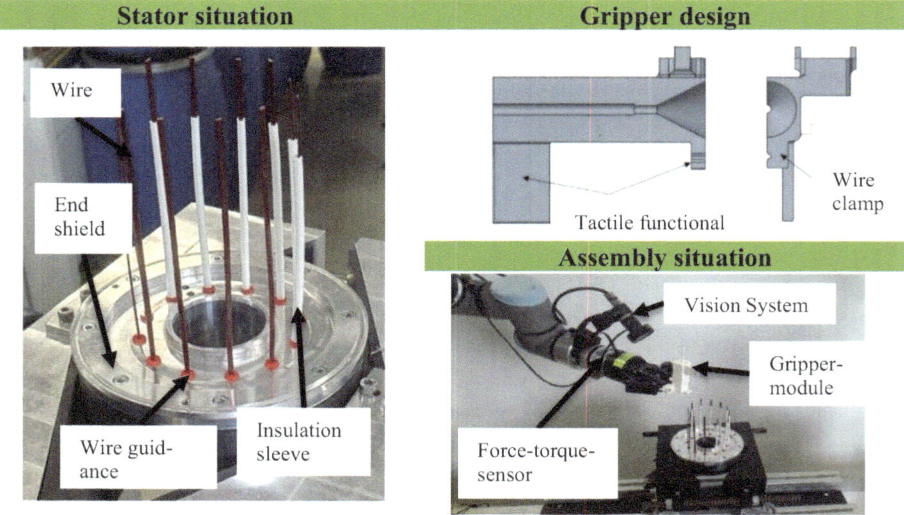

Fig. 7 Stator situation, gripper design and whole scenery

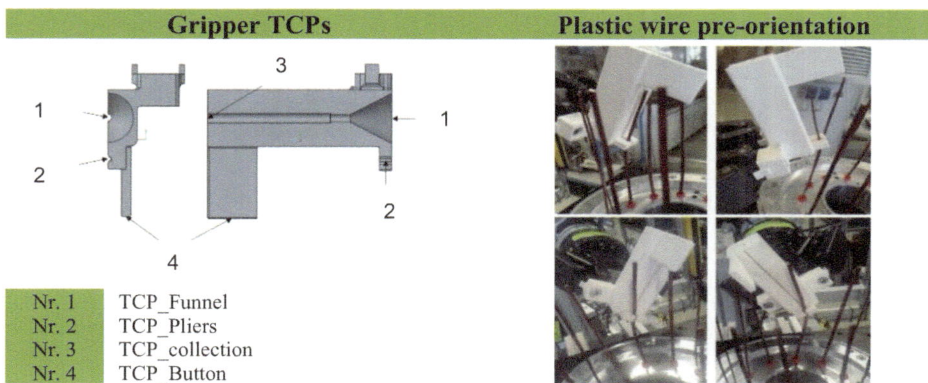

Fig. 8 Gripper-TCP-situation and wire pre-orientation

4.3 Tactile Detection of Wire Ends

In order to insulated the flat wire, at first its position has to be determined. Therefore, various TCPs have been defined on the gripper jaws, which must be changed during the program run. A total of four different TCPs are created and can be seen in Fig. 8 on the left. However, during the tactile detection of the wire, only the TCP_button is used in this use case. First, the robot moves to the approximate wire position based on the geometry of the lamination Then, the robot moves linearly in x-direction and reads the force at the end effector. If the force exceeds a certain threshold value, the wire position in this spatial direction is detected and the x-position is saved. Afterward, the same procedure is repeated for the y-axis as well and its position is also saved.

4.4 Visual Detection of Wire Ends

Another method instead of tactilely detect the wire ends it to visually detect them. The visual detection of the wire end is done with the MIRAI vision system launched by Micropsi. This system is based on machine learning algorithms and uses prerecorded data to fulfill the task. It is essential for the algorithm that both gripper jaws and the object to be detected are in the camera's field of view. Once the system-integrated algorithm has taken enough scenes of the point of interest before, e.g. the flat wire, the current image from the camera is compared with those teached scenes and a robot movement is calculated until the deviation falls below a certain threshold value. Once the robot stops moving, the detection is complete and the wire end position is saved.

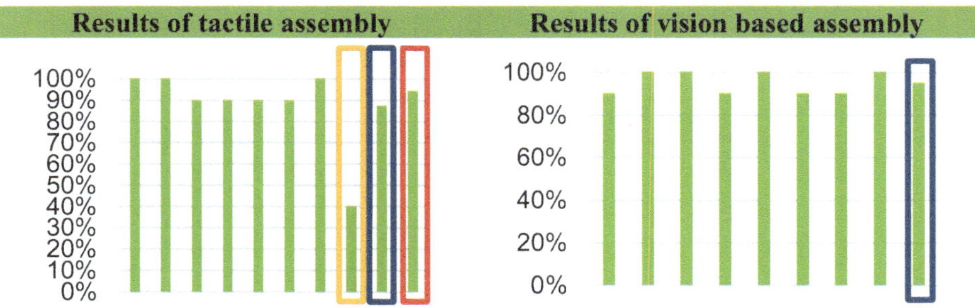

Fig. 9 Process results of both detection approaches on different deformation situations

4.5 Comparison of Both Techniques

Looking at the results in Fig. 9 left sided, it is noticeable that the success rate of several defined deformation situations shown on the x-axis is very high, resulting in an overall success rate of 88% (blue colored). This means that 88% of all placement attempts were successful and the insulation sleeve was joined. Twisted wires are a major problem, highlighted in orange. These only have a success rate of 40%. It should be noted, that even if the insulation sleeves were successfully joined, the wire was still twisted at the end of the process. If this variant is excluded from the overall success rate, the overall success rate is as high as 94% (red colored). For all other variants, apart from twisted wires, the success rate is 90% or higher. For straight wires with a length of 11 cm, which represent the actual standard in the process, the rate is even 100%. Figure 9 right handed shows the placement attempts of the visual insulation sleeve assembly. They are all characterized by a success rate of 90% or higher. With an overall blue highlighted success rate of 95%, the process already delivers very good values. A closer look must be taken at the variant in which the wire was twisted by 90°. Although the placement attempts resulted in a success rate of 100%, the wire was still twisted at the end of each placement process.

4.6 Joining Process

The whole joining process chain is depicted in Fig. 10, beginning with the wire detection. This step differs depending on whether the tactile or visual approach is used. The wire pre-orientation is initially intended to react to rough position deviations and minimize them. Starting from the known wire position in the immediate vicinity of the lamination, the wire is shaped into the process window by plastic forming. Now, since the wire end position is detected and due to the wire pre orientation it is also ensured that the wire lies within a certain process window the insulation process can start. Therefore, the robot picks up an insulation sleeve which is position in the collection slot recognizable as TCP_

Fig. 10 Insulation processes sequence

Collection in Fig. 8. Then, the robot rotates its wrist and moves up to the known wire end position so that the TCP_Funnel is directly above the wire end. Due to the funnel the wire is guided into the collection slot where the insulation sleeve is located. By moving down the wire is the sleeve can be joined safely. The contact force is set to 25 N, as frictional forces arise between the wire and the insulating sleeve during joining. After the gripper has reached the end shield, the gripper is opened and moved upwards. This is necessary so that the gripper can grip the sleeve again to join it up to the stop. Finally, the sleeve is installed and the robot can move away.

4.7 Result Discussion

Both systems show similar success rates in the test series, but the results differ on closer inspection. The tactile approach is much more stable in detecting the flat wire due to the physical contact with the wire than the visually supported approach using virtual image processing. As wire detection is the first step in automated insulating sleeve assembly, the subsequent processes of pre-orienting the wire and joining the sleeve are heavily dependent on the result of the wire detection. Poor or faulty detection results in poorer pre-orientation, which may lead to faulty joining of the insulating sleeve. For this reason, the tactile approach is technically superior to the visual approach, as it ensures a more robust process flow. The precision of the tactile system is higher than that of the visual system. This is again due to the more precise detection of the wire. The internal complexity is also kept to a minimum with tactile insulation sleeve assembly, as the robot controller permanently controls the robot. In the case of visually supported insulation sleeve assembly, the controller of the MIRAI—system and the robot controller alternate in controlling the robot, which can lead to complications.

Acknowledgements The corresponding author thanks Philipp Mathea and Andreas Morello for their revising and consulting. Additionally, the authors thanks Florian Schächinger and Johannes Engelmaier for their support during practical experiments. Furthermore, the author thanks Dr. Alexander Kühl and Prof. Dr. Jörg Franke for the supervision of this work.

References

1. Mahr, A., Mayr, A., Jung, T., Franke, J.: Robot-assisted concept for assembling form coils in laminated stator cores of large electric motors. Procedia Manuf. **38**, 866–875 (2019). https://doi.org/10.1016/j.promfg.2020.01.168
2. Mahr, A., Wurm, M., Bickel, B., Franke, J., Halder, H.: Development of a control system for a universal winding machine based on virtual path planning. In: 2017 7th International Electric Drives Production Conference (EDPC), Würzburg, pp. 1–5 (2017)
3. Mayr, A. et al.: Electric Motor Production 4.0—Application Potentials of Industry 4.0 Technologies in the Manufacturing of Electric Motors, pp. 1–13
4. Kühl, A., Furlan, S., Gutmann, J., Meyer, M., Franke, J.: Technologies and processes for the flexible robotic assembly of electric motor stators. In: 2017 IEEE International Electric Machines and Drives Conference (IEMDC), 21–24 May 2017, Miami, FL, USA, pp. 1–6 (2017)
5. Kuehl, A., Franke, J.: Robot-based forming of hairpin winding. In: 2021 IEEE International Electric Machines and Drives Conference (IEMDC), vol. 2021 (2021). https://doi.org/10.1109/IEMDC47953.2021.9449576
6. Hultman, E.: Introducing robotized stator cable winding to rotating electric machines. Machines **10**(8), 695 (2022). https://doi.org/10.3390/machines10080695
7. Hultman, E., Salar, D., Leijon, M.: Robotized surface mounting of permanent magnets. Machines **2**(4), 219–232 (2014). https://doi.org/10.3390/machines2040219
8. Vogel, A., Morello, A., Mahr, A., Franke, J., Kuehl, A.: Challenges in cost-effective automated detection of enameled flat wires in the context of robot-based stator manipulation. In: 2022 12th International Electric Drives Production Conference (EDPC), Regensburg, Germany, pp. 1–8 (2022)
9. Mahr, A., Mathea, P., Vogel, A., Morello, A., Franke, J., Kühl, A.: Robotic based assembly of insulating sleeves onto winding coil ends of electric drive stators. In: 2023 13th International Electric Drives Production Conference (EDPC), Regensburg, Germany, pp. 1–7 (2023)

Potentials and Challenges of Enabling Industrial and Professional Service Robots to Be Social

Nina Merz, Jörg Franke, and Sebastian Reitelshöfer

Abstract

Humanoid robots that support us in our daily lives as attentive servants and communicate with us in an empathetic, humorous, and competent way are the epitome of many visions of the future. However, most research in the area of social robotics so far has focused on applications for patients in healthcare settings, leaving other applications unexplored. Recently, the research on social robots for other professional applications has gained further attention from the community. This qualitative analysis aims to highlight the relevance of this research area and to provide further research directions. Therefore, it summarizes the current potentials and challenges that arise when industrial and professional service robots are enabled to be social. Positive effects on the health of employees and their satisfaction of social needs are identified as well as benefits for the actual use of robots in companies. Further research is identified in the areas of social concepts, social software and employee acceptance.

Keywords

Social robots • Industrial robots • Professional service robots

N. Merz (✉) · J. Franke · S. Reitelshöfer
Institute for Factory Automation and Production Systems (FAPS), Friedrich-Alexander-Universität Erlangen-Nürnberg, Erlangen, Germany
e-mail: nina.merz@faps.fau.de

M.-C. Wanner et al. (eds.), *Annals of Scientific Society for Assembly, Handling and Industrial Robotics 2024*, https://doi.org/10.1007/978-3-031-91463-8_19

1 Introduction

Robots are considered a core technology in industrial applications such as automotive assembly. Because of their technical flexibility, they have the potential to be the future core technology for many other professional applications, too. In this context, the goal of using robots in professional settings is to digitize repetitive or time-consuming manual tasks, allowing skilled workers to focus on specific and more flexible tasks. Thus they are addressing the current major challenges in industry—demographic change and skills shortages [1].

In this context, the research area of robotics focuses on the continuous improvement of the range of functions, precision and payload of robots. To enable robots for further applications, research in the field of Human–Robot-Interaction (HRI) focuses on the possibilities of cooperation between humans and robots. The continuous research in this area already enabled professional service robots to perform autonomous pick-up and delivery services in spaces shared with humans, as well as simple collaborative tasks together with humans. However, so far robots are still another technology tool to be used in industries. Social skills would enable them to communicate more effectively—including non-verbally—by recognizing and expressing emotions and thus becoming an assistant to employees. For example an automated guided vehicle (AGV) could recognize the fear of an employee it is passing and calming him down by telling him that he recognized him and will move around him. The research field of the so-called "social robotics" explores this ability of automated devices to communicate and interact socially with a human.

Because of the combination of the benefits of a never tired and strong robot with social skills, social robots are one of the most important emerging technologies [2]. Consequently, the number of research activities on social robots has increased in recent years [3]. This trend is enabled by both, the development of low-cost sensors and the development of powerful artificial intelligence (AI) methods that enable social interactions. Although appropriate social components are also of crucial importance in the context of industrial robotics and other professional applications of robots, most current application scenarios focus on healthcare [4]. Consequently, the state of the art lacks an overview of the potentials and current limitations of social robots focusing on the interaction with other professional users. Thus the objective of this study is to address the aforementioned research gap through the implementation of a qualitative analysis.

Therefore, the next chapter of this paper presents the scope of this research by defining industrial, professional service and social robots. In order to postulate the need for social robots an overview of positive effects of enabling robots to be social in professional settings follows. To provide a roadmap for further research, general conceptual challenges in the development of professional social robots are presented. The paper concludes with giving limitations and a short summary of the results.

2 Industrial Robots, Professional Service Robots and Social Robots

Although different definitions exist, a robot in this contribution is defined as an actuated mechanism programmable in two or more axes with a degree of autonomy, moving within its environment, to perform intended tasks. A distinction is made between industrial (industrial automation settings) and service robots (non-industrial), where service robots are further divided into professional (service robots for professional use) and private ones (service robots for private use) [5]. Examples of professional service robots include amongst others transportation systems (e.g. drones, autonomous guided vehicles), and humanoid robots. Due to their versatility, any robot can be used in different industries, such as manufacturing, commerce, or medicine. The classification is undertaken according to the application of the robots [5]. This research focuses on industrial and professional service robots and professional users (employees) as interaction partners to the robots.

The research field of social robotics focuses on the development of intuitive and human-like communication between humans and robots. Consequently, the intention is to provide robots with social capabilities. The social capabilities of a robot are achieved by sensors collecting multimodal data, the robot's software processing this data and the robot performing an acoustic (e.g. verbal speech) or visual (e.g. movement) action. Because of its focus on the interaction of the robot, social robotics is a sub-area of HRI. There are many definitions of social robots in the literature. What they have in common, is that social robots should be able to autonomously perceive and respond to environmental stimuli, interact with humans or robots and understand and follow general social rules [6]. Thereby the robot and its social skills must be appropriate to the situation. For example, people's expectations of the robot's abilities are influenced by its appearance, resulting in people expecting robots to follow general moral rules, when a robot has a humanoid appearance [7]. In this paper a social robot is a physical robot that has the ability to socially communicate and interact with a human counterpart adapted to the situation [8, 9]. Not only humanoid robots are considered social robots, but also any technical system that can interact socially according to the above definition. This therefore also includes, for example, self-driving transportation systems, drones, and stationary and mobile industrial robots.

3 Potentials of Enabling Industrial and Professional Service Robots to Be Social

As mentioned in the introduction, robots address the challenge of labor shortage that companies face today. Labor shortages are caused by demographic changes and result in unpopular positions and professions being unfilled. Robots offer the possibility to take over unpopular parts of physical work or to upgrade disliked positions and professions.

For example, heavy tasks that are physically demanding for the worker, repetitive tasks that are unpopular to the workers, or fetching tasks that are time-consuming can be taken over by a robot [1]. Thus, they support ergonomics and the well-being of the employees [1, 10]. At the same time, robots are still a disruptive innovation to employees in most industries or are only known in a non-collaborative setting behind a fence. As a result, there is a high risk that the workers who need to interact with the robots will not accept them and will refuse to work collaboratively with them. Current reasons include the fear of job loss [11] and safety and security concerns [12]. It is argued that implementing social capabilities in robots will improve HRI and thus increases their acceptance [12, 13]. This leads to the use of robots by employees for the intended tasks.

In addition to the benefits to companies, social robots have the potential to increase the employee job satisfaction, making a company that uses robots a more attractive place to work and therefore attracting more applications from motivated and skilled workers. For example, social robots have the potential to improve employee health by providing preventive health support to employees who (1) face an imbalance between work demands and capabilities, (2) have a temporary reduced work ability, or (3) face a permanently reduced work ability due to chronic illness and/or age [14]. Next, social robots also have the potential to meet the social needs of employees. For example, they can increase the meaningfulness of employees' work, and increase the connection among employees [15].

In summary, enabling industry and professional service robots to be social offers benefits to companies that result in being a better workplace. Employees experience an improved physical and emotional health. Both of these benefits occur when robots are implemented in general, but are multiplied when social capabilities are added.

4 Challenges When Enabling Industrial and Professional Service Robots to Be Social

Research efforts are making robots even more precise, powerful, flexible, sensitive and autonomous. Although these improved technical capabilities enable safe human–machine collaboration in general [16], currently, research in HRI focuses only on applications in the healthcare sector [4]. This results in minimal knowledge about the use of social robots in other professional scenarios and most automated devices operating in the same environment as humans lacking social skills. However, knowledge is needed to be able to quickly adapt and adjust social skills and/or robots to new circumstances (e.g. environment, group of people, tasks). In addition, only a few empathetic robots with a range of functions dedicated to specific interaction scenarios in specific domains are currently available for purchase [17], resulting in a lack of research platforms with open interfaces, which makes research on social skills time-consuming and expensive. To be able to efficiently research on how to enable robots to be social, several basic concepts need to be in place.

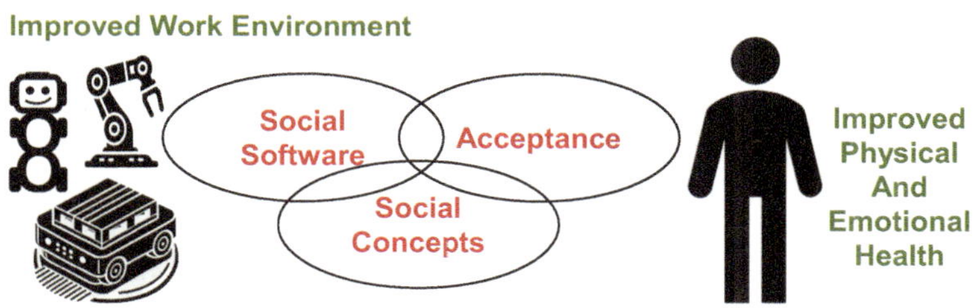

Fig. 1 Potentials (green) and challenges (red) of enabling industrial and professional service robots to be social

The qualitative analysis revealed three interacting clusters—social concepts, social software and employee acceptance—as current conceptual challenges that address the aforementioned general research barriers for enabling industrial and professional service robots to be social. When these challenges, which are explained in the following, are focused on in research, the benefits outlined in Sect. 3 can arise (see Fig. 1).

4.1 Social Concepts

The core of social robots is the interaction with humans. Since the robot may look and act differently depending on the scenario, and the human may have different roles, several taxonomies exist to structure HRI [9, 18]. Onnasch and Roesler [18] derived a taxonomy that builds on existing ones and characterizes each HRI by its interaction context classification (field of application, exposure), robot classification (task specification, morphology, degree of autonomy), and team classification (human role, team composition, communication channel, proximity). However, the sociality of the robot is not included. There exist classifications that analyzed the sociality of a robot, but they do not consider professional applications [9, 19]. Due to the many possibilities of a social industry or professional service robot, it is necessary to extend current HRI classifications with sociality in order to keep the comparability of research activities.

Other unresolved challenges continue to exist in the autonomous HRI and the creation of goal-oriented and accepted holistic social communication and cooperation between robots and humans in a professional environment. On the one hand, it must be ensured that the communication is as natural as possible, but on the other hand that it must not have a frightening effect on humans (keyword: uncanny valley [20]). Next, the interaction must be designed so that the productivity and product quality is not negatively affected. Last, the robot and its social skills must be suitable to the situation, which makes the holistic development of social skills particularly complex. Thus for each HRI scenario it is necessary to identify, which social abilities are necessary and which are not. For the use of a social robot in a private environment, Graaf et al. [21] identified eight factors (reciprocal interaction, showing thoughts and feelings, social awareness, social support, autonomy, comfort, similarity to oneself and mutual respect) that are necessary for social interactions. However, it is possible that in professional applications, due to the forced interaction, different factors are necessary to enable a robot to be social. Thus, appropriate social capabilities of industry and professional service robots need to be identified [14]. In addition it is unclear how many and to what extend a robot must have social capabilities to be considered social [9].

In summary, research in social robotics lacks social concepts that (1) allow a structured view on social HRI and (2) show which social capabilities of a robot are necessary for a social HRI in a professional environment.

4.2 Social Software

Programming by persons of a broad variety of knowledge domains or end users may be of interest for social robots [22]. Therefore, there is a need for suitable frameworks the implementation of social capabilities for robotic systems in scientific settings as well as in the development of commercial products with social capabilities. In addition to known taxonomies for characterizing specific HRI scenarios [18], further studies, for example on moral frameworks for social robots [23], can be used as a basis for a social robot framework. In terms of software frameworks, the Robot Operating System (ROS) is an established middleware for implementing a wide range of robot applications that has become the global quasi-standard as a framework for research projects. Up to now, software packages or interfaces that specifically and comprehensively address sociality functionalities are not yet available. However, various projects exist to bundle elements for a wide range of perception tasks based on deep learning approaches [24], for recognizing the actions and intentions of human users [25] or for various components of social interaction such as eye movements [26] body language [27, 28] or dialogs in the form of text or spoken language [29]. Independent form concrete software frameworks, various architectures for social robot systems have already been proposed for example, the use of social agents for multi-robot systems [30]. A frequently described application scenario for deriving software architectures for social robots addresses their interaction with older people. There are cloud-based services with relevant functions to interact with older people [31]. An architecture with four components for navigation, person recognition and tracking, localization and interaction for a specific service robot as an assistant for older people is presented in [32]. A service-based architecture with over 30 subcomponents is also presented in [33] for social robots in elderly care. Another system for carrying out exercises with elderly people, which uses six modules is presented in [34]. A robot coach architecture for elderly care based on multi-user engagement models is presented in [35]. In [36], a rule engine is used to execute certain robot actions on an NAO robot based on collected medical data of elderly persons. In [37], a layer-based architecture for a specific robot for dealing with elderly people with five layers and over 20 subcomponents is described.

For the therapeutic context several examples are described, mapping the behavior to the morphology of the robot [38], screening people with autism spectrum disorder [39] or for cardio-rehab [40]. A more complex architecture again with over 20 subcomponents is also presented for the therapeutic context in [41]. A controller architecture for mapping user states to robot activities is also presented in [42] in order to guide users through various tasks and exercises in a task-controlled manner. An architectural proposal with very complex levels is shown in [43]. An approach to accompany planned activities of users throughout the day with a service robot and to trigger adapted behavior patterns is described in [44]. A collection of software modules for using a specific robot platform for games with humans is presented in [45]. An emotion-based approach for social robots in general, which triggers certain behavior patterns based on detected emotions using a complex hierarchical behavior control, is described in [46]. Another complex architecture for the implementation of social robots in different application scenarios is described in [47], which is based on layers of 15 components that build on each other.

To summarize, it can be stated that numerous concepts for architectures are described in literature. As described above those architectures are complex system developed with specific settings for specific settings.

4.3 Employee Acceptance

In the third section, reasons are given why social robots have a positive impact on the work environment, which only apply if the robots are implemented by the companies and used by the employees. A crucial factor for the use of innovative technologies is the acceptance by the future user. Vice versa, in this research acceptance is defined as the actual use of a technology [48]. The identification of the level of acceptance of a technology (acceptance object; here: social robot) by the subsequent user (acceptance subject; here: employee) under certain conditions (acceptance context; here: HRI) is complex [49], because of numerous individual influences, which change as soon as the characteristics of the scenario (robot, employee, HRI) change. In detail, each of the applications a robot can work in and shapes a robot could have trigger different expectations by the employee that have to be met by the robot to be accepted.

For a structured analysis of acceptance, different Technology Acceptance Models (TAM) are used [48, 50, 51]. Influences analyzed within the models are, for example the perceived ease of use, performance expectancy, social influence and voluntariness. Demographics such as age, gender and former experience can act as moderators [51, 52]. However, traditional acceptance models are not considered adequate for interaction-oriented technologies such as social robots [51]. In addition, developing only one acceptance model for social robots in general is not adequate. This is reasoned since they are, as already stated, a multifunctional tool for a variety of purposes and can have different embodiments and since the characteristics of the interaction scenario mutational interdepend each other, it results in different influences on the acceptance of different social robots in different applications. This assumption is reflected in literature, since different study designs arise [50, 53]. In addition, as stated in Sect. 2.1, no aligned concept on social robots exist, which makes comparisons between the studies and general conclusions difficult and time intensive [50, 53].

In summary, acceptance is, next to social concepts and social software, another crucial concept to be looked after when developing social skills. However, currently, due to different study designs, it is difficult to draw general conclusions based on existing research and developing models for acceptance analysis of social industrial robots and professional social service robots is time intensive and requires skilled researchers.

5 Conclusion

Robots are multifunctional tools that are used in many different applications. To make them available for further applications and to increase the user satisfaction in a HRI, they are currently being enabled to act socially. However, the research focus in social robotics in professional applications currently lies in healthcare scenarios. This research paper gives a short but structured summary of the potentials of the implementation and use of social industrial and social professional service robots as well as displaying current open topics for further research.

Although this qualitative research aims to provide a complete synthesis of existing research, some limitations must be acknowledged. The research field of social robots is interdisciplinary by nature. Despite the efforts of the authors to consider different disciplines, it is possible that relevant studies are missing. Next, qualitative reviews are subject to selection, interpretation and reviewer bias by nature. Although the research team objectively reviewed the identified literature, it cannot be excluded that these biases are present in this study.

In summary, implications that social robots increase work satisfaction and extinct employees' fears of robots have been revealed. Open research topics exist in the descriptive area of social concepts, the enabler for professional social robots social software, and the analytical topic of the acceptance of the professional robots by employees. It is recommended that further research is conducted with the objective of closing these research gaps, thereby enabling more robots to act social and allowing companies and employees to benefit from the advantages associated with this technology.

Acknowledgements This work has been supported by a grant of the Bavarian Research Society under the project "FORSocialRobots" AZ-1594-23.

References

1. International Federation of Robotics: Robots in Daily Life: The Positive Impact of Robots on Wellbeing. Frankfurt (2021)
2. Lathan, C.E., Ling, G.: Top 10 Emerging Technologies of 2019: Engineering: Social Robots Play Nicely with Others. https://www.scientificamerican.com/article/top-10-emerging-technologies-of-2019/. Last accessed 08 April 2024
3. Mejia, C., Kajikawa, Y.: Bibliometric analysis of social robotics research: identifying research trends and knowledgebase. Appl. Sci. **7**, 1316 (2017)
4. Lambert, A., Norouzi, N., Bruder, G., et al.: A systematic review of ten years of research on human interaction with social robots. Int. J. Hum.-Comput. Interact. **36**, 1804–1817 (2020)
5. International Organization for Standardization: Robots and Robotic Devices—Vocabulary (8373:2012) (2012). https://www.iso.org/obp/ui/#iso:std:iso:8373:ed-2:v1:en. Accessed 30 April 2022
6. Sarrica, M., Brondi, S., Fortunati, L.: How many facets does a "social robot" have? A review of scientific and popular definitions online. Inf. Technol. People **33**, 1–21 (2020)
7. Roesler, E., Naendrup-Poell, L., Manzey, D., et al.: Why context matters: the influence of application domain on preferred degree of anthropomorphism and gender attribution in human-robot interaction. Int. J. Soc. Robot. **14**, 1155–1166 (2022)
8. Duffy, B.R.: Anthropomorphism and the social robot. Robot. Auton. Syst. **42**, 177–190 (2003)
9. Fong, T., Nourbakhsh, I., Dautenhahn, K.: A survey of socially interactive robots. Robot. Auton. Syst. **42**, 143–166 (2003)
10. Smids, J., Nyholm, S., Berkers, H.: Robots in the workplace: a threat to—or opportunity for—meaningful work? Philos. Technol. **33**, 503–522 (2020)
11. Yam, K.C., Tang, P.M., Jackson, J.C. et al.: The rise of robots increases job insecurity and maladaptive workplace behaviors: multimethod evidence. J. Appl. Psychol. (2022)
12. Naneva, S., Sarda, G.M., Webb, T.L., et al.: A systematic review of attitudes, anxiety, acceptance, and trust towards social robots. Int. J. Soc. Robot. **12**, 1179–1201 (2020)
13. Broadbent, E.: Interactions with robots: the truths we reveal about ourselves. Annu. Rev. Psychol. **68**, 627–652 (2017)
14. Vänni, K.J., Korpela, A.K.: Role of social robotics in supporting employees and advancing productivity. In: Tapus, A. (ed.) Social Robotics: 7th International Conference, ICSR 2015, Paris, France, October 26–30, 2015; Proceedings, pp. 674–683. Springer, Cham (2015)
15. Sartore, M., Ocnarescu, I., Joly, L.-R. et al.: Ikigai robotics: how could robots satisfy social needs in a professional context? A positioning from social psychology for inspiring the design of the future robots. In: Cavallo, F., Cabibihan, J.-J., Fiorini, L. et al. (eds.) Social Robotics: 14th International Conference, ICSR 2022, Florence, Italy, December 13–16, 2022, Proceedings, vol. 13818, pp. 701–709. Springer, Cham (2022)
16. Müller, R., Franke, J., Henrich, D., et al. (eds.): Handbuch Mensch-Roboter-Kollaboration. Hanser, München (2019)
17. Mahdi, H., Akgun, S.A., Saleh, S., et al.: A survey on the design and evolution of social robots—past, present and future. Robot. Auton. Syst. **156**, 104193 (2022)
18. Onnasch, L., Roesler, E.: A taxonomy to structure and analyze human-robot interaction. Int. J. Soc. Robot. **13**, 833–849 (2021)
19. Bartneck, C., Forlizzi, J.: A design-centred framework for social human-robot interaction. In: RO-MAN 2004. 13th IEEE International Workshop on Robot and Human Interactive Communication (IEEE Catalog No. 04TH8759), pp. 591–594. IEEE (2004)
20. Mori, M.: The Uncanny Valley: The Original Essay by Masahiro Mori. IEEE Spectrum (2012)

21. Graaf, M.M.A. de, Ben Allouch, S., van Dijk, J.A.G.M.: What makes robots social?: A user's perspective on characteristics for social human-robot interaction. In: Tapus, A., André, E., Martin, J.-C. et al. (eds.) Social Robotics, vol. 9388, pp. 184–193. Springer International Publishing, Cham (2015)
22. Coronado, E., Mastrogiovanni, F., Indurkhya, B., et al.: Visual programming environments for end-user development of intelligent and social robots, a systematic review. J. Comput. Lang. **58**, 100970 (2020)
23. Malle, B.F., Scheutz, M.: Moral competence in social robots. In: 2014 IEEE International Symposium on Ethics in Science, Technology and Engineering (ETHICS 2014), Chicago, Illinois, USA, 23–24 May 2014, pp. 1–6. IEEE, Piscataway, NJ (2014)
24. Passalis, N., Pedrazzi, S., Babuska, R. et al.: OpenDR: An Open Toolkit for Enabling High Performance, Low Footprint Deep Learning for Robotics. In: 2022 IEEE/RSJ International Conference on Intelligent Robots and Systems (IROS). IEEE (2022)
25. Sarabia, M., Ros, R., Demiris, Y.: Towards an open-source social middleware for humanoid robots. In: 2011 11th IEEE-RAS International Conference on Humanoid Robots. IEEE (2011)
26. Ruhland, K., Peters, C.E., Andrist, S., et al.: A review of eye gaze in virtual agents, social robotics and HCI: behaviour generation, user interaction and perception. Comput. Graph. Forum **34**, 299–326 (2015)
27. McColl, D., Nejat, G.: Recognizing emotional body language displayed by a human-like social robot. Int. J. Soc. Robot. **6**, 261–280 (2014)
28. Marmpena, M., Garcia, F., Lim, A. et al.: Data-Driven Emotional Body Language Generation for Social Robotics (2022)
29. Jokinen, K., Wilcock, G.: Dialogues with Social Robots: Enablements, Analyses, and Evaluation. Springer, Singapore (2017)
30. O'Hare, G.M.P., Duffy, B.R., Collier, R. et al.: Agent Factory: Towards Social Robots (1999)
31. Bonaccorsi, M., Fiorini, L., Cavallo, F., et al.: A cloud robotics solution to improve social assistive robots for active and healthy aging. Int. J. Soc. Robot. **8**, 393–408 (2016)
32. Coşar, S., Fernandez-Carmona, M., Agrigoroaie, R., et al.: ENRICHME: perception and interaction of an assistive robot for the elderly at home. Int. J. Soc. Robot. **12**, 779–805 (2020)
33. Portugal, D., Alvito, P., Christodoulou, E., et al.: A study on the deployment of a service robot in an elderly care center. Int. J. Soc. Robot. **11**, 317–341 (2019)
34. Fasola, J., Mataric, M.A.: Socially assistive robot exercise coach for the elderly. J. Hum.-Robot Interact. **2** (2013)
35. Fan, J., Bian, D., Zheng, Z., et al.: A robotic coach architecture for elder care (ROCARE) based on multi-user engagement models. IEEE Trans. Neural Syst. Rehabil. Eng. **25**, 1153–1163 (2017)
36. Torta, E., Werner, F., Johnson, D.O., et al.: Evaluation of a small socially-assistive humanoid robot in intelligent homes for the care of the elderly. J. Intell. Robot Syst. **76**, 57–71 (2014)
37. Gross, H.-M., Schroeter, C., Mueller, S. et al.: I'll keep an eye on you: home robot companion for elderly people with cognitive impairment. In: IEEE International Conference on Systems, Man, and Cybernetics (SMC), 2011, 9–12 Oct. 2011, Anchorage, Alaska, USA; conference proceedings; [including workshop papers], pp. 2481–2488. IEEE, Piscataway, NJ (2011)
38. Cao, H.-L., Esteban, P.G., de Beir, A. et al.: A Platform-Independent Robot Control Architecture for Multiple Therapeutic Scenarios (2016)
39. Dehkordi, P.S., Moradi, H., Mahmoudi, M., et al.: The design, development, and deployment of RoboParrot for screening autistic children. Int. J. Soc. Robot. **7**, 513–522 (2015)
40. Casas, J., Gomez, N.C., Senft, E. et al.: Architecture for a social assistive robot in cardiac rehabilitation. In: 2018 IEEE 2nd Colombian Conference on Robotics and Automation (CCRA). IEEE (2018)

41. Hoang-Long, C., Van De Greet, P., James, K., et al.: A personalized and platform-independent behavior control system for social robots in therapy: development and applications. IEEE Trans. Cogn. Dev. Syst. **11**, 334–346 (2019)
42. Mead, R., Wade, E., Johnson, P. et al.: An architecture for rehabilitation task practice in socially assistive human-robot interaction. In: 19th International Symposium in Robot and Human Interactive Communication. IEEE (2010)
43. Kim, J.-H., Jeong, I.-B., Park, I.-W., et al.: Multi-layer architecture of ubiquitous robot system for integrated services. Int. J. Soc. Robot. **1**, 19–28 (2009)
44. Louie, W.-Y.G., Vaquero, T., Nejat, G. et al.: An autonomous assistive robot for planning, scheduling and facilitating multi-user activities. In: 2014 IEEE International Conference on Robotics and Automation (ICRA). IEEE (2014)
45. Gonzalez-Pacheco, V., Ramey, A., Alonso-Martin, F., et al.: Maggie: a social robot as a gaming platform. Int. J. Soc. Robot. **3**, 371–381 (2011)
46. Hirth, J., Schmitz, N., Berns, K.: Towards social robots: designing an emotion-based architecture. Int. J. Soc. Robot. **3**, 273–290 (2011)
47. Asprino, L., Ciancarini, P., Nuzzolese, A.G., et al.: A reference architecture for social robots. J. Web Semant. **72**, 100683 (2022)
48. Davis, F.D.: Perceived usefulness, perceived ease of use, and user acceptance of information technology. MIS Q. **13**, 319 (1989)
49. Lucke, D.: Akzeptanz: Legitimität in der "Abstimmungsgesellschaft". VS Verlag für Sozialwissenschaften, Wiesbaden, s.l. (1995)
50. Merz, N., Franke, J., Bodendorf, F.: Are we prepared for the rise of service robots?—A review on acceptance measurement. In: The Human Side of Service Engineering. AHFE International (2022)
51. Heerink, M., Kröse, B., Evers, V., et al.: Assessing acceptance of assistive social agent technology by older adults: the Almere model. Int. J. Soc. Robot. **2**, 361–375 (2010)
52. Venkatesh, V., Davis, F.D.: A theoretical extension of the technology acceptance model: four longitudinal field studies. Manag. Sci. **46**, 186–204 (2000)
53. Savela, N., Turja, T., Oksanen: A social acceptance of robots in different occupational fields: a systematic literature review. Int. J. Soc. Robot. **10**, 493–502 (2018)

ROS-Driven Simulation Framework for Exoskeletons in a Robot-Based Development Environment

Ramazan Gökay, Rajal Nagwekar, Seyed Milad Mir Latifi, Julian Öltjen, Alexander Stark, and Robert Weidner

Abstract

One of the biggest challenges in developing exoskeletons is that they need a long development process. Emulation of exoskeleton behaviour can be a noticeable advantage for exoskeleton development and evaluation, e.g., by reducing the number of necessary physical exoskeleton prototypes and thus accelerating system optimization and evaluation. This paper proposes a framework to emulate the behaviour of exoskeletons by using an approach with collaborative robots (cobots), motion tracking system

R. Gökay (✉) · R. Nagwekar · S. M. M. Latifi · R. Weidner
Laboratory of Manufacturing Technology, Helmut-Schmidt-University/University of the Federal Armed Forces Hamburg, Hamburg, Germany
e-mail: ramazan.goekay@hsu-hh.de

R. Nagwekar
e-mail: rajal.nagwekar@hsu-hh.de

S. M. M. Latifi
e-mail: milad.mirlatifi@hsu-hh.de

R. Weidner
e-mail: robert.weidner@hsu-hh.de; Robert.Weidner@aas.tu-freiberg.de

J. Öltjen · A. Stark
voraus robotik GmbH, Hannover, Germany
e-mail: julian.oeltjen@vorausrobotik.com

A. Stark
e-mail: alexander.stark@vorausrobotik.com

R. Weidner
Chair for Automated and Autonomous Systems, TU Bergakademie Freiberg, Freiberg, Germany

M.-C. Wanner et al. (eds.), *Annals of Scientific Society for Assembly, Handling and Industrial Robotics 2024*, https://doi.org/10.1007/978-3-031-91463-8_20

and Robot Operating System 2 (ROS2). The proposed approach offers an ability to see the exoskeleton-human interaction in physical environment while the exoskeleton is configured in digital environment. Within this ability, it differentiates from existing approaches. According to system testing, it is seen that the measured force profile and calculated force profile are close to each other. Therefore, the proposed approach gives promising results about that the physical simulation of exoskeleton is possible by using motion capture system, exoskeleton simulation framework and cobots. In future, the developed system will be further confirmed with more experimentation to improve the settings.

Keywords

Exoskeleton • Physical and virtual exoskeleton simulation • ROS • Motion tracking • Robot control

1 Introduction

Exoskeletons are wearable, mechanical or robotic systems which aim to support and enhance human physical capabilities [1]. They are used in different sectors, e.g., in rehabilitation of impaired people, in military to enhance strength and endurance or support workers during long-term and repetitive industrial tasks [2, 3]. Depending on the use cases, exoskeletons have different functionalities and dimensions and thus different requirements, coming from human beings, activity and support.

Due to the direct interaction between the user and exoskeleton, the development of exoskeletons requires the consideration of human-exoskeleton movement synchronization. For developing such systems, interdisciplinary knowledge as well as a human-centred approach including methods for describing and modelling properties are helpful.

In a typical process of exoskeleton development, firstly the use case is analysed, and requirements are figured out. Next, draft ideas are derived, and the preliminary design stage is started [4]. After that, the detailed design stage is moved on basis of the preliminary design, so that early stage protypes or later prototypes with increased maturity are built up. These prototypes are normally tested first in laboratory and then in field environment on different criteria to get feedback and use these results for optimizing the system [5]. Traditional development approaches include some tasks which require a significant amount of effort, time and budget. Since these ways are mostly based on "trial and error" procedure, they require frequent prototype iteration and field studies [6] and cause long development procedures.

In scope of this paper, a ROS2 simulation framework for emulating the exoskeleton behaviour is proposed. In Sect. 2, the existed approaches in the literature are mentioned. In Sect. 3, the proposed method and system components are introduced. In Sect. 4, the developed system architecture is explained and then the integration of components is showed. In Sect. 5, the experimental setup and limitations are presented and then the experiment results are analysed. In Sect. 6, the presented work is concluded and then ideas for the future work are discussed.

2 Related Works

Within technological developments, digital solutions have become widespread so that exoskeleton development process is shortened [7]. Various design and evaluation ideas are realized quickly by using simulation frameworks and digital models. Biomechanical model-based approaches are included in the exoskeleton development process [8]. Multi-body simulation methodologies are used to evaluate exoskeletons [8, 9]. A co-simulation model which includes musculoskeletal human model, and the models of the technical systems are proposed to see the impact of technical systems on the human body [10]. Robotic-based exoskeleton testing methods are also proposed [11]. With the use of robotics simulators, researchers and engineers can develop, test, and evaluate robotic systems in virtual settings [12]. One of the most popular platforms used by robotics researchers for a wide range of applications, including motion planning, control, sensing, localization, and mapping, is ROS [13].

As an alternative exoskeleton development approach, a development environment is proposed in scope of the dtec.bw (Digitalization and Technology Research Center of the Bundeswehr) project EVO-MTI [14]. This development environment aims to simulate exoskeletons and users with physical environment [5]. The physical simulation of exoskeletons including support characteristic should be realized by using several cobots [5]. Therefore, ROS2 environment is used to model all relevant properties.

3 Methodology and System Components

In scope of emulation of the exoskeleton behaviours, a simulation environment which includes physical mock-ups is prepared. Since exoskeletons are worn by humans, there is a direct interaction between both humans and exoskeleton [15]. This requires a perfect matching between human kinematics and exoskeleton workspace with respect to task.

To simulate the exoskeleton behaviour, actuated technical systems which can be adjustable for different exoskeleton functionalities, use case and human anthropometry are needed. This situation pops out two main needs: following the human motion and applying respective support force. To meet these needs, a development environment which includes motion capture system and cobots is proposed. A robot-based approach for shoulder exoskeleton simulation is shown as an example in Fig. 1. The connection between motion capture system and cobot is provided via the developed simulation framework.

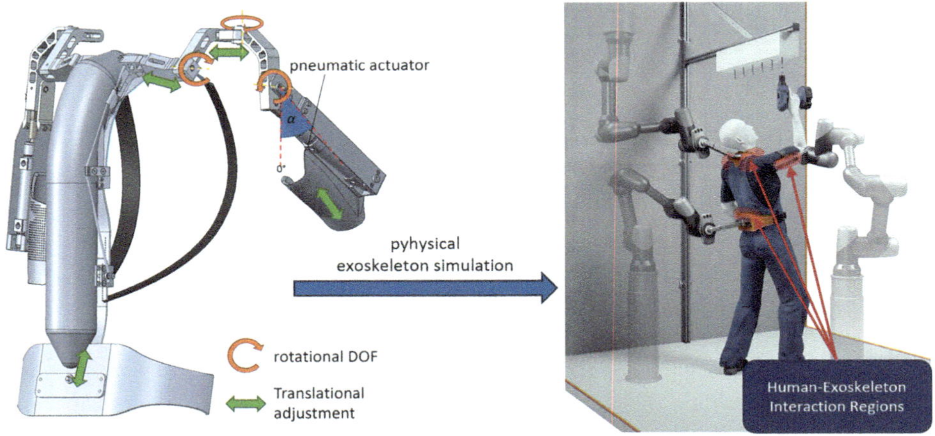

Fig. 1 Robot-based approach for simulating shoulder exoskeleton as an example

3.1 Hardware Components

As a first example, the shoulder exoskeleton "Lucy" [16] is used to implement and confirm the simulation environment. Thereby, further developments and later evaluation can be helped within a standardized framework. With the employed control system and the pneumatic actuator, the exoskeleton effectively supports the shoulder joint during tasks at head level or above [16].

To model physical interaction and implement the properties of support systems in real-world environments, cobots are used in scope of this paper. These cobots play a crucial role in simulating the kinematic chain of exoskeletons and the interaction forces appearing at the interfaces. In proposed approach, YU5 (Industrial cobot, Agile Robots AG, Germany) cobots are used as they meet the necessary criteria. These cobots boast a maximum support force of 70 N at a working range of 850 mm. To ensure the requisite velocities for accurately tracking human movement, motion capture data from various applications were recorded tracked using the end effector of cobot.

The Vicon Bonita motion capture system, a marker-based approach, is used for obtaining the position and orientation of the human upper body. It uses a set of infrared cameras and markers. The system used for this approach consists of eight near-infrared cameras that run at 240 frames per second (fps) and have a recording frequency of 100 Hz. Each camera consists of a lens and strobe. The strobe is an array of near-infrared light signals that when emitted, reflects by the markers. The accuracy of this system is estimated to be around 0.5 mm.

3.2 Software Components

The developed simulation framework is constructed on ROS2 environment. ROS2 is an open-source middleware that is widely used for robotic applications [17]. Since it holds various libraries, development and monitoring tools, it has become a de facto standard in robotic applications. In ROS2, the structure is built on nodes. Nodes are defined as independent computational processes.

The software used to visualize the motion capture data is Vicon Nexus 2.8. It is widely used as a modelling and processing tool for motion analysis. It acts as a user interface (UI) to aid in calibrating the camera setup as well as setting the origin of the workspace. A set of markers were used at different points of the human body to extract the motion data. The UI enabled in creating subjects by setting the origin and axis for every marker set, real-time data stream enables to receive global position and orientation of the subject in the workspace.

4 Implementation

The developed framework consists of three main steps which are motion tracking, exoskeleton simulation and cobot control. In the proposed idea, firstly human motion is tracked by motion capture system. Next, the required support force is calculated via simulation of exoskeletal behaviour according to tracked human motion. In this implementation, simulation of shoulder exoskeleton is aimed. And finally, the cobot is moved with respect to calculated support profile and human motion.

4.1 System Architecture

In context of efficient, modular, sustainable system development, a ROS2-based simulation framework is proposed. The system is constructed on node architecture as illustrated in Fig. 2.

The *Vicon system* includes the test setup with the human in the Vicon workspace. There are sets of markers attached to the human on the arm, shoulder and pelvis to receive the position and orientation of the respective body motion. These marker sets are given the subject name of *arm_marker*, *shoulder_marker* and *pelvis_marker*. The *Vicon node* acts as a real-time data stream that publishes the global position and orientation of the captured markers. The *Body node* subscribes to the pose information of the marker, calculates the arm angle with respect to the marker pose and publishes the shoulder arm angle. The *Exo node* is a node that simulates exoskeletal behaviours. It therefore needs to subscribe to the arm angle topic to give a resulting support force output from the control loop. This support force is then published for the other existing nodes. The *Target node* plans the motion of

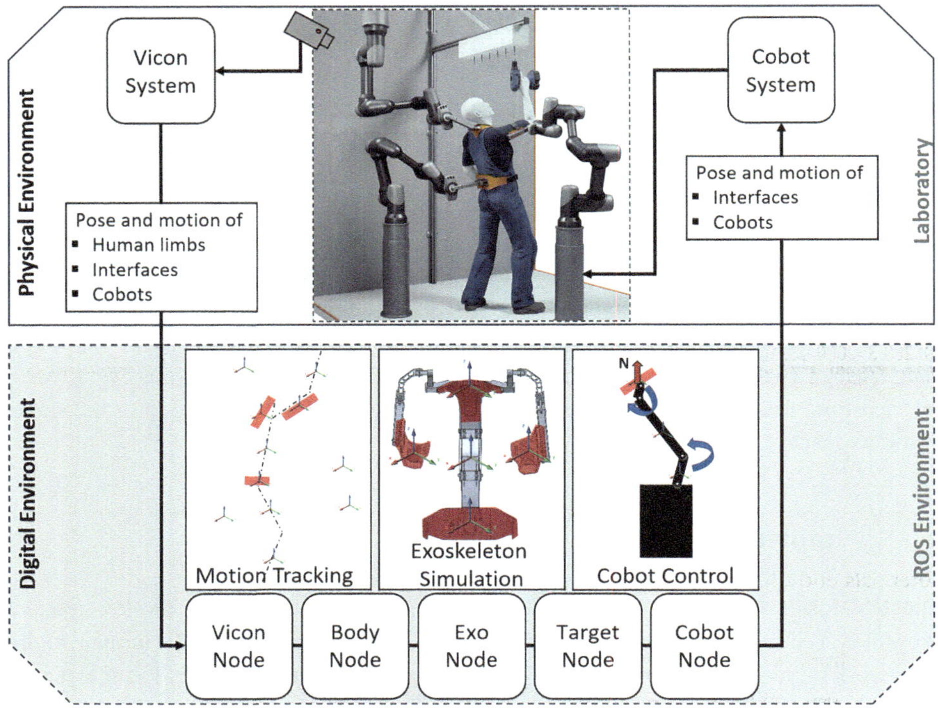

Fig. 2 Proposed system architecture and signal flow with respect to physical and digital environment

the cobot. It subscribes to the support force topic and the target motion trajectory is planned. This target motion information is then published as a topic. The *Cobot node* acts as a command center of the cobot. It moves the cobot with respect to the target trajectory information it receives after subscribing to the target generator topic. In addition, it helps in controlling the cobot via a user interface. This UI is used for resetting errors, setting cobot impedance parameter and requesting cobot operation. This node publishes the joint states of the cobot. The *Cobot system* represents the cobot. It moves according to the subscribed joint state topic. The cobot applies the resulting support force depending on the arm angle of the human.

4.2 System Integration

The system has a heterogeneous structure which includes detection, tracking, simulation and control modules on different platforms as shown in Fig. 3. The markers that are attached to human body are detected by Nexus software on Windows OS based Vicon

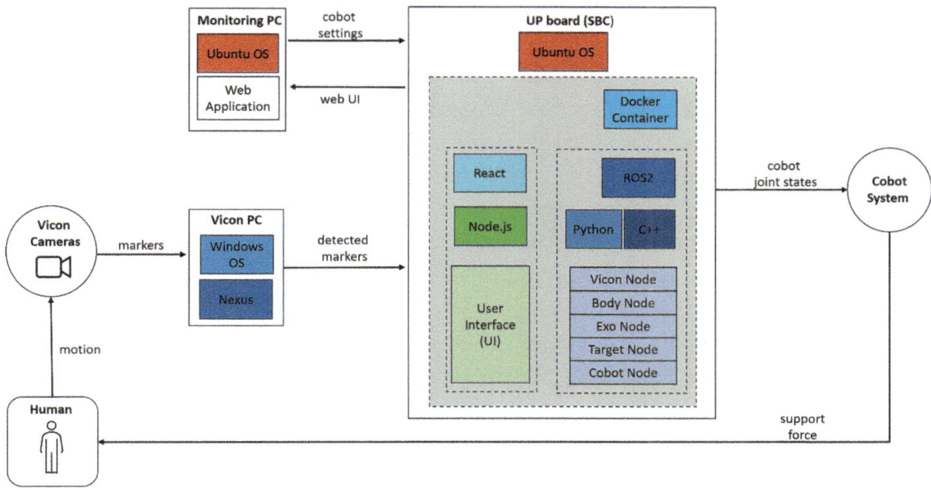

Fig. 3 System integration with respect to hardware and software components

computer. Pose information of the detected markers are sent to UP board. UP board is the brain of the system. It is a single board computer (SBC) which runs on Ubuntu OS. All developed modules are run on UP board. These modules are developed by using various frameworks, libraries and programming languages. The motion tracking, exoskeleton simulation and cobot control modules are developed in ROS2 environment by using C++ and Python programming languages. The user interface module is developed by using React library and Node.js runtime environment. The system is watched via web application through monitoring computer.

To make the development and deployment of the system manageable level, Docker container [18] technology is used. Thus, the dependencies of the system can be installed easily, the deployment of the system is realized without problem and the system is executed as independent from the host machine.

5 System Testing

The system testing aims to justify the proposed system simulates the behaviour of the shoulder exoskeleton through developed ROS2-based simulation framework.

5.1 Experimental Setup and Limitations

In the experimentation procedure, the use case of drilling including the resulting human motion is simulated as a task. This task is repeated several times. A person wears a Vicon-suit, and three marker-sets are placed on the arm, shoulder and pelvis regions to obtain the arm angle as shown in Fig. 4. While the person moves his right arm continuously between 0° and 90° in the Vicon workspace, the cobot applies emulated support force on the table according to arm angle. The applied force is measured by an external force-torque sensor, which is mounted on the robot flange. This measured force simulates the upper-arm support of shoulder exoskeleton in case of drilling task.

In this setup, the robot is operated in impedance control mode and configured for "zero-gravity free-drive", which means the gravitation forces of the robot due to its own mass are compensated, while the robot can be moved freely in the complete workspace

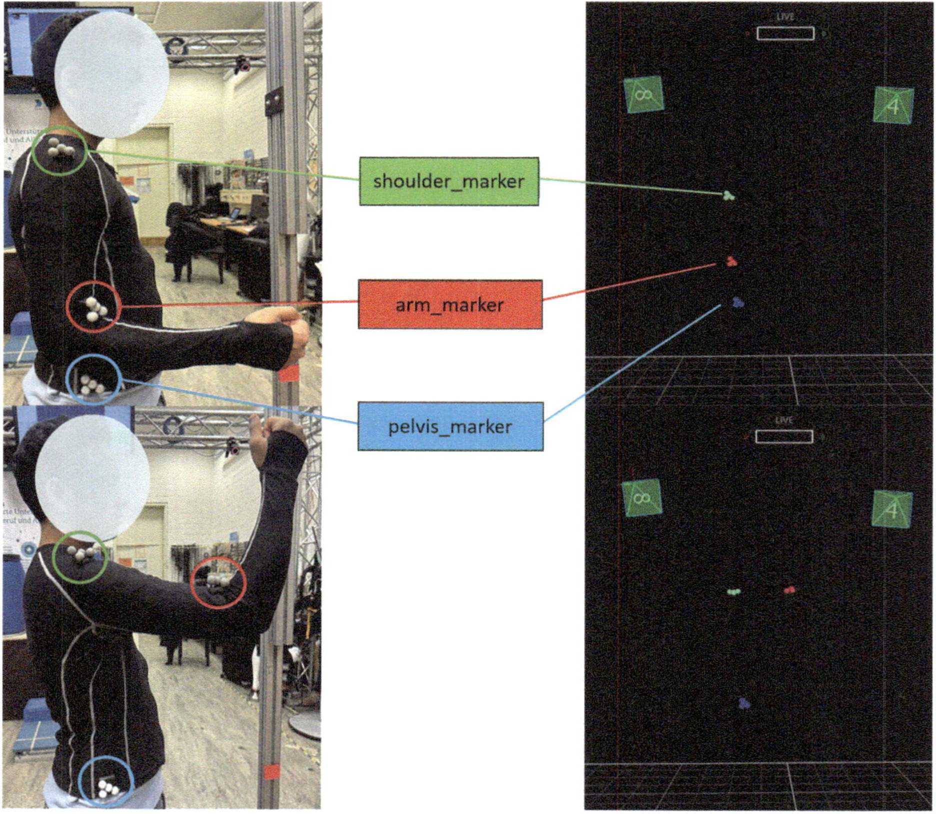

Fig. 4 Marker placement on human body (on left) and detected markers on Nexus (on right)

and apply forces to its environment. The upper arm motion of a person is tracked within three tracking marker sets. A desired force is decided by the *Exo node* and adapted by *Target node* based on the angle between upper arm and torso.

As experimentation limitations, firstly the direct human-cobot interaction is avoided because of safety precautions. Another limitation is the limited workspace because of optical motion capture system. The detection of three marker-set is crucial to calculate the arm angle, so that it requires specific workspace.

5.2 Results and Discussion

The experiment results where a person is tracked while raising and lowering the arm at different speeds are shown in Fig. 5. The plot shows the tracked arm angle (magenta colour in lower plot), the calculated desired force (blue colour in upper plot) and the measured force of the external force-torque sensor (red colour in upper plot), as well as the difference (yellow colour in upper plot) between the desired and measured force over time.

When the arm moves from 0° to 90°, the required support force and the measured force increase. At 90° arm angle, the required support force and measured force reaches maximum value (70 N). When the arm moves from 90° to 0°, the required support force and the measured force decrease.

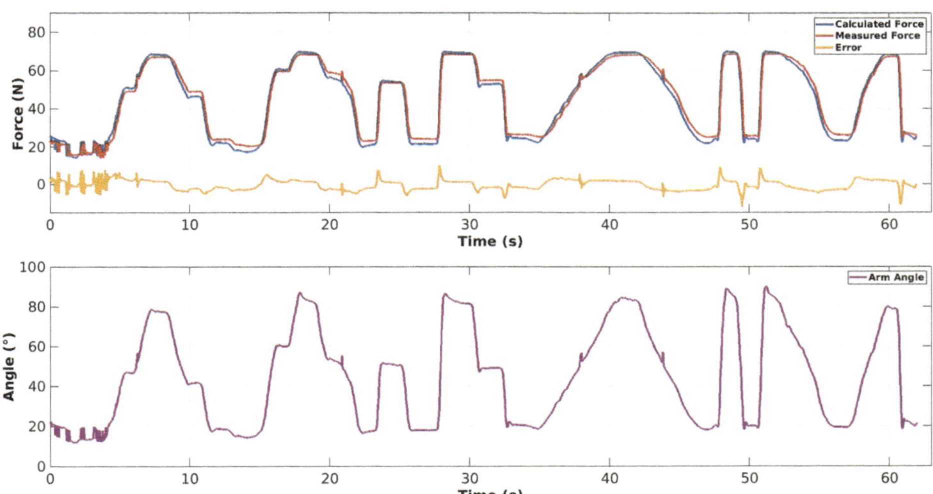

Fig. 5 Force profiles that simulate upper-arm support force with respect to arm movement at different speed

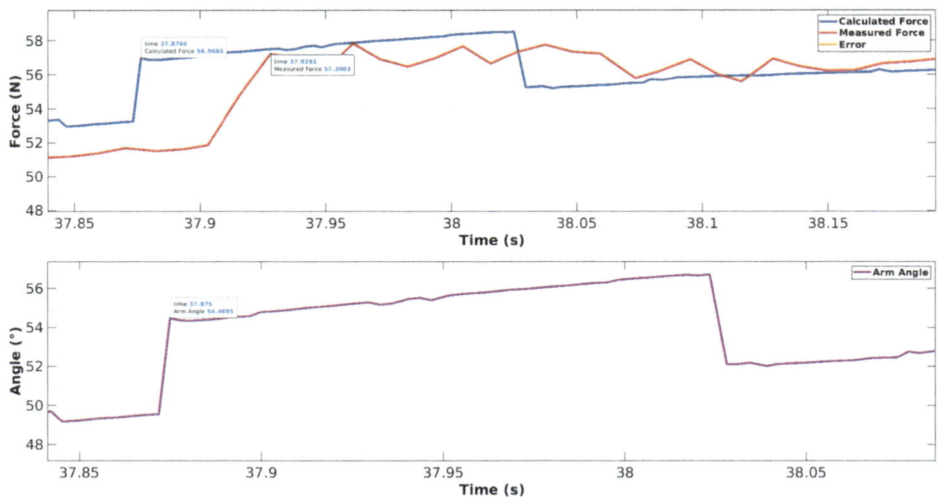

Fig. 6 Latency analysis on arm angle detection, commanded force and applied force

For increasing force characteristics, the measurements show a following behaviour with an error between desired and measured force is less than 5 N (force error is 7.14%). With decreasing force, a delayed response of the real force application is observed. Further, a remaining deviation between the desired and applied force is less than 10 N (force error is 14.28%) stays.

The higher following error and remaining error at decreasing force characteristics is attributed to disturbance effects due to breakout torque and stick friction in the robot drive trains. These effects have a particular impact for decreasing forces, since the robot applies the force to the sensor in the direction of gravity, i.e., negative z-direction of the world coordinate system. These effects can be compensated by adding a force feedback loop using an extra force-torque sensor on the robot flange. The added sensor provides more accurate feedback on the forces and torques applied by the robot to its environment.

The measurements show that the latency between commanded force and applied force is 51.5 ms while the latency between arm angle detection and commanded force is 1.6 ms as shown in Fig. 6. To decrease the latency, better processors can be used, and more optimized algorithms can be developed to control the system.

6 Conclusion

In scope of this paper, a ROS2-based simulation framework is proposed for exoskeleton development environment which consists of motion capture system and cobot. Traditional approaches require a long product development process, and full-simulation-based approaches lack a realistic physical human-exoskeleton interaction [5]. On the other hand, the proposed framework offers an ability to see the exoskeleton-human interaction in a physical environment while the exoskeleton is configured in digital environment. Within this ability, it differentiates from both full-simulation approaches and traditional approaches. In addition, a modular system architecture is developed via this method so that it offers some notable abilities such as adding new features or adapting to new exoskeleton comfortably. Since it drops physical prototype for each iteration, it shortens the exoskeleton development and evaluation process.

As a proof-of-concept, shoulder exoskeleton behaviour is modelled, and the required support force is applied through one cobot which is assumed to attach to human upper arm. The system testing shows promising results to simulate the behaviour of the exoskeleton by using motion capture system, exoskeleton simulation framework and cobot. In future, the developed system will be further confirmed with more experimentation to improve the settings. A more advanced exoskeleton modelling is aimed within increased number of cobots.

Acknowledgements This research is funded by dtec.bw which we gratefully acknowledge [project EVO-MTI]. dtec.bw is funded by the European Union—NextGenerationEU.

References

1. Pons, J.L.: Wearable Robots: Biomechatronic Exoskeletons. John Wiley & Sons (2008)
2. Weidner, R., Otten, B., Argubi-Wollesen, A., Yao, Z.: Support technologies for industrial production. In: Developing Support Technologies: Integrating Multiple Perspectives to Create Assistance that People Really Want, pp. 149–156 (2018)
3. Monica, L., Draicchio, F., Ortiz, J., Chini, G., Toxiri, S., Anastasi, S.: Occupational Exoskeletons: A New Challenge for Human Factors, Ergonomics and Safety Disciplines in the Workplace of the Future. Springer International Publishing, Cham (2021)
4. Weidner, R., Argubi-Wollesen, A., Karafillidis, A., Otten, B.: Human-Machine Integration as Support Relation: Individual and Task-Related Hybrid Systems in Industrial Production, i-com, vol. 16, no., 2, pp. 143–152 (2017)
5. Hoffmann, N., Latifi, S.M.M., Nagwekar, R., Weidner, R.: Towards a robotic-based development environment for designing and evaluating exoskeletons. In: ISR Europe 2023; 56th International Symposium on Robotics, pp. 354–360. VDE (2023)
6. Bengler, K., Harbauer, C.M., Fleischer, M.: Exoskeletons: A Challenge for Development. Wearable Technol. **4**, e1 (2023)
7. Zheng, L., Lowe, B., Hawke, A.L., Wu, J.Z.: Evaluation and test methods of industrial exoskeletons in vitro, in vivo, and in silico: A Critical Review. Crit. Rev. Biomed. Eng. **49**(4), 1–13 (2021)
8. Tröster, M., Wagner, D., Müller-Graf, F., Maufroy, C., Schneider, U., Bauernhansl, T.: Biomechanical model-based development of an active occupational upper-limb exoskeleton to support healthcare workers in the surgery waiting room. Int. J. Environ. Res. Public. Health. **17**(14), 5140 (2020)

9. Fritzsche, L., Gärtner, C., Spitzhirn, M., Galibarov, P.E., Damsgaard, M., Maurice, P., Babič, J.: Assessing the efficiency of industrial exoskeletons with biomechanical modelling—comparison of experimental and simulation results. In: Proceedings of the 21st Congress of the International Ergonomics Association (2021)
10. Molz, C., Scherb, D., Löffelmann, C., Sänger, J., Yao, Z., Lindenmann, A., Miehling, J.: A co-simulation model integrating a musculoskeletal human model with exoskeleton and power tool model. Appl. Sci. **14**(6), 2573 (2024)
11. Ito, T., Ayusawa, K., Yoshida, E., Kobayashi, H.: Evaluation of active wearable assistive devices with human posture reproduction using a humanoid robot. Adv. Robot. **32**(12), 635–645 (2018)
12. Asad, U., Khan, M., Khalid, A., Lughmani, W.A.: Human-centric digital twins in industry: A Comprehensive Review of Enabling Technologies and Implementation Strategies. Sensors **23**(8), 3938 (2023)
13. Macenski, S., Foote, T., Gerkey, B., Lalancette, C., Woodall, W.: Robot operating system 2: Design, Architecture, and Uses in the Wild. Sci. Robot. **7** (2022)
14. Dtec.bw: EVO-MTI—Umgebung zur Entwicklung für phys. Unterstützungssysteme. https://dtecbw.de/home/forschung/hsu/projekt-evo-mti. Last accessed 15 Feb 2024
15. Ajayi, M.O., Djouani, K., Hamam, Y.: Interaction control for human-exoskeletons. J. Control Sci. Eng. (2020)
16. Otten, B.M., Weidner, R., Argubi-Wollesen, A.: Evaluation of a novel active exoskeleton for tasks at or above head level. IEEE Robot. Autom. Lett. (2018)
17. Ros Wiki Website: https://wiki.ros.org/Documentation. Last accessed 14 March 2024
18. Bashari Rad, B., Bhatti, H., Ahmadi, M.: An introduction to Docker and analysis of its performance. IJCSNS Int. J. Comput. Sci. Netw. Secur. (2017)

Artificial Intelligence

Towards Explainable AI-Based In-Line Inspection and Quality Assurance for Drilling in Aircraft Manufacturing

Ole Stüven, Felix Geiger, Keno Moenck, and Thorsten Schüppstuhl

Abstract

In-line inspection and quality assurance are essential activities in manufacturing high-value and safety–critical products. In the scope of manufacturing aircraft fuselages, the high-quantity operation drilling is one of the critical steps, in which even slight improvements concerning quality or lot times are majorly impactful. Data recorded during drilling can be used to infer information about, e.g., the quality of the hole. Then, Artificial Intelligence (AI) methods can infer information about the quality of the process. More complex AI models have the advantage of more advanced predictive capabilities, which comes at the cost of being opaque even to experts. That hinders the deployment of the models, especially if the human must still be part of the process. Here, methods of Explainable AI (XAI) can augment AI predictions with explanations to enable humans to make more informed decisions. In this work, we propose a process framework that uses XAI to enable efficient in-line inspection of drill holes.

Keywords

Drilling • Aircraft manufacturing • In-line inspection • Quality assurance • XAI

O. Stüven (✉) · F. Geiger · K. Moenck · T. Schüppstuhl
Institute of Aircraft Production Technology, Hamburg University of Technology, Hamburg, Germany
e-mail: ole.stueven@tuhh.de
URL: https://www.tuhh.de/ifpt

M.-C. Wanner et al. (eds.), *Annals of Scientific Society for Assembly, Handling and Industrial Robotics 2024*, https://doi.org/10.1007/978-3-031-91463-8_21

1 Introduction

Drilling and, subsequently, riveting is an essential process within the production and assembly of aircraft structures, including the connection of the fuselage elements, e.g., frames, stringer, and skin. It can be estimated that about half of the assembly cycle consists of drilling and riveting tasks [1]. Figure 1 displays a section of an aircraft fuselage with an expectable distribution of drill holes and riveting joints. The major challenge that arises from drilling is poor hole quality, which can lead to cracks within the airframe and reduce its reliability [2]. However, because of the high-security requirements, high reliability and process quality are necessary. Increasing the process reliability includes, e.g., premature tool change or the insertion of additional connections to counteract the uncertain drill hole quality. The drilling process for riveted joints in aircraft production consists of drilling and countersinking the hole. Therefore, highly specialized automatic drilling machines or robots with drilling tools are used [3].

Inspection and quality assurance are essential for successful automation. The use of AI to monitor, secure, and eventually optimize processes has been a popular field of research for years. Subsequently, a variety of AI-based applications have been developed and successfully implemented within a wide range of manufacturing processes. However, AI models are usually challenging to comprehend and trust due to their black-box nature, which makes effective use in sensitive areas such as the aircraft industry very difficult from a regulatory perspective. Thus, XAI has become a popular research subject within AI in recent years [4].

In this work, we analyze and evaluate the possibility of using methods of XAI to in-line optimize the drilling process for the production and assembly of aircraft structures while providing transparency, interpretability, and explainability.

The rest of this work is structured as follows: First, in Sect. 2, we briefly summarize state-of-the-art measuring methods for drill hole inspection, the usage of AI in process

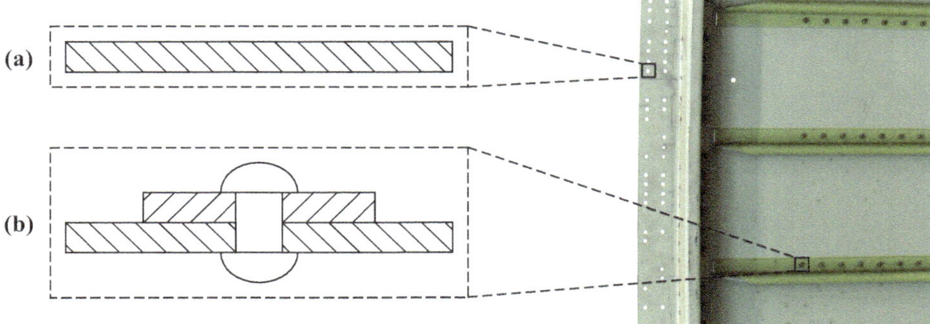

Fig. 1 Drill holes **a** and riveted joints **b** on an aircraft fuselages

monitoring and assurance, and common XAI methods. In Sect. 3, we propose our process framework and, in Sect. 4, a concept of an experimental setup to repeatedly drill and accurately measure boreholes to generate process data to evaluate XAI methods. Finally, we discuss and conclude this work in Sect. 5.

2 Related Work

2.1 Drill Hole Inspection

Drill hole inspection targets the assessment of parameters like diameter, roundness, angular deviation from the planned drilling axis, or surface quality. Furthermore, defects like cracks and, e.g., delamination in composite materials are relevant features. Process and tool data like forces, torques, vibrations, or noise [5] can be used solely or combined with additional process knowledge to estimate tool conditions and infer drill hole qualities. The named parameters can be derived from measuring quantities of different measuring principles, which we present in the following differentiating between tactile and non-tactile methods.

Non-complex tactile measurement methods can measure drill holes employing a probe that collects point coordinates from predefined spots within the drill hole through touching. A 3D point cloud can be derived from a multitude of measuring points that provide highly accurate information about parameters like geometry, straightness, or evenness. A disadvantage is the linear correlation between measuring time and point cloud resolution. Also, surface defects, e.g., cracks or blowholes, small in size, typically fall under the probing resolution, such as not being measurable.

There are several non-tactical measuring principles like laser scanning, ultrasonic measurement, and White Light Interferometry (WLI). Out of these, WLI offers the ability to produce point clouds of drill holes with the highest accuracy and resolution within a reasonable time. This detailed digitalization/representation of the drill hole is an important component of the proposed process framework of this paper since a plethora of quality information can be derived from it. Therefore, we focus on the usage of WLI for our use case.

Like conventional microscopes, systems based on the physical measuring principle of interference consist of an illumination unit, imaging optics, and a camera. The beam is split into a measurement beam and a reference beam. The measurement beam hits the surface to be measured, and the reference beam hits a reference mirror whose position is known. Light interference between the two paths allows for deriving changes in distances on the surface as measuring points. White light interferometry has already been successfully used for applications like crack inspection of engine combustion chambers [6], investigating pitting corrosion [7], automated assessment for grinding spots on aircraft landing gear components [8], or robot-based inspection of freeform components [9]. WLI systems for inspecting the drill hole wall are still subject to research and development, in which the authors are involved in concurrent work.

2.2 AI in Process Monitoring and Assurance

In order to ensure and increase the quality and reliability of automated manufacturing, the process must be monitored to post- or in-process optimize products, processes, or resources. In the scope of drilling, monitoring enables drawing conclusions about the process and the condition of the tool and, if necessary, process parameters to be corrected. Monitoring machining systems using AI process models includes several topics, such as sensor systems selection, sensor fusion methods, signal processing, feature selection/extraction, design of experiments, and AI models [10]. Compared to analytical process models, which require a deep understanding of the complex process, (end-to-end) learnable AI methods offer the potential for rapid evaluation of diverse, non-linear process and sensor data and process assessment. AI enables adaptive process control by, e.g., continuously monitoring parameters during the process and correcting them if necessary. As of today, a variety of modern AI-based methods are already being used to monitor different manufacturing processes. A holistic overview of AI in process monitoring is adopted from [11] and depicted in Table 1.

While inductive learning-based methods better generalize as well as depend on less expert knowledge, they have in common that decision-making and the generation of output resemble non-transparent black-box behavior so that human comprehensibility, acceptance, and trust are impaired, and effective use in sensitive areas such as medicine or aviation is problematic. Methods and approaches of XAI and associated aspects such as transparency, interpretability, and explainability are therefore playing an increasingly important role [12].

2.3 Methods of XAI

The term XAI is often used interchangeably with Interpretable AI (IAI). There is a wide range of methods that can be used for XAI. A possible categorization is adopted from [19] and depicted in Fig. 2.

On the top level, the methods can be divided into inherently interpretable models, model-agnostic, and model-specific methods. Interpretable models contain conventional statistical methods like linear regression or methods of classical machine learning like decision trees or k-nearest neighbors. These models have in common that humans can directly interpret the learned parameters. The weights of, e.g., a linear regression represent each feature's impact on the target value. The major disadvantage is that their predictive capabilities are not as good as those of more complex models containing more

Table 1 Overview of AI methods in process monitoring. Structure adopted from [11]

Area of application	AI-methods	Example of use
Assistance and learning systems	Neural networks	Supporting and enhancing the capabilities of employees by providing guidance on individual learning processes and controlling competence saturation [13]
Quality assurance	Neural networks, Decision tree, Support Vector machine, Bayesian network	Optical quality detection and defect classification with the possibility of identifying the root cause for quality deviations [14, 15]
Predictive maintenance	Decision tree, Principal, Component analysis	Fault detection and estimation of the remaining useful life for industrial equipment [16]
Process control	Artificial neural networks, Fuzzy logic, Particle swarm optimization, Response surface method	Modeling and adaptively controlling relevant process parameters [17]
Tool condition monitoring	Artificial neural networks, Back propagation, DNA-based computing	Monitoring and predicting the degree of tool wear as a function of machining time [18]

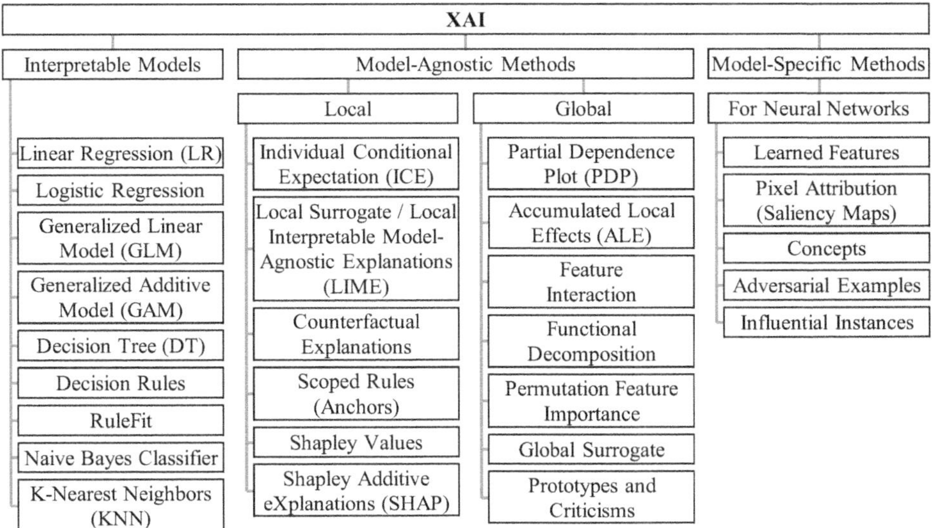

Fig. 2 Overview of XAI methods. Structure adopted from [19]

parameters. Model-agnostic methods are more complex and classifiable into local and global methods. Local means the methods provide explanations for single instances of a dataset, while global methods provide explanations for an entire model. A major advantage is the characteristic of model-agnostic, which facilitates substituting or comparing the deployed model. In contrast, model-specific XAI methods are suited for ANNs. Since ANNs perform better in non-linear and complex settings than inherently interpretable models, model-specific XAI methods have the potential to deliver explainable results even for complex correlations.

A large amount of literature exists on XAI, wherefore we build upon a recent and extensive survey paper. Alexander et al. gathered a large number of papers concerning IAI and XAI, selected the most relevant, and categorized them [20]. Of all listed papers, solely [21] examines XAI for a drilling process by predicting the state of drills from top-view RGB images of the holes in pieces of wood. The used method is dependent on visual features appearing on the surface of the working piece, which is not given for all use cases, e.g., the necessity of observing drill hole walls. Besides, the terms XAI and drilling appear together in the field of geothermal drilling, e.g., in [22]—unrelated and barely adaptable to drilling in manufacturing. There is no published research that directly focuses on using XAI for drilling processes that lie within the scope of this work. There are, however, many papers that propose and test XAI methods for manufacturing use cases. The application, adaption, further development, and evaluation of these methods for a drilling process that enables in-line inspection and quality assurance is, therefore, still an open research question.

3 Process Framework

For the automated inspection of a drill hole, a system must gather information about the hole, from which it can infer the drill hole's quality and further conclusions, e.g., how to influence and rectify the hole's quality. This information can either be gathered explicitly using a device that scans the hole or implicitly by evaluating the data collected during the drilling process. These explicit methods represent the actual state, while the implicit methods predict this state, which is why they are inherently less certain. The explicit method's disadvantage is that, depending on the device used, the inspection process is time-consuming, which renders it practically infeasible for every hole in a high-quantity application. The proposed process framework combines both methods to provide a solution for reliable and fast inspection of drill holes. The idea is to collect and correlate explicit and implicit data in the preliminary stage and use this gained knowledge in the production stage to infer hole quality from implicit data while collecting explicit data only sporadically to assure process stability. Methods of XAI are deployed on top of AI models to increase transparency, interpretability, and explainability for the operator, which then can decide how to influence/change the drilling process. This fundamental process framework is depicted in Fig. 3. In the following section, we elaborate on the steps of the process framework more closely.

Implicit data and controllable parameters that can be collected during the drilling process are listed in Table 2.

There are several methods and devices to scan drill holes that produce explicit data, e.g., 2D or 2,5D/3D data. Independent of the data type, the quality and failure mode of the hole has to be quantified and classified based on the data. Possible measuring principles and systems are listed in Sect. 2.1. The WLI, as a 2,5D sensor system, which can be extended to a 3D measuring system with additional rotational and linear axes, offers the

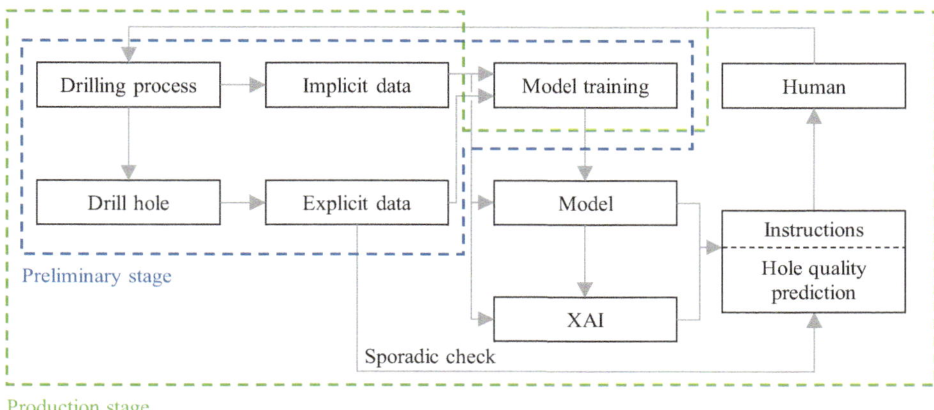

Fig. 3 Fundamental framework of the proposed process (the preliminary and production stages are highlighted with dashed lines)

Table 2 Possible process parameters to control and data to collect during the process

Process parameter	Data		
	Direct data	Metadata	Ambient data
Drill rotational speed	Reaction forces	Drill bit wear	Temperature
Feed profile	Reaction moments	Workpiece material	Humidity
Feed speed	Spindle current	Drill bit operation time	Room acoustic emission
Feed force	Spindle moment		
Drill bit	Spindle voltage		
Coolant flow rate	Spindle speed		
	Contact acoustic emission		

necessary accuracy and resolution for the use case compared to tactile systems or other non-tactile systems.

The prediction of the model should contain information about the quality of the hole and instructions on how to improve the process to rectify insufficient quality by adjusting process parameters. This instruction is created by correlating the process parameters with the hole quality in the preliminary stage. In the production stage, correction of the process parameters can be inferred based on the predicted hole quality. Potentially, multiple models or ensemble learning, which combines the predictions of multiple models, have to be utilized.

As mentioned in Sect. 2.2, there are various different approaches to using AI methods within different areas of application in process monitoring and assurance. In principle, all XAI methods listed in Sect. 2.3 are conceivable to be applied to the use case. Naturally, the method has to be compatible with the chosen AI model. The XAI method can augment the prediction of the model by, e.g., specifying which features in the underlying data led the model to its prediction. If, e.g., specific periods during the drilling operation are crucial for the prediction of a model, XAI methods can point out these periods and present humans with examples of holes with different qualities. Furthermore, a metric that represents the confidence of the prediction is helpful for a human to make a decision based on the prediction.

Since the input data listed in Table 2 is almost entirely time series data, the relevant fields of methods are time series classification and regression. A state-of-the-art method regarding accuracy in time series classification is HIVE-COTE 2.0 [23], an ensemble learner. Therefore, the method can serve as a first baseline approach. Since it is an ensemble learner, model-agnostic XAI methods are best suited to augment its predictions.

4 Conceptual Test Design

The framework proposed in Sect. 3 is used to develop an experimental setup to test and validate the process. In Fig. 4, a draft of the setup is depicted. The setup consists of an xy-stage to position the coupon in the horizontal plane. The spindle and WLI are mounted in such a way that they can access the whole horizontal coupon area and are vertically actuated. All listed data and process parameters from Table 2 are chosen to be collected within the setup due to their accessibility and potential information concerning the quality of the hole. A 6-axis force/torque load cell is used to measure reaction forces and moments. Measurement of spindle speed and moment is directly integrated into the spindle. A contact microphone is used to capture acoustic emissions on the spindle. Temperature and humidity sensors are installed in the drill's vicinity. Metadata is generated by an application that runs on a control system and keeps track of the workpiece material and the drill bit operation time. If known, the wear of a newly inserted drill bit can be optionally input by the operator. All available process parameters are also captured.

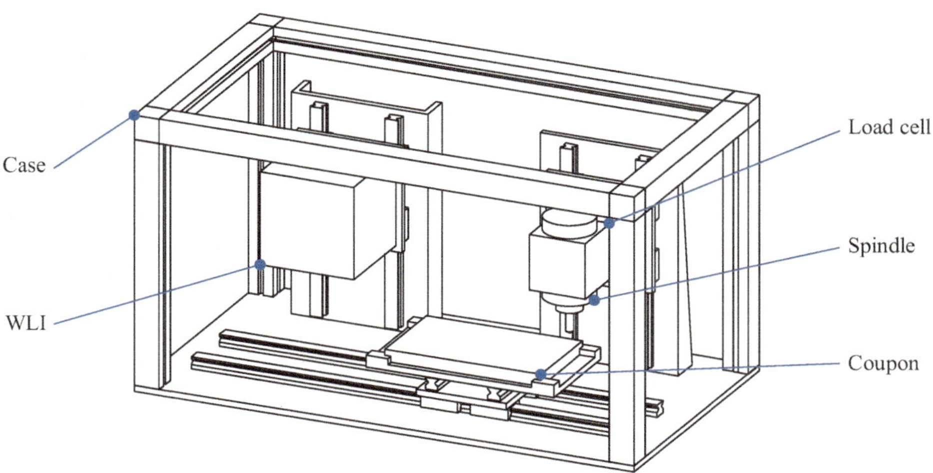

Fig. 4 Draft of an experimental setup to test and validate the proposed process framework

Explicit data is captured with the WLI, which generates high-accuracy and resolution 3D point clouds. Different AI models can be examined to extract quality data from the point cloud and correlate explicit and implicit data.

In the preliminary stage, data has to be generated by drilling holes in coupons and collecting all measurements. To make the model sensitive to all relevant failure modes, these have to occur in the data with sufficient quantity. Therefore, expert knowledge is needed to adjust the setup in such a way that the relevant failure modes occur. We can, e.g., achieve it by purposefully using a damaged or worn drill bit, but identifying ways to reproduce all relevant failure modes is one of the major challenges in the preliminary stage. Afterward, an algorithm has to be developed that is able to classify and quantify a hole based on the explicit data regarding the failure modes and further parameters, e.g., roundness or diameter. This serves as the ground truth, which is correlated with the implicit data. Here, different models will be evaluated, and the best ones will be chosen to be augmented with XAI methods.

The production stage is tested on the setup by artificially introducing sources of errors. Firstly, it is evaluated to see if the model is able to accurately predict the failure modes of the holes from the implicit data by comparing it to the explicit data. Furthermore, it can be evaluated if the model is able to rectify failure modes by adjusting the process parameters according to the model's recommendation.

5 Conclusion

In this work, a novel general process framework to enable XAI-based in-line drilling inspection is proposed. By combining explicit data from scanning devices and implicit data collected during the process, we estimate that a fast and reliable process can be established. We also presented an overview of the relevant topics: drill hole inspection, AI in process monitoring, and XAI. Furthermore, we presented an experimental setup that can be used in future work to develop and evaluate the process further. The motivation for the proposed process framework comes from aircraft production. Nevertheless, it can probably be adapted to other drilling tasks or technologies.

Acknowledgements This work is part of the research project *Ganzheitliches Datenmodell und XAI-basierte Analysen zur Bohrprozessadaption* (HoleListic, grant number: *20D2205C*), supported by the *Federal Ministry for Economic Affairs and Climate Action* (BMWK) as part of the *Federal Aeronautical Research Programme LuFo VI-3.*

References

1. Zhang, L., Gao, W., Lu, D., Zeng, D., Lei, P., Yu, J., Tang, M.: Robotic simple and fast drilling system for automated aircraft assembly. Int. J. Adv. Manuf. Technol. **122**(1), 411–426 (2022). https://doi.org/10.1007/s00170-022-09789-7
2. Aamir, M., Giasin, K., Tolouei-Rad, M., Vafadar, A.: A review: drilling performance and hole quality of aluminium alloys for aerospace applications. J. Mater. Res. Technol. **9**(6), 12484–12500 (2020). https://doi.org/10.1016/j.jmrt.2020.09.003
3. Zhao, H., Xi, J., Zheng, K., Shi, Z., Lin, J., Nikbin, K., Duan, S., Wang, B.: A review on solid riveting techniques in aircraft assembling. Manuf. Rev. **7**, 40 (2020). https://doi.org/10.1051/mfreview/2020036
4. Ali, S., Abuhmed, T., El-Sappagh, S., Muhammad, K., Alonso-Moral, J.M., Confalonieri, R., Guidotti, R., Del Ser, J., Díaz-Rodríguez, N., Herrera, F.: Explainable artificial intelligence (XAI): What we know and what is left to attain trustworthy artificial intelligence. Inf. Fusion **99**, 101805 (2023). https://doi.org/10.1016/j.inffus.2023.101805
5. Koch, J., Schoepflin, D., Venkatanarasimhan, A., Schüppstuhl, T.: Tool wear classification in automated drilling operations of aircraft structure components using artificial intelligence methods. SAE Int. J. Adv. Curr. Prac. Mobil. **4**(4), 1072–1081 (2022). https://doi.org/10.4271/2022-01-0040
6. Domaschke, T., Schueppstuhl, T., Otto, M.: Robot guided white light interferometry for crack inspection on airplane engine components. In: ISR/Robotik 2014; 41st International Symposium on Robotics, pp. 1–7 (2014)
7. Holme, B., Lunder, O.: Characterisation of pitting corrosion by white light interferometry. Corros. Sci. **49**(2), 391–401 (2007). https://doi.org/10.1016/j.corsci.2006.04.022
8. Ehrbar, J., Kähler, F., Schüppstuhl, T.: Automated assessment for grinding spots on aircraft landing gear components using robotic white light interferometry. In: Lehmann, P., Osten, W., Albertazzi Gonçalves, A. (eds.) Optical Measurement Systems for Industrial Inspection XIII.

Optical Measurement Systems for Industrial Inspection XIII, Munich, Germany, 26.06.2023–30.06.2023, p. 96. SPIE (2023–2023). https://doi.org/10.1117/12.2673757

9. Ehrbar, J., Schoepflin, D., Schüppstuhl, T.: Robot-based inspection of freeform components: process analysis and challenges in using a lateral scanning WLI. In: Silva, F.J.G., Pereira, A.B., Campilho, R.D.S.G. (eds.) Flexible Automation and Intelligent Manufacturing: Establishing Bridges for More Sustainable Manufacturing Systems. Lecture Notes in Mechanical Engineering, pp. 21–28. Springer Nature Switzerland, Cham (2024)
10. Abellan-Nebot, J.V., Romero Subirón, F.: A review of machining monitoring systems based on artificial intelligence process models. Int. J. Adv. Manuf. Technol. **47**(1–4), 237–257 (2010). https://doi.org/10.1007/s00170-009-2191-8
11. Fahle, S., Prinz, C., Kuhlenkötter, B.: Systematic review on machine learning (ML) methods for manufacturing processes—identifying artificial intelligence (AI) methods for field application. Procedia CIRP **93**, 413–418 (2020). https://doi.org/10.1016/j.procir.2020.04.109
12. Gilpin, L.H., Bau, D., Yuan, B.Z., Bajwa, A., Specter, M., Kagal, L.: Explaining explanations: An overview of interpretability of machine learning. In: 2018 IEEE 5th International Conference on Data Science and Advanced Analytics (DSAA). 2018 IEEE 5th International Conference on Data Science and Advanced Analytics (DSAA), Turin, Italy, 01.10.2018–03.10.2018, pp. 80–89. IEEE (2018). https://doi.org/10.1109/DSAA.2018.00018
13. Haslgrubler, M., Gollan, B., Ferscha, A.: A cognitive assistance framework for supporting human workers in industrial tasks. IT Prof. **20**(5), 48–56 (2018). https://doi.org/10.1109/MITP.2018.053891337
14. Huber, J., Tammer, C., Krotil, S., Waidmann, S., Hao, X., Seidel, C., Reinhart, G.: Method for classification of battery separator defects using optical inspection. Procedia CIRP **57**, 585–590 (2016). https://doi.org/10.1016/j.procir.2016.11.101
15. Lokrantz, A., Gustavsson, E., Jirstrand, M.: Root cause analysis of failures and quality deviations in manufacturing using machine learning. Procedia CIRP **72**, 1057–1062 (2018). https://doi.org/10.1016/j.procir.2018.03.229
16. Pinto, R., Cerquitelli, T.: Robot fault detection and remaining life estimation for predictive maintenance. Procedia Comput. Sci. **151**, 709–716 (2019). https://doi.org/10.1016/j.procs.2019.04.094
17. Zuperl, U., Cus, F., Reibenschuh, M.: Modeling and adaptive force control of milling by using artificial techniques. J. Intell. Manuf. **23**(5), 1805–1815 (2012). https://doi.org/10.1007/s10845-010-0487-z
18. D'Addona, D.M., Ullah, A.M.M.S., Matarazzo, D.: Tool-wear prediction and pattern-recognition using artificial neural network and DNA-based computing. J. Intell. Manuf. **28**(6), 1285–1301 (2017). https://doi.org/10.1007/s10845-015-1155-0
19. Molnar, C.: Interpretable machine learning. In: A Guide for Making Black Box Models Explainable, 2nd edn. (2022)
20. Alexander, Z., Chau, D.H., Saldaña, C.: An interrogative survey of explainable AI in manufacturing. IEEE Trans. Ind. Inf. 1–13 (2024). https://doi.org/10.1109/TII.2024.3361489
21. Bukowski, M., Kurek, J., Antoniuk, I., Jegorowa, A.: Decision confidence assessment in multi-class classification. Sensors (Basel, Switzerland) **21**(11), 3834 (2021). https://doi.org/10.3390/s21113834
22. Feng, Z., Gani, H., Damayanti, A.D., Gani, H.: An explainable ensemble machine learning model to elucidate the influential drilling parameters based on rate of penetration prediction. Geoenergy Sci. Eng. **231**, 212231 (2023). https://doi.org/10.1016/j.geoen.2023.212231
23. Middlehurst, M., Large, J., Flynn, M., Lines, J., Bostrom, A., Bagnall, A.: HIVE-COTE 2.0: a new meta ensemble for time series classification (2021)

BY

Efficiently Linking Real Scenes with Synthetic Data Generation for AI-Based Cognitive Robotics and Computer Vision Applications

Paul Koch, Vivek Chavan, André Sers, Adem Karakurt, Paul Hofmann, Mohamad Zaher Ziadeh and Jörg Krüger

Abstract

AI vision models are a driving factor for the potential use case scenarios of cognitive robotics within in the industry and household applications. A large array of methods from semantic environment analysis towards 6D and grasping pose estimation have been proposed based on the latest AI achievements. However, such advancements require further strong and efficient methods w.r.t. training data and AI-architectures, which are capable in synergy to tackle current challenges, precision limits, and scalability beyond domain gaps. In this paper, we discuss these current limits and trends in the related state-of-the-art which are challenging those. Further we discuss our current work in progress on bridging the domain gap between simulations and real world applications by linking those in the training data generation.

Keywords

Artificial intelligence • Cognitive robotics • Data generation

Website: www.ipk.fraunhofer.de.

P. Koch (✉) · V. Chavan · A. Sers · A. Karakurt · P. Hofmann · M. Z. Ziadeh · J. Krüger
Fraunhofer IPK, Berlin, Germany
e-mail: paul.koch@ipk.fraunhofer.de

J. Krüger
TU Berlin, Berlin, Germany

M.-C. Wanner et al. (eds.), *Annals of Scientific Society for Assembly, Handling and Industrial Robotics 2024*, https://doi.org/10.1007/978-3-031-91463-8_22

1 Introduction

In recent years, AI-methods have shown significant advancements in various data-driven applications, surpassing traditional computer science methods. Initially successful in computer vision in 2012 [1], AI has now made significant progress in robotic-related vision applications. Many methods have been proposed to detect objects [2–6], including pixel-wise panoptic segmentation of the environment and objects [2, 4, 6, 7]. Additionally, AI has made advancements in 6D pose estimation and grasping pose estimation of partially occluded objects [8–12]. Overall, AI has demonstrated its potential to generalize with self-supervised learning towards various data-driven downstream applications with optional task specific fine-tuning [13–15].

However, the success of AI-methods in robotic-related applications still heavily relies on the availability of well-annotated large datasets with high diversity to ensure generalizability within a certain task and domain. To address this need, numerous datasets and challenges for robotic-related applications have been introduced. Among these, notable pioneers for end2end AI-models in their respective challenges are YCB-Video [8] and Linemod [16] for 6D pose estimation, and GraspNet [17] for two-finger grasping pose estimation. Other robotic manipulation centered challenges such as SuchtionNet [18], 6D pose estimation of unseen objects [19], and Grasping transparent objects [20] have been proposed. Additionally, some work further focuses on incorporating simulations [8, 21–26] and robots [21, 27–29] for unlimited synthetic data generation to tackle the high demands of precise annotated training data. Though they come with issues of the domain gap [30], execution speed [17], and implementation costs.

W.r.t robot control policy learning various methods have been proposed. Driess et al. [31] created with PalmE a multi modal model for language controlled robot action, decision and motion planning, which is capable to grasp objects on request. This innovation is based on heavy parallel robotic training data collection, where hard coded robot actions are observed based on a natural language text requests. Similarly, Chi et al. [32] observe use human controlled robot actions in order to learn end2end manipulation policies. W.r.t reinforcement learning, further control policy learning are proposed for simulated data [33–36]. Anyhow, these methods, datasets and challenges play a crucial role in advancing the capabilities of AI in robotic vision applications by providing standardized evaluation benchmarks and fostering research innovations.

Ever since various AI-architectures [9–12, 21, 37–39] with hard-coded engineering methods (tailored methods for their specific problems) have been proposed, which are further increasing precision on their related downstream tasks and benchmarks. One can increasing consider in particular Bin-Picking challenges to be solved [12, 39]. Here AnyGrasp [12] shows a high resilience on bin-picking various objects which can be scaled towards unknown and dynamic objects. Although AnyGrasp is versatile, it is still limited to a semi-2D top-down grasping setup with a two finger gripper or suction end-effector [18, 39] and not proven in many industrial setting. Moreover, these current grasping specific methods lack generally knowledge about the environment and the objects within the scene, making them powerful modular tools but not scalable towards full AI end-to-end reasoning-based manipulation planning—rather they are a streamlined end-to-end mapping of the statistical source domain knowledge to the target domain of; grasping poses [17, 38, 39], grasping rectangle areas [22, 40–42], grasping affordance maps [37, 43, 44], 6D pose estimation attached with predefined grasping poses [8, 10, 21, 29], or robot control policies [31, 32]. One may argue that those robot control policies [31, 32] are reasoning-based methods, but they are still limited towards their training domain and lack explicit physical world knowledge such as "what is object mass", "whats a center of mass", "whats stiffness", "whats friction", etc. Following this logic, Huang et al. [45] improved grasping of deformable objects based on a physic-reasoning enhanced backbone.

In this paper we continue this logic and argue that achieving a generalized and physics grounded reasoning capable system would require a more "cognitive" AI robotic model rather then many individually engineering tailored service modules—and in order to achieve such a AI-system extensive multi modal well annotated data is required. Building upon pioneers related work, we propose a work-in-progress data collection and AI-training loop, which links real world scenes with simulations for unlimited and well annotated real data.

2 Related Work

Hand held data generation: The most strait forward method for vision-based cognitive robotic centered data generation is to use single images and videos for data collection. This works fine for classical vision problems such as classification [46], detection [47] and segmentation [47, 48]. However, annotations such as poses for cognitive robotic tasks are much harder to annotate. Therefore, related work often use video streams with camera trajectory tracking, such that poses only need to be annotated once per scene [8, 16, 49, 50]. This creates a dense data collection but is limited to a few scenes.

Data generation with robots: In addition to hand held data collection, other work uses robotic arms in order to roam the work space and track the camera positions [17, 21, 29]. This method gives a marker free scene with precise camera trajectory tracking within the target robot workspace. However, this setup is also limited to a few scenes. Therefore, Deng et al. [21] proposed a data collection loop within which the robot can iteratively learn to move objects in the scenery. Hence, generation many constellations to refine the performance, but with a limited object variety. Furthermore, these methods are further time restricted and scalable only with a large invest. In contrast to pose estimation datasets, other work proposes to observe the hard-coded or human controlled robot actions in order to directly learn robot control and manipulation policies [31, 32]. However, these methods are also bound by extensive investment costs and their training domain.

Data generation with simulations: In order to create well annotated training data with scene and object diversity, related work opted to employ simulations for unlimited synthetic data generation [8, 16, 21, 49]. Other work uses simulators as a source for annotating grasping poses [17]. However, these methods suffer from the domain gap [30] when shifted to the target domain [21]. Although methods such as domain randomization [51] and style transfer [52, 53] are trying to bridge the domain gap, synthetic data remains to be a supporting data source along real data [21, 26].

Grasping Annotations: Unlike other computer-vision annotations such as classification labels, bounding boxes, segmentation masks, and none symmetric 6D object poses are grasping poses not of discrete nature. The realm of possible grasping poses are unlimited in the continuous cartesian space [27]. Hence, learning regression methods for estimation is an ill defined problem. Nevertheless, related work pursed this avenue with success [17]. In order to deal with this discrete vs. continuous problem some work focuses on introducing grasping affordance [43, 44], none-discrete matrices [8], or offset tolerance based loss functions [17]. Learning from demonstration [31, 32] and reinforcement learning [33–36] is circumventing this issue by directly learning the robot control policy. But here learning from demonstration is limited to the discrete number of demonstrations, while reinforcement learning is mostly bound to simulations and has a hard time to converge for complex problems without constrains or human feedback [36].

3 From Real Scenes to Simulations and Back

We outline our main data generation pipeline in Fig. 1.

(1) Scanning the Scenery: Creating simulations requires many assets and textures in order to represent the real world scenery. Therefore, scanning the scenery assets has been proposed by the research community rather then using time consuming and expert crafting required CAD modelling. Recent advancements in AI research further improved scanning methods, such that no expensive hardware is required. Here Nerfs have been found to be well suited to reconstruct 3D scenes [54] (see Fig. 2), but it is hard to convert them into mesh and texture assets of individual objects for simulations. In contrast other scanning methods such as Nvdiffrec [55, 56] allow us to directly get those assets disentangled from the background Fig. 3. For the scenery itself nvdiffrec can also be used, but this is not yet thoroughly investigated.

(2) Generate Simulations: From the 3D assets we can now build novel simulated scenarios with randomly placed objects. However the scene generation needs to incorporate prior knowledge about the possible constellations of things, such that e.g. the objects are reasonably placed on tables and in shelves rather then on the floor. While it is fairly easy to hardcode this on a small scale, it becomes quite a challenge scaling to generate well structured simulations without human controlled design in order to cut costs. Here one needs assets which are connected to meta information on how a simulation can use this assets within a full autonomous scene generation–a table asset has object placement areas and object assets require such area and are graspable. A full logic needs to be developed here.

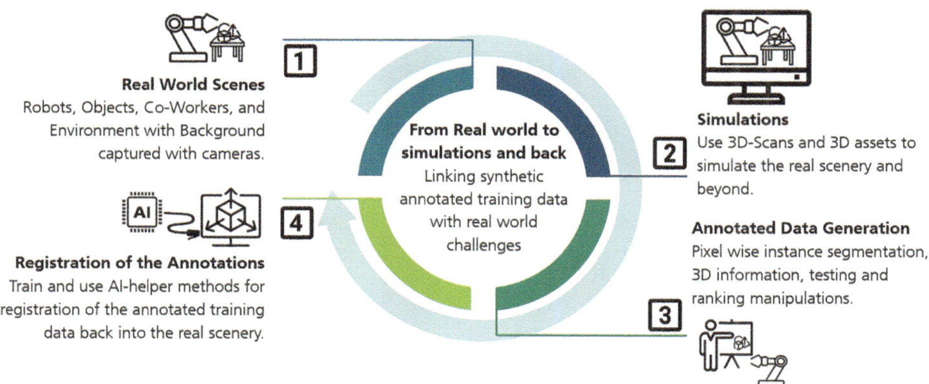

Fig. 1 **Linking real robot workspace scenes with simulations**: In a continuous loop we propose to scan real scenes (1) and transform them into simulations (2). Here we can conduct many experiments, find grasping candidates, train control policies and annotate training data (3). Eventually, we can now train further AI methods to help to transform the annotations and control policies back into the real scenery (4). This process is looped in order to refine AI-helpers and deploy methods

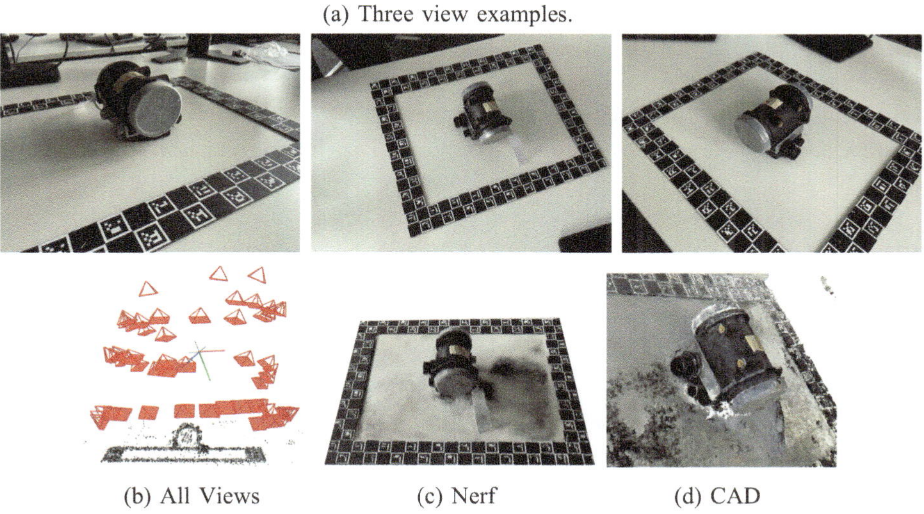

Fig. 2 With Nerfs [54] we can create good looking 3D representations (some issues with the shade), but we cant yet export high quality masked 3D metric assets and textures

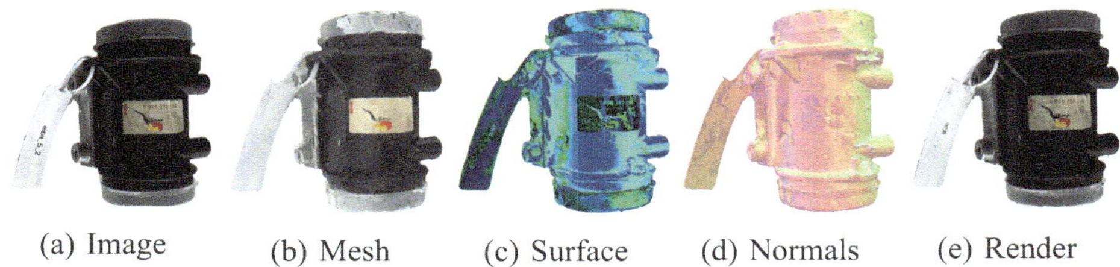

Fig. 3 3D assets with textures from Nvdiffrec [55]

(3) Annotating Data: Modern simulators such as Omniverse™ and Blender are capable to fully randomize their assets and create photorealistic images from their generated scenes. Moreover, these simulators are capable to extract segmentation annotations and other meta information such as point clouds and 6D poses from their scenes. In addition to logical placements of assets in the scene, a simulation also need to incorporate reasonable lighting and physics (e.g. gravity). Nvidia Issac Sim™ is a recent development within simulators, which allows users to create and generate robotic scenes. Building on top of Omniverse™ this simulator delivers photorealistic and physically accurate virtual environments. Additionally, the Issac Sim™ simulator can be used to carry out many robotic experiments. Hence, one can use it in order to generate for a given scene e.g. possible grasping poses with force requirements or vision based trajectory planning. This allows us to create annotations (positive and negative samples) for many different tasks, which are exhaustive and expensive in reality.

(4) Train AI-Helpers: Rather then training AI-models to solve a given downstream task from the simulation data alone, it is has been found to be very beneficial to incorporate real data along site of the synthetic data, in order to close the sim2real gap [21, 26]. Annotating the real world data can be at times very exhausting, especially for robotic related downstream task. Therefore, we propose to train the AI models iterative and step-by-step in order to help to link the simulated data back to the real world, where the 3D assets originated from. In particular we aim e.g. to use 6D pose estimation models trained from simulated data given the 3D assets from real world scenes in order to initialize poses of those real assets in their real world scenes. This in turn allows us to deploy further registration onto the scenery scanning in order to increase the precision of the annotated the real world data. From this linkage we can then e.g. update the training data for the pose estimation and further importantly bring the real world scene into the simulation in order to find possible grasping poses or execute and validate planning strategies. Once the simulated experiments are conducted we have a annotated real world scene with real world data from which we can train novel end2end solutions (grasping pose estimation or planning). Other possible AI-helpers which we can train in this scenery are style transfers [52] and texture asset augmentations Fig. 4. Moreover, we aim to combine simulation trained segmentation models with generalized models [13,

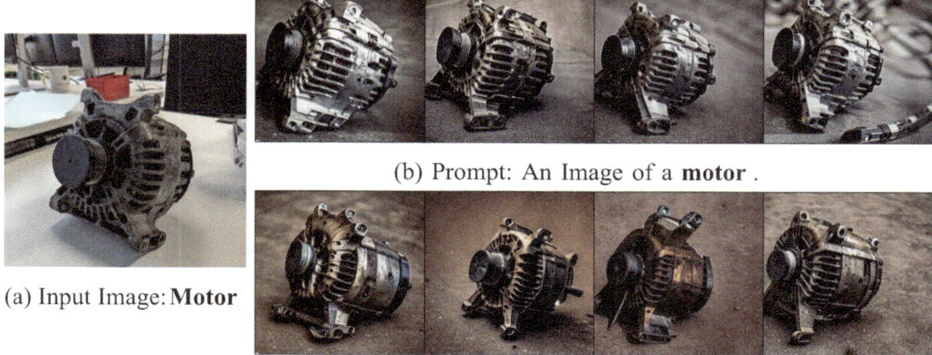

(a) Input Image: **Motor**

(b) Prompt: An Image of a **motor** .

(c) Prompt: An Image of a **motor with a rusty surface**

Fig. 4 Image augmentation with Perfusion [57]. First experiments for natural language based editing of the visual appearance of assets (before or after simulation)

58] for identifying overlaps for masking annotations with a linkage to the asset knowledge (e.g. the placement area).

4 Conclusion

In this paper we outlined the current trends in vision-based cognitive robotics for object manipulation. Based on those trends we identified the need for a large amount of well annotated training data from real and simulated environments. With our proposed linkage of real world scenes with simulations we aim to provide such data, iterative bridge the sim2real domain gap, and create a training foundation for many AI-tool for cognitive robotics.

References

1. Krizhevsky, A., Sutskever, I., Hinton, G.E.: Imagenet classification with deep convolutional neural networks. In: Pereira, F., Burges, C., Bottou, L., Weinberger, K. (eds.) Advances in Neural Information Processing Systems, vol. 25. Curran Associates, Inc. (2012)
2. Ren, S., He, K., Girshick, R.B., Sun, J.: Faster R-CNN: towards real-time object detection with region proposal networks (2015). CoRR arXiv:1506.01497
3. Redmon, J., Divvala, S.K., Girshick, R.B., Farhadi, A.: You only look once: unified, real-time object detection (2015). CoRR arXiv:1506.02640

4. Carion, N., Massa, F., Synnaeve, G., Usunier, N., Kirillov, A., Zagoruyko, S.: End-to-end object detection with transformers (2020). CoRR arXiv:2005.12872
5. Zhu, X., Su, W., Lu, L., Li, B., Wang, X., Dai, J.: Deformable DETR: deformable transformers for end-to-end object detection (2020). CoRR arXiv:2010.04159
6. Zong, Z., Song, G., Liu, Y.: Detrs with collaborative hybrid assignments training (2023)
7. Cheng, B., Misra, I., Schwing, A.G., Kirillov, A., Girdhar, R.: Masked-attention mask transformer for universal image segmentation (2021). CoRR arXiv:2112.01527
8. Xiang, Y., Schmidt, T., Narayanan, V., Fox, D.: Posecnn: a convolutional neural network for 6d object pose estimation in cluttered scenes (2017). CoRR arXiv:1711.00199
9. Rünz, M., Agapito, L.: Maskfusion: Real-time recognition, tracking and reconstruction of multiple moving objects (2018). CoRR arXiv:1804.09194
10. Wang, C., Xu, D., Zhu, Y., Martín-Martín, R., Lu, C., Fei-Fei, L., Savarese, S.: Densefusion: 6d object pose estimation by iterative dense fusion (2019). CoRR arXiv:1901.04780
11. Su, Y., Saleh, M., Fetzer, T., Rambach, J., Navab, N., Busam, B., Stricker, D., Tombari, F.: Zebrapose: coarse to fine surface encoding for 6dof object pose estimation (2022)
12. Fang, H.-S., Wang, C., Fang, H., Gou, M., Liu, J., Yan, H., Liu, W., Xie, Y., Lu, C.: Anygrasp: robust and efficient grasp perception in spatial and temporal domains. IEEE Trans. Robot. (T-RO) (2023)
13. Oquab, M., Darcet, T., Moutakanni, T., Vo, H., Szafraniec, M., Khalidov, V., Fernandez, P., Haziza, D., Massa, F., El-Nouby, A., Assran, M., Ballas, N., Galuba, W., Howes, R., Huang, P.-Y., Li, S.-W., Misra, I., Rabbat, M., Sharma, V., Synnaeve, G., Xu, H., Jegou, H., Mairal, J., Labatut, P., Joulin, A., Bojanowski, P.: Dinov2: learning robust visual features without supervision (2023)
14. Radford, A., Kim, J.W., Hallacy, C., Ramesh, A., Goh, G., Agarwal, S., Sastry, G., Askell, A., Mishkin, P., Clark, J., Krueger, G., Sutskever, I.: Learning transferable visual models from natural language supervision (2021). CoRR arXiv:2103.00020
15. Brown, T.B., Mann, B., Ryder, N., Subbiah, M., Kaplan, J., Dhariwal, P., Neelakantan, A., Shyam, P., Sastry, G., Askell, A., Agarwal, S., Herbert-Voss, A., Krueger, G., Henighan, T., Child, R., Ramesh, A., Ziegler, D.M., Wu, J., Winter, C., Hesse, C., Chen, M., Sigler, E., Litwin, M., Gray, S., Chess, B., Clark, J., Berner, C., McCandlish, S., Radford, A., Sutskever, I., Amodei, D.: Language models are few-shot learners (2020). CoRR arXiv:2005.14165
16. Hinterstoisser, S., Lepetit, V., Ilic, S., Holzer, S., Bradski, G., Konolige, K., Navab, N.: Model based training, detection and pose estimation of texture-less 3d objects in heavily cluttered scenes. In: Lee, K.M., Matsushita, Y., Rehg, J.M., Hu, Z. (eds.) Computer Vision–ACCV 2012, pp. 548–562. Springer, Berlin, Heidelberg (2013)
17. Fang, H.-S., Wang, C., Gou, M., Lu, C.: Graspnet-1billion: a large-scale benchmark for general object grasping. In: Proceedings of the IEEE/CVF Conference on Computer Vision and Pattern Recognition, pp. 11444–11453 (2020)
18. Cao, H., Fang, H.-S., Liu, W., Lu, C.: Suctionnet-1billion: a large-scale benchmark for suction grasping. IEEE Robot. Autom. Lett. **6**(4), 8718–8725 (2021)
19. Gou, M., Pan, H., Fang, H.-S., Liu, Z., Lu, C., Tan, P.: Unseen object 6d pose estimation: a benchmark and baselines (2022)
20. Fang, H., Fang, H.-S., Xu, S., Lu, C.: Transcg: a large-scale real-world dataset for transparent object depth completion and a grasping baseline. IEEE Robot. Autom. Lett. 1–8 (2022)
21. Deng, X., Xiang, Y., Mousavian, A., Eppner, C., Bretl, T., Fox, D.: Self-supervised 6d object pose estimation for robot manipulation (2019). CoRR arXiv:1909.10159
22. Mahler, J., Liang, J., Niyaz, S., Laskey, M., Doan, R., Liu, X., Ojea, J.A., Goldberg, K.: Dex-net 2.0: Deep learning to plan robust grasps with synthetic point clouds and analytic grasp metrics (2017). CoRR arXiv:1703.09312
23. Mousavian, A., Eppner, C., Fox, D.: 6-dof graspnet: Variational grasp generation for object manipulation (2019). CoRR arXiv:1905.10520

24. Depierre, A., Dellandréa, E., Chen, L.: Jacquard: a large scale dataset for robotic grasp detection (2018). CoRR arXiv:1803.11469
25. Yan, X., Hsu, J., Khansari, M., Bai, Y., Pathak, A., Gupta, A., Davidson, J., Lee, H.: Learning 6-dof grasping interaction via deep geometry-aware 3d representations. In: 2018 IEEE International Conference on Robotics and Automation (ICRA), pp. 3766–3773 (2018)
26. Koch, P., Schlüter, M., Krüger, J.: With synthetic data towards part recognition generalized beyond the training instances. In: AIP Conference Proceedings, vol. 2989, p. 020007 (2024)
27. Pinto, L., Gupta, A.: Supersizing self-supervision: learning to grasp from 50k tries and 700 robot hours (2015). CoRR arXiv:1509.06825
28. Levine, S., Pastor, P., Krizhevsky, A., Quillen, D.: Learning hand-eye coordination for robotic grasping with deep learning and large-scale data collection (2016). CoRR arXiv:1603.02199
29. Koch, P., Schlüter, M., Thill, S., Krüger, J.: Towards robot-assisted data generation with minimal user interaction for autonomously training 6d pose estimation in operational environments. Procedia CIRP **120**, 249–254 (2023). 56th CIRP International Conference on Manufacturing Systems 2023
30. Jakobi, N., Husbands, P., Harvey, I.: Noise and the reality gap: the use of simulation in evolutionary robotics, vol. 929, pp. 704–720 (1995)
31. Driess, D., Xia, F., Sajjadi, M.S.M., Lynch, C., Chowdhery, A., Ichter, B., Wahid, A., Tompson, J., Vuong, Q., Yu, T., Huang, W., Chebotar, Y., Sermanet, P., Duckworth, D., Levine, S., Vanhoucke, V., Hausman, K., Toussaint, M., Greff, K., Zeng, A., Mordatch, I., Florence, P.: Palm-e: an embodied multimodal language model (2023)
32. Chi, C., Feng, S., Du, Y., Xu, Z., Cousineau, E., Burchfiel, B., Song, S.: Diffusion policy: visuomotor policy learning via action diffusion (2023)
33. Yu, T., Quillen, D., He, Z., Julian, R., Hausman, K., Finn, C., Levine, S.: Meta-world: a benchmark and evaluation for multi-task and meta reinforcement learning (2019). CoRR arXiv:1910.10897
34. Yu, T., Kumar, S., Gupta, A., Levine, S., Hausman, K., Finn, C.: Gradient surgery for multi-task learning (2020). CoRR arXiv:2001.06782
35. Sodhani, S., Zhang, A., Pineau, J.: Multi-task reinforcement learning with context-based representations (2021)
36. Hejna, J., Rafailov, R., Sikchi, H., Finn, C., Niekum, S., Knox, W.B., Sadigh, D.: Contrastive preference learning: learning from human feedback without rl (2023)
37. Wang, C., Fang, H.-S., Gou, M., Fang, H., Gao, J., Lu, C.: Graspness discovery in clutters for fast and accurate grasp detection. In: Proceedings of the IEEE/CVF International Conference on Computer Vision (ICCV), pp. 15964–15973 (2021)
38. Liu, J., Zhang, R., Fang, H.-S., Gou, M., Fang, H., Wang, C., Xu, S., Yan, H., Lu, C.: Target-referenced reactive grasping for dynamic objects, pp. 8824–8833 (2023)
39. Mahler, J., Matl, M., Satish, V., Danielczuk, M., DeRose, B., McKinley, S., Goldberg, K.: Learning ambidextrous robot grasping policies. Sci. Robot. **4**(26), eaau4984 (2019)
40. Jiang, Y., Moseson, S., Saxena, A.: Efficient grasping from rgbd images: Learning using a new rectangle representation. In: 2011 IEEE International Conference on Robotics and Automation, pp. 3304–3311 (2011)
41. Lenz, I., Lee, H., Saxena, A.: Deep learning for detecting robotic grasps. Int. J. Robot. Res. **34**(4–5), 705–724 (2015)
42. Zhang, H., Lan, X., Zhou, X., Zheng, N.: Roi-based robotic grasp detection in object overlapping scenes using convolutional neural network (2018). CoRR arXiv:1808.10313
43. Ardón, P., Pairet, È., Petrick, R.P.A., Ramamoorthy, S., Lohan, K.S.: Learning grasp affordance reasoning through semantic relations (2019). CoRR arXiv:1906.09836
44. Bamford, L., Klassen, N., Karl, J.: Faster recognition of graspable targets defined by orientation in a visual search task. Exp. Brain Res. **238**, 04 (2020)

45. Huang, I., Narang, Y., Eppner, C., Sundaralingam, B., Macklin, M., Bajcsy, R., Hermans, T., Fox, D.: Defgraspsim: physics-based simulation of grasp outcomes for 3d deformable objects. IEEE Robot. Autom. Lett. **7**, 6274–6281 (2022)
46. Deng, J., Dong, W., Socher, R., Li, L.-J., Li, K., Fei-Fei, L.: Imagenet: a large-scale hierarchical image database. In: 2009 IEEE Conference on Computer Vision and Pattern Recognition, pp. 248–255 (2009)
47. Lin, T., Maire, M., Belongie, S.J., Bourdev, L.D., Girshick, R.B., Hays, J., Perona, P., Ramanan, D., Dollár, P., Zitnick, C.L.: Microsoft COCO: common objects in context (2014). CoRR arXiv:1405.0312
48. Cordts, M., Omran, M., Ramos, S., Rehfeld, T., Enzweiler, M., Benenson, R., Franke, U., Roth, S., Schiele, B.: The cityscapes dataset for semantic urban scene understanding (2016). CoRR arXiv:1604.01685
49. Kaskman, R., Zakharov, S., Shugurov, I., Ilic, S.: Homebreweddb: RGB-D dataset for 6d pose estimation of 3d objects (2019). CoRR arXiv:1904.03167
50. Sundermeyer, M., Hodan, T., Labbe, Y., Wang, G., Brachmann, E., Drost, B., Rother, C., Matas, J.: Bop challenge 2022 on detection, segmentation and pose estimation of specific rigid objects (2023)
51. Tobin, J., Fong, R., Ray, A., Schneider, J., Zaremba, W., Abbeel, P.: Domain randomization for transferring deep neural networks from simulation to the real world (2017). CoRR arXiv:1703.06907
52. Gatys, L.A., Ecker, A.S., Bethge, M.: A neural algorithm of artistic style 2015. CoRR arXiv:1508.06576
53. Zhang, Y., Tang, F., Dong, W., Huang, H., Ma, C., Lee, T.-Y., Xu, C.: Domain enhanced arbitrary image style transfer via contrastive learning. In: Special Interest Group on Computer Graphics and Interactive Techniques Conference Proceedings, SIGGRAPH '22. ACM (2022)
54. Müller, T., Evans, A., Schied, C., Keller, A.: Instant neural graphics primitives with a multiresolution hash encoding (2022). CoRR arXiv:2201.05989
55. Munkberg, J., Hasselgren, J., Shen, T., Gao, J., Chen, W., Evans, A., Müller, T., Fidler, S.: Extracting triangular 3d models, materials, and lighting from images. In: Proceedings of the IEEE/CVF Conference on Computer Vision and Pattern Recognition (CVPR), pp. 8280–8290 (2022)
56. Hasselgren, J., Hofmann, N., Munkberg, J.: Shape, Light, and Material Decomposition from Images using Monte Carlo Rendering and Denoising (2022). arXiv:2206.03380
57. Tewel, Y., Gal, R., Chechik, G., Atzmon, Y.: Key-locked rank one editing for text-to-image personalization (2023)
58. Kirillov, A., Mintun, E., Ravi, N., Mao, H., Rolland, C., Gustafson, L., Xiao, T., Whitehead, S., Berg, A.C., Lo, W.-Y., Dollár, P., Girshick, R.: Segment anything (2023)

Handling

Adaptive Preload Control of Cable-Driven Parallel Robots for Handling Task

Thomas Reichenbach, Johannes Clar, Andreas Pott and Alexander Verl

Abstract

This paper presents a method for dynamic adjustment of cable preloads based on the actuation redundancy of cable-driven parallel robot (CDPRs), which allows increasing or decreasing the platform stiffness depending on task requirements. This is achieved by computing preload parameters with an extended nullspace formulation of the kinematics. The method facilitates the operator's ability to specify a defined preload within the operation space. The algorithms are implemented in a real-time environment, allowing for the use of optimization in hybrid position-force control. To validate the effectiveness of this approach, a simulation study is performed, and the obtained results are compared to existing methods. Furthermore, the method is investigated experimentally and compared with the conventional position-controlled operation of a cable robot. The results demonstrate the feasibility of adaptively adjusting cable preloads during platform motion and manipulation of additional objects.

Keywords

Adaptive • Nullspace • Control • Cable-driven • Parallel • Robot

T. Reichenbach (✉) · J. Clar · A. Pott · A. Verl
Institute for Control Engineering of Machine Tools and Manufacturing Units (ISW), University of Stuttgart, Stuttgart, Germany
e-mail: thomas.reichenbach@isw.uni-stuttgart.de
URL: https://www.isw.uni-stuttgart.de/

M.-C. Wanner et al. (eds.), *Annals of Scientific Society for Assembly, Handling and Industrial Robotics 2024*, https://doi.org/10.1007/978-3-031-91463-8_23

1 Introduction

The use of serial manipulators is well established for handling tasks in automation and production environments. However, for large-scale manipulation the workspace and payload of serial robots are limited. These disadvantages can be overcome by using cable-driven parallel robot (short: cable robots) that use cables instead of rigid prismatic actuators to control an end-effector. Due to their flexibility and low weight, cables can be stored compactly on drums so that high maximum actuation length and acceleration can be reached. Thus, a large geometric workspace can be made possible. A well-known example for a large-scale cable robot is FAST [8], which is a spherical radio telescope with a span width of 600 m and a payload of approx. 30 t. Other examples are the Robotic Seabed Cleaning Platform (RSCP) [2] for removing marine litter and high rack warehouse solutions [4]. Efficient operation is crucial for handling tasks of industrial robots, requiring high accuracy for loading or unloading objects and low energy consumption for intermediate movements. Cable robots offering high range and speed but lower repeatability compared to conventional industrial robots. For cable robots, increasing the platform stiffness leads to a more precise operation and higher energy consumption, whereas decreasing leads to a low energy consumption due to lower cable force and also lower accuracy [9]. High energy consumption can be assumed by applying high cable forces on the platform, which require high motor torques and therefore also high motor currents. Conversely, low energy consumption can be achieved with low cable forces.

Parallel cable robots are redundant mechanisms with number of actuated cables $m \geq n + 1$, whereby the platform can be manipulated with n degrees of freedom. Verhoeven [16] already describes that cable robots have the potential to increase or decrease the cable preload and consequently adjusting the platform stiffness by exploiting the nullspace. Later, Kraus et al. [6] presented an energy efficient computation method of the force distribution of a cable robot. Therefore, a gradient vector is used to compute a reference force and subsequently solve the static equation of a cable robot with the Moore-Penrose pseudo inverse. Disturbances due to changes during handling tasks, i.e. loading or unloading an object, and the pose error of the platform were not investigated. Lamaury et al. [7] develop an adaptive control system for a redundantly restrained suspended cable robot (CoGiRo). Here, 26 parameters are estimated during operation (online) and continuously adapted in the feedforward or closed-loop control and transferred to the drives as torque target values. The adaptation of dynamic and controller parameters to increase accuracy is also investigated by Godbole et al. [1], Harandi et al. [3], and Zhang et al. [17].

In these works, adaptation is limited to the adjustment of the dynamics and controller parameters. However, the aim of the presented control concepts is not to control the cable preload in nullspace and thus adaptively adjust the platform stiffness, but to reduce disturbing effects caused by the cable actuation system or the environment. Hence, in this paper, an adaptive preload control (APC) for redundantly restrained cable robots which allows adjustment of the cable preload during motion and during object manipulation is presented.

2 Advanced Kinematics

2.1 Geometric Relations

In Fig. 1, the geometric relations of a cable robot are deduced from the algebraic loop of the inverse kinematics with the example of the ith cable. This leads to

$$l_i = \|\boldsymbol{l}_i\|_2 = \|\boldsymbol{a}_i - \boldsymbol{r} - \boldsymbol{R}\boldsymbol{b}_i\|_2 , \tag{1}$$

where $\boldsymbol{l}_i$ is the vector of the ith actuated cable with l_i as its length, $\boldsymbol{a}_i$ the frame anchor points, $\boldsymbol{r}$ the platform position, $\boldsymbol{R}$ the platform rotation matrix, and $\boldsymbol{b}_i$ the platform anchor points within the local platform coordinate system.

The so-called structure matrix $\boldsymbol{A}^\top$ is computed by normalizing inverse kinematics for each cable i with given platform pose $(\boldsymbol{r}, \boldsymbol{R}) \in \mathrm{SE}(3)$ with the Euclidean norm $\| \cdot \|_2$. The structure matrix can also described as the negative transposed Jacobian of direct kinematics $\boldsymbol{A}^\top = -\boldsymbol{J}_{\mathrm{DK}}^\top$ which has the cable direction $\boldsymbol{u}_i = \frac{\boldsymbol{l}_i}{\|\boldsymbol{l}_i\|_2}$ and twist with respect to a reference coordinate system $\boldsymbol{R}\,\boldsymbol{b}_i \times \boldsymbol{u}_i$ as matrix entries,

$$\boldsymbol{A}^\top = -\boldsymbol{J}_{\mathrm{DK}}^\top = \begin{bmatrix} \boldsymbol{u}_1 & \cdots & \boldsymbol{u}_m \\ \boldsymbol{R}\,\boldsymbol{b}_1 \times \boldsymbol{u}_1 & \cdots & \boldsymbol{R}\,\boldsymbol{b}_m \times \boldsymbol{u}_m \end{bmatrix} . \tag{2}$$

For the technical use of running cables, the use of deflection pulleys are common, so that the inverse kinematics can be extended with pulley kinematics, as described in [10]. This extended pulley kinematics is used for the simulative and experimental investigations in this paper.

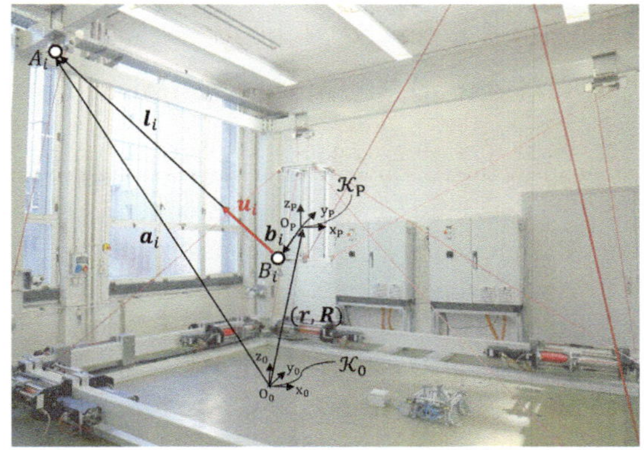

Fig. 1 Algebraic loop of inverse kinematics of *COPacabana* robot at *ISW* [15]. With $A_i \in \mathcal{K}_0$ as frame anchor points and $B_i \in \mathcal{K}_\mathrm{P}$ as platform anchor points

2.2 Nullspace Extension

The dynamics of the mobile platform is summarized to wrench $\boldsymbol{w}_\mathrm{c}$ such that $\boldsymbol{A}^\top \boldsymbol{f} = \boldsymbol{w}_\mathrm{c}$ is fulfilled, where $\boldsymbol{f} = [f_1, \ldots, f_m]$ are the applied cable forces due to joint-space actuation. Because the cable robot platform is actuated with at least $m \geq n + 1$ cables, the structure matrix $\boldsymbol{A}^\top$ is non-quadratic and not invertible. Thus, the kernel of the structure matrix $\ker(\boldsymbol{A}^\top) = \{\boldsymbol{v} \in \mathbb{R}^m \mid \boldsymbol{A}^\top \boldsymbol{v}_j = 0\}$ is not empty, assuming that the robot is not in a singular pose. For example, a redundancy of $\rho = m - n$ leads to a pose dependent nullspace with size $(m \times \rho)$ such that $\boldsymbol{N} = [\boldsymbol{v}_1, \ldots, \boldsymbol{v}_\rho]$. The main approach of the APC which is presented in this paper is adjusting the geometric stiffness of the platform on the platform and identify a set of preload parameters such that a solution of force distribution is found. To do so, cable force are measured on the frame (i.e. with pulleys) and converted to the applied wrench $\boldsymbol{w}_\mathrm{c}$. The stiffness of a cable robot is represented by two parts: the pose-dependent stiffness applied due to cable elongation and the geometric stiffness which depends on the applied cable forces and pose (see Kraus [5]). To adjust the geometric stiffness, the non-quadratic structure matrix $\boldsymbol{A}^\top$ will be extended by its transposed nullspace, such that

$$\boldsymbol{A}_\mathrm{adv}^\top = \begin{bmatrix} \boldsymbol{A}^\top \\ \boldsymbol{N}^\top \end{bmatrix}. \tag{3}$$

Hence, $\boldsymbol{A}_\mathrm{adv}^\top$ is an invertible quadratic matrix which can be used to compute a unique force distribution for a given platform wrench with

$$(\boldsymbol{A}_\mathrm{adv}^\top)^{-1} \boldsymbol{w}_\mathrm{c,adv} = \boldsymbol{f}\,, \tag{4}$$

where $\boldsymbol{w}_\mathrm{c,adv} = [\boldsymbol{w}_\mathrm{c}, \boldsymbol{\lambda}]^\top$, with preload parameters $\boldsymbol{\lambda} = [\lambda_1, \ldots, \lambda_\rho]^\top$.

2.3 Preload Parameter Optimization

Depending on the platform pose $(\boldsymbol{r}, \boldsymbol{R})$ and the orientation of the nullspace, feasible preload parameters $\boldsymbol{\lambda}$ must be determined. Furthermore, the cable force should neither drop below a specified minimum nor exceed a maximum limit. The cable force limitations result from the mechanical properties of the cable robot. For example, depending on the size of the robot and cable type a specific minimum force $\mathbf{f}_\mathrm{min}$ must be applied to the cable to avoid uncontrolled motion of the platform due to lag of stiffness. The maximum cable force $\mathbf{f}_\mathrm{max}$ should not be exceeded because of cable damage or overload of the cable actuation system. For handling tasks, the platform is moved with or without objects, whereby high stiffness is required during the loading or unloading process. The movement along the planned trajectory between the two operations must be efficient, whereas a high stiffness is not necessary. The geometrical stiffness can be seen as a measure of the resistance of the platform, which influences the positioning accuracy [9]. Hence, it is necessary to adaptively change the preload of the platform with the force distribution of the cables. Therefore, the preload control parameter η_c is introduced, such that the platform is specifically preloaded for each pose of a tajectory. Consequently, these conditions can be formulated as the following minimization problem

$$\min_{\boldsymbol{\lambda}} \quad e = \| \underbrace{\boldsymbol{N\lambda}}_{\boldsymbol{f}_{\text{tar}}} - (\eta_c \mathbf{f}_{\text{max}} + (1 - \eta_c)\mathbf{f}_{\text{min}}) \|_2 \tag{5}$$

$$\text{subject to:} \quad \underbrace{(\boldsymbol{A}_{\text{adv}}^{\top})^{-1} \boldsymbol{w}_{\text{c,adv}}}_{\boldsymbol{f}_{\text{tar}}} - \mathbf{f}_{\text{max}} \leq 0 \tag{6}$$

$$\mathbf{f}_{\text{min}} - \underbrace{(\boldsymbol{A}_{\text{adv}}^{\top})^{-1} \boldsymbol{w}_{\text{c,adv}}}_{\boldsymbol{f}_{\text{tar}}} \leq 0 \tag{7}$$

$$0 < \eta_c < 1 \quad , \tag{8}$$

where $\boldsymbol{f}_{\text{tar}}$ describes the target force distribution for the force controller. The preload control parameter η_c can be adapted between 0% and 100% during motion. Thus, a real-time capable optimization algorithm with nonlinear constraints is used to compute feasible preload parameters $\boldsymbol{\lambda}$.

3 Adaptive Preload Controller

To give an overview, Fig. 2 shows a schematic block diagram of the APC and the relation between the platform and drive systems of the cable robot. The basic structure of the control system comes from Reichenbach et al. [13], and is now adapted for the integration of the APC. The above-mentioned work also includes a detailed description of the modeling of drive controller, winch mechanics, cable dynamics, and platform dynamics, which can be assumed as rigid compared to the flexible cables. Hence, the following section focus on APC.

To achieve linear actuation using a cable, the target pose $\boldsymbol{y}_{\text{tar}} = [x, y, z, \alpha, \beta, \gamma]^{\top}$ of the platform, which includes Cartesian position $\boldsymbol{r}_{\text{tar}}$ and transformed orientation $\boldsymbol{R}_{\text{tar}}$ specified in Euler angles, requires the calculation and transformation of the target cable length $l_{\text{tar},i}$ into angular positions $\varphi_{\text{tar},i}$ for the drive systems. To do so, the inverse kinematics (IK), as defined in (1), is used with pulley kinematics [10], taking into account the transmission ratio of the planetary gear t_G and the winch $\text{t}_\text{W} = 2\pi\text{R}_\text{D}$ with R_D as nominal drum radius. Thus, the position controller output is the angular velocity target value

$$\omega_{\text{q},i} = \text{K}_\text{v}\text{t}_\text{W}\text{t}_\text{G}(l_{\text{d},i} - l_i) = \text{K}_\text{v}(\varphi_{\text{d},i} - \varphi_i), \tag{9}$$

with φ_i for the actual motor position, K_v as position controller gain. For preload parameter optimization, the actual wrench on the platform is observed by transforming the actual cable forces with the structure matrix, such that $\boldsymbol{w}_{\text{c,obs}} = \boldsymbol{A}^{\top}\boldsymbol{f}$. Consequently, friction forces, i.e. within the cable guidance system, are also transformed into the operational space. The

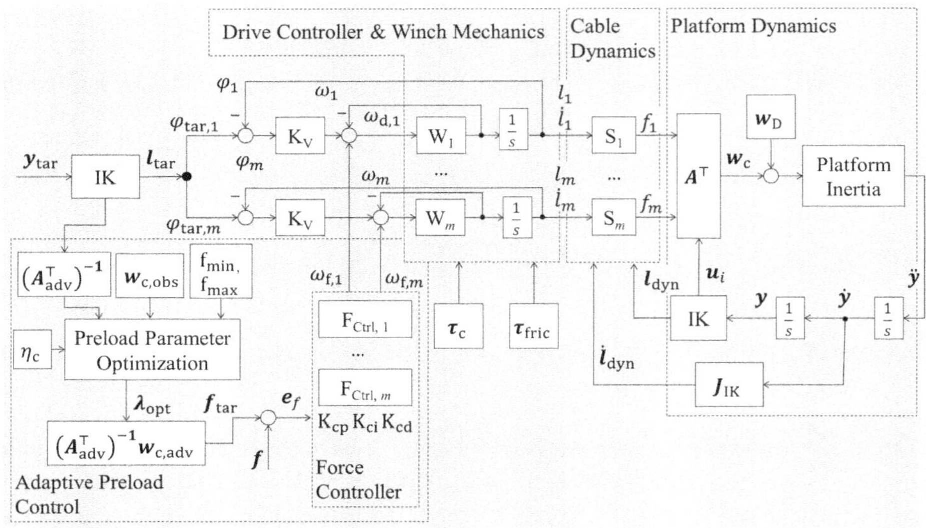

Fig. 2 Schematic overview of the implementation of APC within the cable robot system (adapted from [13])

inverted advanced structure matrix $(\boldsymbol{A}_\text{adv}^\top)^{-1}$ is computed with the target pose $\boldsymbol{y}_\text{tar}$ according to (2) and (3). Both, the actual wrench and the inverted advance structure matrix are computed in real-time for each target pose along a planned trajectory. Additionally, the adaptive preload control parameter η_c and cable force limits $\text{f}_\text{lim} = [\text{f}_\text{min}, \text{f}_\text{max}]^\top$ are specified. To preload the cables within the cable robots nullspace, a force controller is implemented. Using the optimized preload parameters $\boldsymbol{\lambda}_\text{opt}$ and the advanced structure matrix $\boldsymbol{A}_\text{adv}^\top$, target cable forces can be computed with (4) and controlled with a PID-controller $\text{F}_{\text{Ctrl},i}$ for each motor. The output of the force controller is an additive angular speed value $\omega_{\text{f},i}$. Thus, the angular speed target for the drive control system consists of the position controller part and the force controller part which leads to $\omega_{\text{tar},i} = \omega_{\text{q},i} + \omega_{\text{f},i}$.

Simulation Study To show that the algorithms of the APC are correctly implemented, a simulation study is performed. Therefore, the cable force computation of the presented APC is compared with the CF method [12]. The simulation study was performed with the geometrical data of the cable robot *COPacabana* which is presented by Trautwein et al. [15]. However, modifications were made to the frame and platform geometry due to reconstruction. Thus, the actual data of *COPacabana* is published in [14] as XML file including geometry, drive and control parameters of the cable robot. For the computation of the force distributions, the Python integration WiPy of WireX is used which is an open-source software environment to develop cable robots and presented by Pott [11]. To solve Eq. (5), optimization algorithms COBYLA and SLSQP are investigated. COBYLA uses local derivative-free optimization and SLSQP uses local gradient-based optimization. Both algorithms produce similar results, and only minor changes can be observed due to numerical noise and different treatment of the termination criteria. The feasibility of the algorithms is calculated for the force distribution at poses along a path with $r_z = [0, 3]$ m and a discretization of 1 mm, which is an upward movement of the platform without rotation. Thereby, the algorithms are verified whether the solutions of the computed target forces show discontinuities and whether the force distribution of the respective pose is valid.

The results of the simulation study in Fig. 3 show, that the approach using a preload control parameter η_c to increase or decrease the geometric stiffness of the platform is feasible. It can be seen that despite different parameters, the optimization algorithm converges until the pose $z_0 = 2.4$ m is reached, marked with a black line in the figure. In addition, the results show that a preload control parameter of $\eta_c = 50\%$ correspond to the results of the CF method.

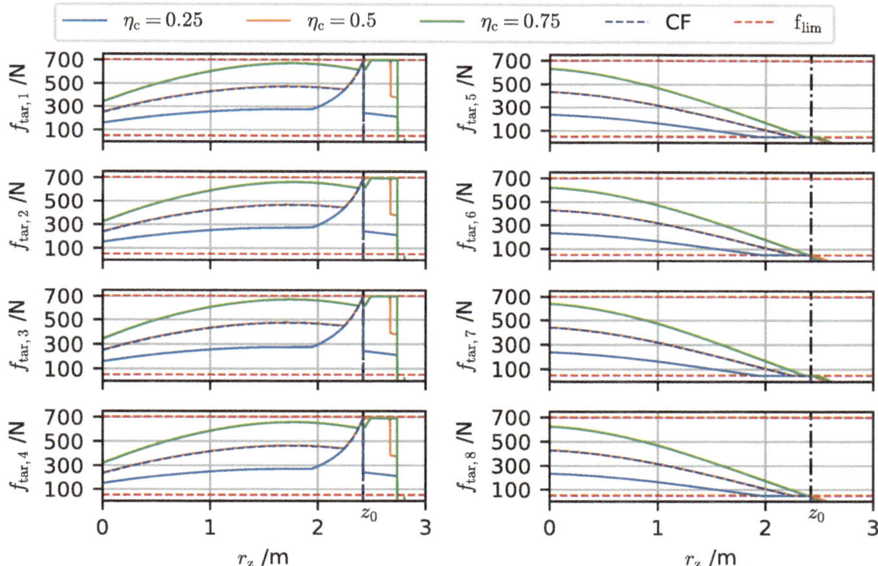

Fig. 3 Computed cable force distributions of APC compared to CF method with a motion along $r_z = [0, 3]$ m and cable force limits $f_{lim} = [50, 700]^T$ N

4 Experimental Investigation

The experimental investigation is performed on the cable robot *COPacabana* as described in Sect. 3 and thus the same parameters are used. To compute cable force target values for the drive control system, the SLSQP algorithm is converted into real-time capable driver object (TcCOM) in Beckhoff TwinCAT 3 (3.1.4024.54) environment. This setup is able to compute the nullspace, the optimization, and the force control algorithms within 20 ms (with a single core of an Intel Core i7-7700 processor). In order to validate the APC for a handling task experimentally, a weight with $m_{\mathrm{Obj}} = 15.2$ kg is loaded via a load hook onto the cable robot, moved along a trajectory with different rotations and then unloaded again. The overall weight is $m_{\mathrm{P}} = m_{\mathrm{EE}} + m_{\mathrm{Obj}}$ with platform mass $m_{\mathrm{EE}} = 13.9$ kg. The procedure is performed with conventional position controlled operation using IK (STD) and repeated with activated APC. The experimental setup and the platform trajectory are shown in Fig. 4 and video recordings of the two experiments are also published in [14]. The trajectory during manipulation was performed with a target path velocity of $\dot{\boldsymbol{y}}_{\mathrm{tar}} = 0.34\ \mathrm{ms}^{-1}$.

Results and Discussion The results of the experiments are shown in Fig. 5. On the left, the cable forces of the experiments are shown, where the cable force target values $f_{\mathrm{tar},i}$ are computed with Eq. (4). The cable force limits are $\mathrm{f}_{\mathrm{lim}} = [50, 700]^{\mathsf{T}}$ N. The measured cable forces during position controlled operation are $f_{\mathrm{STD},i}$, whereas cable forces of APC operation are $f_{\mathrm{APC},i}$. On the right, position and rotation errors of the platform are shown to validate that changes in geometric stiffness do not lead to large changes in position and rotation, and stay within the tolerance STD operation. The platform pose $\boldsymbol{y}_{\mathrm{DK}}$ was determined by solving forward kinematics using measured motor encoder position φ. Thus, pose errors due to cable

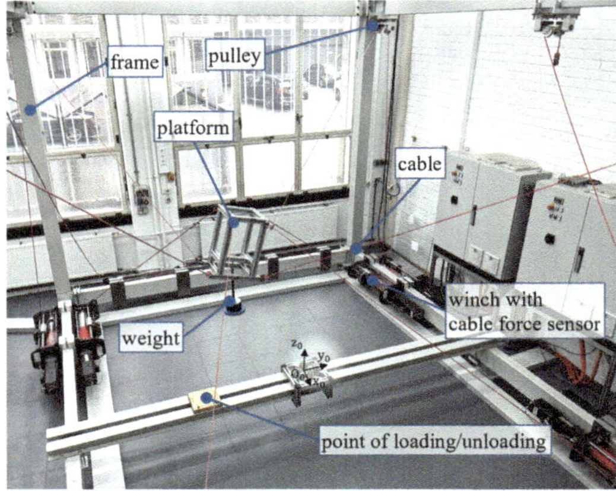

	r /m	$R(\alpha)$ /°	$R(\beta)$ /°	η_c /%	m_P /kg
Load weight:					
1	$[0.00\ \ 0.00\ \ 0.60]^{\mathsf{T}}$	0	0	25	13.9
2	$[0.00\ \ -0.87\ \ 0.38]^{\mathsf{T}}$	20	0	75	13.9
3	$[0.00\ \ -0.93\ \ 0.38]^{\mathsf{T}}$	0	0	75	29.1
Motion along trajectory:					
4	$[0.00\ \ -0.93\ \ 0.80]^{\mathsf{T}}$	0	0	50	29.1
5	$[-0.75\ \ -0.50\ \ 0.80]^{\mathsf{T}}$	-15	0	50	29.1
6	$[-0.75\ \ 0.50\ \ 0.40]^{\mathsf{T}}$	-15	0	50	29.1
7	$[0.75\ \ 0.50\ \ 0.80]^{\mathsf{T}}$	0	0	50	29.1
8	$[0.75\ \ -0.50\ \ 0.40]^{\mathsf{T}}$	0	-15	50	29.1
9	$[0.00\ \ -0.75\ \ 0.80]^{\mathsf{T}}$	-15	15	50	29.1
10	$[0.00\ \ 0.00\ \ 0.80]^{\mathsf{T}}$	0	0	50	29.1
Unload weight:					
11	$[0.00\ \ -0.93\ \ 0.38]^{\mathsf{T}}$	0	0	75	29.1
12	$[0.00\ \ -0.87\ \ 0.38]^{\mathsf{T}}$	20	0	75	29.1
13	$[0.00\ \ 0.00\ \ 0.60]^{\mathsf{T}}$	0	0	25	13.9

Fig. 4 Experimental setup on the left and schematic representation of the performed motion on the right

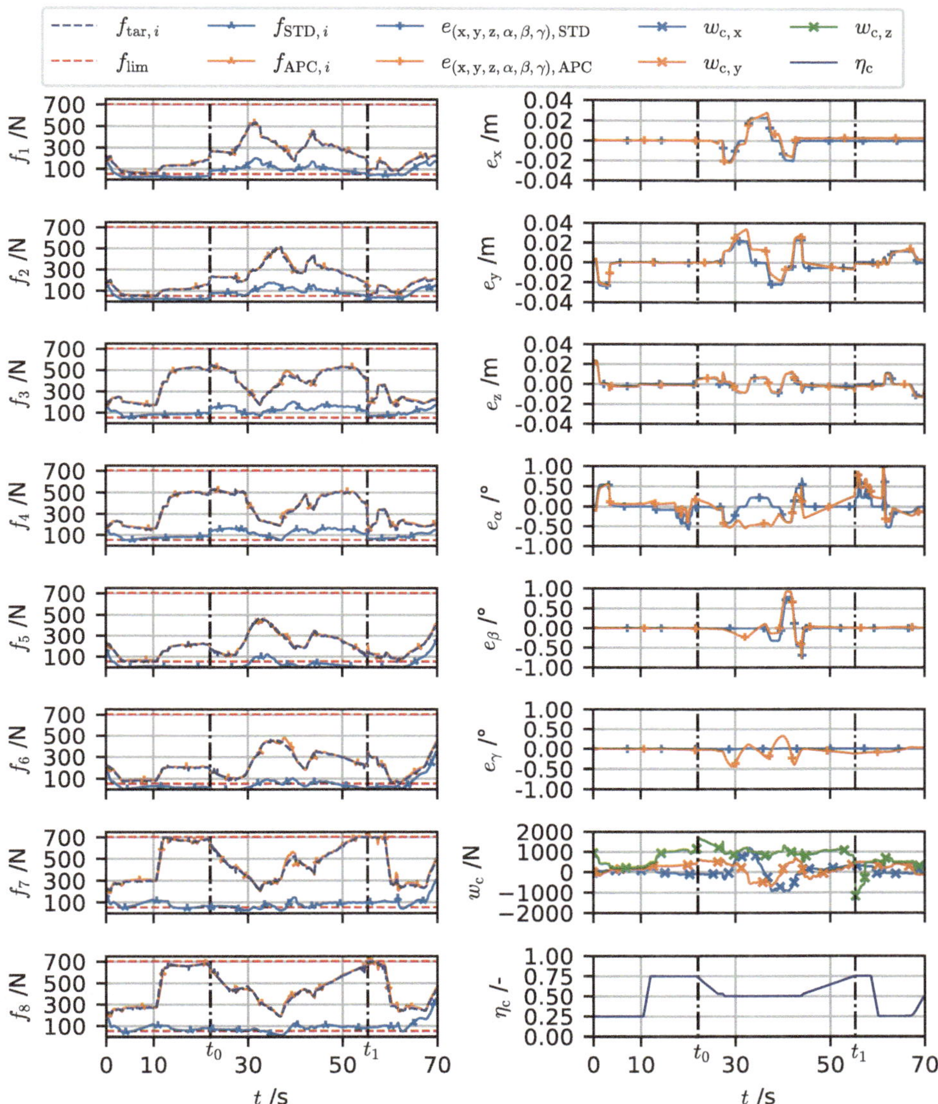

Fig. 5 Experimental results with cable forces on the left and pose errors on the right. In addition, measured translational wrenches and preload control parameters during motion are shown

elongation between frame and platform anchors cannot be investigated. The pose errors $\boldsymbol{e} = \boldsymbol{y}_{\text{tar}} - \boldsymbol{y}_{\text{DK}}(\boldsymbol{\varphi})$ are computed for platform position and rotation $\boldsymbol{e} = [e_{\text{x}}, e_{\text{y}}, e_{\text{z}}, e, e_{\text{fi}}, e]^{\top}$ for both experiments, STD and APC operation. At $t_0 = 22$ s the weight is loaded by the platform and at $t_1 = 55.4$ s the weight is unloaded.

For APC, the measured cable forces follow the target cable forces continuously and stay within force limits, even if the preload control parameter η_{c} is changed during motion. While unloading the weight at t_1 the cable forces of the lower cables $i = [5, 6, 7, 8]$ exceed the maximum cable force limit for about 0.5 s, as the weight was jammed when it was unloaded. Furthermore, minor control errors of the cable force occur during the movement between t_0 and t_1, which can be explained by the incomplete identification of the wrench $\boldsymbol{w}_{\text{c}}$ from the force sensors on winches and the swinging weight. This effect can also be observed in the minor increased rotation error e. Comparing the cable forces of STD and APC operation, the cable forces during STD operation often fall below the minimum force, which leads to a loss of platform stiffness and reduced position accuracy. Additionally, there only small differences found between STD and APC operation of the computed pose errors. Hence, the geometrical stiffness of the platform can be adaptively adjusted without loss of pose accuracy within the tolerance of STD operation. The point in time of loading and unloading the weight can be observed by identified wrench $w_{\text{c,z}}$. The swinging effect of the weight during the trajectory can also be observed with the wrenches $w_{\text{c,(x,y)}}$.

5 Conclusion

In summary, a novel preload control for CDPRs was investigated and validated in this work. For this purpose, the transposed Jacobian (structure matrix) is extended with the nullspace due to the drive redundancy and combined with the online identification of the acting wrenches on the platform. This allows an optimization problem formulation with which the cable preload can be adaptively changed with a solely preload control parameter. The results were evaluated and analyzed simulatively and experimentally. This method can be used for redundantly-restrained cable robots as long as kinematic relationships have been sufficiently formulated, because the nullspace and optimization problem formulation is generic. In the future, the accuracy of the cable robot will be investigated in detail with an absolute coordinate measuring system (laser tracker) to show the advantages of APC with hybrid position-force control of cable robots. Additionally, friction disturbances within the drive system will be further investigated in order to save costs on additional cable force sensors. Furthermore, the developed algorithms can be used for the design and control of highly redundant reconfigurable cable robots.

Acknowledgements This work was supported by the German Research Foundation (DFG-project numbers: 470674716 and 317440765) at the University of Stuttgart.

References

1. Godbole, H.A., Caverly, R.J., Forbes, J.R.: Dynamic modeling and adaptive control of a single degree-of-freedom flexible cable-driven parallel robot. J. Dyn. Syst. Meas. Control **141**(10) (2019). https://doi.org/10.1115/1.4043427
2. Gouttefarde, M., Rodriguez, M., Barrelet, C., Hervé, P.E., Creuze, V., Gorrotxategi, J., Oyarzabal, A., Culla, D., Sallé, D., Tempier, O., Ferrari, N., Chaumont, M., Subsol, G.: The robotic seabed cleaning platform: An underwater cable-driven parallel robot for marine litter removal. In: Caro, S., Pott, A., Bruckmann, T. (eds) Cable-Driven Parallel Robots, vol. 132, Springer Nature Switzerland, Cham, pp. 430–441 (2023). https://doi.org/10.1007/978-3-031-32322-5_35
3. Harandi, J., MR, Khalilpour SA, Taghirad HD, Romero JG,: Adaptive control of parallel robots with uncertain kinematics and dynamics. Mech. Syst. Signal Process. **157**(107), 693 (2021). https://doi.org/10.1016/j.ymssp.2021.107693
4. Khajepour A, Mendez ST, Rushton M, Jamshidianfar H, Qi R, Pazooki A, Durali L, Soltani A (2023) A warehousing robot: From concept to reality. In: Caro S, Pott A, Bruckmann T (eds) Cable-Driven Parallel Robots, vol 132, Springer Nature Switzerland, Cham, pp 397–406, https://doi.org/10.1007/978-3-031-32322-5_32
5. Kraus W (2015) Force control of cable-driven parallel robots. phdthesis, University of Stuttgart, Stuttgart, Germany, https://doi.org/10.18419/opus-6899
6. Kraus, W., Spiller, A., Pott, A.: Energieeffizienz von parallelen seilrobotern. In: Frey, G., Schumacher, W., Verl, A. (eds.) Elektrische Automatisierung - Systeme und Komponenten (SPS/IPC/DRIVES Kongress 2013). VDE-Verlag and VDE Verlag, Berlin (2013)
7. Lamaury J, Gouttefarde M, Chemori A, Hervé PE (2013) Dual-space adaptive control of redundantly actuated cable-driven parallel robots. In: Intelligent Robots and Systems (IROS 2013), https://doi.org/10.1109/IROS.2013.6697060
8. Li H, Li MZ (2019) An experimental study on control accuracy of fast cable robot following zigzag astronomical trajectory. In: Pott A, Bruckmann T (eds) Cable-Driven Parallel Robots (CableCon 2019), Springer, Cham, Mechanisms and Machine Science, vol 74, pp 245–253, https://doi.org/10.1007/978-3-030-20751-9_21
9. Nguyen DQ, Gouttefarde M (2014) Study of reconfigurable suspended cable-driven parallel robots for airplane maintenance. In: Intelligent Robots and Systems (IROS 2014), IEEE, pp 1682–1689
10. Pott A (2012) Influence of pulley kinematics on cable-driven parallel robots. In: Lenarčič J, Husty ML (eds) Latest Advances in Robot Kinematics (ARK 2012), Springer, Dordrecht, pp 197–204, https://doi.org/10.1007/978-94-007-4620-6_25
11. Pott A (2019) Wirex. https://doi.org/10.13140/RG.2.2.25754.70088, https://gitlab.cc-asp.fraunhofer.de/wek/wirex
12. Pott A, Bruckmann T, Mikelsons L (2009) Closed-form force distribution for parallel wire robots. In: Kecskeméthy A, Müller A (eds) Computational Kinematics, Springer, Berlin and Heidelberg, pp 25–34, https://doi.org/10.1007/978-3-642-01947-0_4

13. Reichenbach T, Rausch K, Trautwein F, Pott A, Verl A (2021) Velocity based hybrid position-force control of cable robots and experimental workspace analysis. In: Gouttefarde M, Bruckmann T, Pott A (eds) Cable-Driven Parallel Robots, Mechanisms and Machine Science, vol 104, Springer International Publishing, Cham, pp 230–242, https://doi.org/10.1007/978-3-030-75789-2_19
14. Reichenbach, T., Clar, J., Pott, A., Verl, A.: Replication data for: Adaptive preload control of cable-driven parallel robots for handling task. (2024). https://doi.org/10.18419/darus-4075
15. Trautwein F, Reichenbach T, Tempel P, Pott A, Verl A (2020) Copacabana. In: Pfurner M, Dohnal F (eds) Sechste IFToMM D-A-CH Konferenz, https://doi.org/10.17185/duepublico/71189
16. Verhoeven R (2004) Analysis of the workspace of tendon-based stewart platforms. phdthesis, University of Duisburg-Essen, Duisburg, Germany
17. Zhang, B., Shang, W., Deng, B., Cong, S., Li, Z.: High-precision adaptive control of cable-driven parallel robots with convergence guarantee. IEEE Trans. Industr. Electron. **71**(7), 7370–7380 (2024). https://doi.org/10.1109/TIE.2023.3310012

Examination of Ultrasonic Non-Destructive Testing for Tailored-Forming-Applications

Caner-Veli Ince, Alexej Verschinin, Jan Peter Niestroj, René Gansel, Sebastian Barton, and Annika Raatz

Abstract

The novel manufacturing process, Tailored Forming, sets new requirements for handling and non-destructive testing. The Tailored-Forming-Process aims to manufacture hybrid workpieces, with a key point being the joining zone where an intermetallic phase may form. This phase directly affects the strength of the workpiece, making it crucial to monitor and maintain its development during the manufacturing process. The monitoring is carried out using the non-destructive testing method of ultrasonic testing, which is performed during handling. This paper explores the requirements for automated testing of Tailored-Forming-Components to effectively monitor the bonding quality of the joining zone. Therefore, Tailored Forming is briefly introduced and the current manual testing and signal processing is presented. Then, a semi-automated workpiece examination with planar and spherical joining zone geometries is performed. Variations in sensor positioning and its tilt angle relative to the specimen have shown significant fluctuations in the measured signals. Therefore, for a fully automated process, accurate handling of both the sensor and the specimen is essential.

Keywords

Automation • Handling • Ultrasonic Testing • Hybrid Workpieces

C.-V. Ince (✉) · J. P. Niestroj · A. Raatz
Institute of Assembly Technology and Robotics, Leibniz University Hannover, An der Universität 2, Garbsen, Germany
e-mail: ince@match.uni-hannover.de

A. Verschinin · R. Gansel · S. Barton
Institut für Werkstoffkunde (Materials Science), Leibniz University Hannover, An der Universität 2, Garbsen, Germany

M.-C. Wanner et al. (eds.), *Annals of Scientific Society for Assembly, Handling and Industrial Robotics 2024*, https://doi.org/10.1007/978-3-031-91463-8_24

1 Introduction

Growing environmental protection requirements have led to a need for novel manufacturing methods that enhance existing technologies by increasing efficiency. One example is the automotive industry, which is implementing lightweight structures and materials to achieve weight reduction, resulting in more fuel-efficient vehicles. The challenge is to achieve such a structure without compromising mechanical properties, while maintaining a competitive price. Lightweight structures require materials with low density and high mechanical strength, such as titanium, which is more expensive than steel [20].

To overcome the mentioned challenge, the development of specially tailored hybrid components could be a solution. Hybrid workpieces allow to combine the positive mechanical properties of different materials and to reduce the cost for lightweight construction. Such a novel process is researched in the central research centre (CRC) 1153 'Tailored Forming' at the Leibniz University [3]. The CRC aims to develop a process chain that enables the manufacture of tailored hybrid components for specific applications. An exemplary component is a shaft made of steel and aluminium in an axial configuration. The steel is located in an area with high stresses, such as a bearing seat, while the aluminium is located in an area with low stresses. This design results in a lighter shaft compared to a mono-material steel shaft, while still being cost-effective compared to a titanium shaft.

In contrast to conventional hybrid manufacturing methods, the Tailored-Forming-Process achieves material joining during the fabrication of the semi-finished workpiece, rather than during or after the forming stage. In the case of the shaft, friction welding joins the materials and afterwards the hybrid workpiece is forged, as shown in Fig. 1. The joint forging massively improves the bonding strength in the joining zone while establishing an intermetallic phase [5]. During the manufacturing process, the intermetallic phase is affected by various parameters such as changing temperature and deformation, which alters the mechanical properties [4].

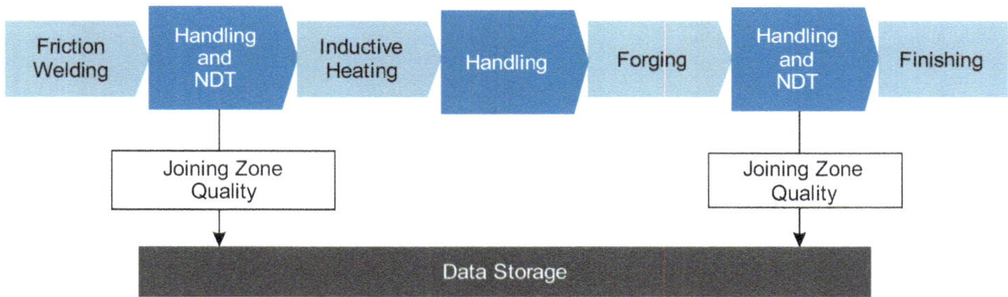

Fig. 1 Tailored-Forming-Process with integrated NDT and data flow

The CRC researches not only the hybrid components, but also the entire process chain [11], including workpiece manufacturing, forging, machining, heat treatment, testing, handling, automation, and data processing, as depicted in Fig. 1. In order to ensure and improve the joining zone quality, it is essential to know how to positively influence the properties of the intermetallic phase during the manufacturing process. Therefore, it is essential to achieve complete bonding in the joining zone of the different materials and to monitor the bonding quality during the manufacturing process, which can be achieved through non-destructive testing (NDT). The main challenge is to identify strong bonding with a limited number of measurements per workpiece due to the high deviation of the feedback signal caused by inadequate manual repeatability. This paper explores the requirements for automated testing of Tailored-Forming-Components to effectively monitor the bonding quality of the joining zone with ultrasonic testing. Firstly, a brief overview of how NDT determines the bonding quality of the joining zone will be provided, followed by a presentation of automated NDT processes. This is followed by an examination of how automation affects the NDT of hybrid Tailored Forming components with varying joint zone geometries.

2 Related Work

In this section, the joining process and the non-destructive testing are briefly introduced. Afterwards, current automated testing methods are presented.

2.1 Joining Process

The joining process of the different materials is crucial for the quality of the resulting components. The CRC 1153 has developed several methods for joining, including deposition welding, laser beam welding, co-extrusion, and friction welding [3]. This paper will focus on hybrid shafts that are manufactured in a friction welding process.

Behrens et al. showed that the geometry of the welding surfaces influences the final bond quality and bond strength compared to plane welding surfaces due to the contact area and the temperature generated during the friction welding process [3]. Therefore, they experimentally examined different friction welding surfaces with the aim of improving bonding quality through shrinkage, undercuts, and surface enlargement. Furthermore, the results indicate that the friction welding parameters have a significant influence on the formation of defects such as air inclusions [5]. Therefore, NDT of the joining zone is necessary to enhance and guarantee consistent bonding quality over time.

2.2 Non-destructive Testing of the Joining Zone

One widespread NDT Method is ultrasonic testing (UT), which is commonly used to examine thickness, defects and bonding properties of coatings [19], composite structurse [2, 18] and (friction-) welded joints [7, 14]. The main principle is the reflection of the ultrasonic waves at surfaces and interfaces, which are represented in the ultrasonic signals by amplitude peaks along the travel time or distance of the wave [13].

In the present case, the ultrasound would behave as follows in the absence of defects such as cracks or pores in the monomaterials. The first reflection caused by the surface to which the ultrasonic waves are applied, is referred as the surface echo (SE) (Fig. 2). It emerges near zero if the ultrasound is applied through a coupling medium on the surface. The second peak is the reflection of the bonding surface of the two materials or rather the debonded interfaces, here referred to as bonding echo (BE). If there is an adhesive or material bond, a part of the ultrasonic waves can pass the bonded areas and be reflected by the back surface of the underlying composite partner, which would emerge in a third amplitude peak, here referred to as back surface echo (BSE) (Fig. 3a). The quality of the bonding decreases the higher the bond echo and the lower the back surface echo. If the ultrasonic waves are applied to an area with no bonding, the signals would only show a damping of the repeated bonding echo (Fig. 3b).

As previously stated, the amplitudes of bonding and back surface echoes are related to the joint quality, but are also affected by factors like the geometry of the probe, the damping property of the materials and the applied ultrasonic frequency. One way to validate the measured value is using a ratio between two subsequent echo peaks, either between the surface and bonding echo or the bonding and back surface echo, which will be used in this study [1, 13].

In the process of friction welding of dissimilar metals, the development of intermetallic phases between the welding partners has a crucial influence on bonding strength. While an insular appearance of intermetallic phases (IMP) is disadvantageous, a consistent layer

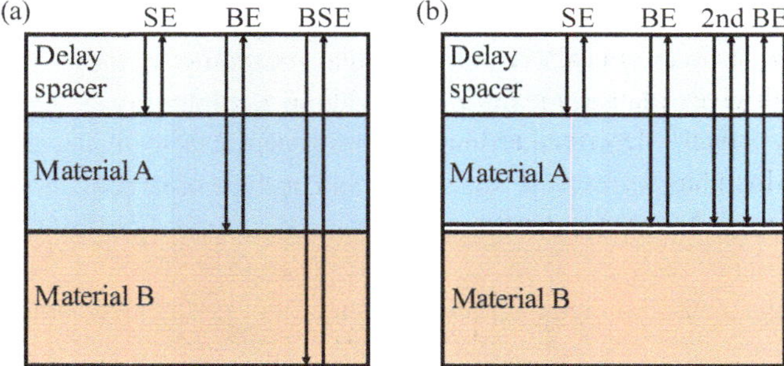

Fig. 2 Ultrasonic waves travelling in layered structures with (**a**) and without bonding (**b**), based on [1]

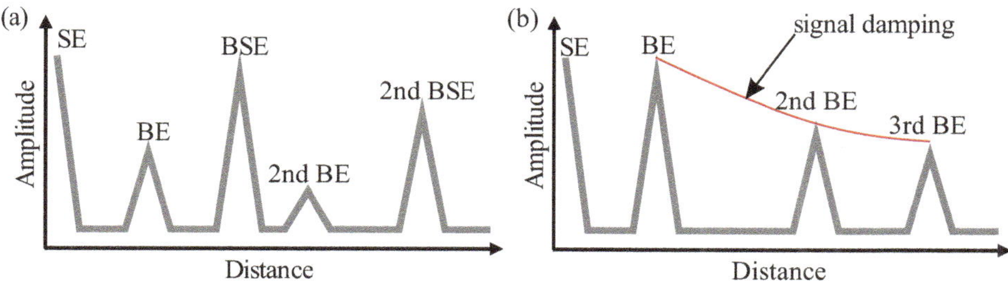

Fig. 3 Schematic ultrasonic signals of bonded (**a**) and debonded areas (**b**), adapted from [13]

of a certain thickness is crucial for high joint strength. However, if the IMP is too thick, this will have a negative effect on the bonding strength. One reason for this behaviour are the differences in the coefficients of thermal expansion between the base materials, which can lead to internal stresses. Additionally, the low solubility of iron in aluminium may result in the formation of brittle IMPs that tend to have higher hardness but lower ductility compared to the base materials [9, 15, 17].

Besides impacting joint strength, the occurrence of IMPs indicates the formation of a material bond between the two materials, which means that a verification of the presence of the IMPs via ultrasonic testing is possible. Investigations on this topic have been carried out as part of the CRC research using co-extruded aluminium-steel profiles as specimens. Although no direct connection between the thickness or type of the IMPs and the ultrasonic signals could be found, a qualitative correlation with the shear strength of the bonding could be verified. That implies that further research on this topic should be performed [8].

Adapting these qualitative results in an automated application in order to develop a quantitative IMP thickness or bonding quality measuring method is rather challenging and requires a suitable regression analysis with an appropriate specimen quantity. Therefore, it is important to first ensure the proper and reliable automated handling of the workpieces and measuring equipment. There are several different handling approaches researched in current studies, which will be briefly presented: Wang et al. employs a robot with six degrees of freedom (6-DoF) where the sensor is used as an end-effector for flexible automated UT [16]. In this way, the robot is able to perform various measurements without moving the sample. In this case, a turntable is used to extend the measurements by rotating the sample. To ensure proper coupling for UT, the sample is immersed in a water bath. However, this solution requires a stationary immersion bath that limits the degree of freedom. A concept for the inspection of spot welds is presented in [12]. An elastic membrane filled with water is attached to the sensor for the coupling between sample and sensor, which allows for usage during movement as opposed to immersion baths. This concept is more versatile, but still requires a stationary system for sample fixation. A more adaptive automation is shown in [10]. They use two robots, one equipped with an ultrasonic transmitter unit and the other with a receiver unit which is required for the ultrasonic transmission method. This dual robot setup enables testing of curved surfaces. The review indicates that 6-DoF robots can be used, but they require separate sample fixation. This must first be solved in the design of the current handling system, since for Tailored Forming, a fully integrated solution into the handling system is desired. This means that the sample is gripped, positioned, and tested in the handling system. As a result, measurement can be performed during handling and the number of required process steps is reduced.

3 Experiments

The experiments aim to determine the requirements for automated testing of rotary friction welded Tailored Forming workpieces. Based on the idea that it is possible to assess the quality of the whole joining surface by conducting the ultrasonic measurement only at one position, given a known typical distribution of joint quality across the specimen cross section and its rotational symmetry, the main objective is to investigate this possibility. Hybrid workpieces will be tested with three different joining zone geometries, ultrasonic sensor angle and surface position parameters. Five workpieces have a planar joining zone, one has a spherical and one has a cylindrical pin emerging from the steel part. Apart from that, all samples are made of EN AW 6082 aluminium alloy and 100Cr6 steel. A 6-DoF collaborative robot is used to manipulate the ultrasonic sensor during the testing phase (Fig. 4). The robot has sensors that monitor the contact force and angle between the ultrasonic sensor and the sample. The ultrasonic measurement was performed utilizing a USLT 2000 device manufactured by Krautkramer. A 7.5 mm diameter impulse-echo sensor was used and water served as the medium for coupling.

Firstly, the effect of sensor placement on the workpiece surface was investigated. For that purpose, each specimen was measured, finely resolved around the central point, at distances of 0, 1, 2, 3, 4 and 5 mm, with additional measurements at 9 and 13 mm to

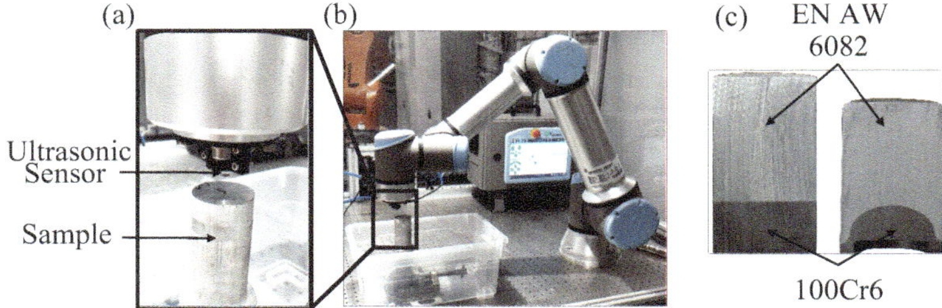

Fig. 4 Experimental set up (**a**), (**b**) and investigated samples (**c**) with planar and spherical joining zone

fully characterise the joining surface. This resulted in eight distinct measurement locations of which each underwent eight measurements at 45° radian intervals, which can be considered equivalent due to the rotation symmetry of the specimen.

The second experimental step was to investigate the impact of the sensor's tilt angle. One specimen, characterized by a planar joining zone, was analysed at the central point using tilt angles of 0°, 1°, 2°, 3°, and 4° between the central axes of the sensor and the specimen. Additionally, the same workpiece was evaluated using the approach previously applied to investigate the influence of sensor position, but with the sensor tilted at a 1° angle.

4 Results

As already mentioned above, the experiments focussed on the influence of the position and the tilt angle of the sensor on the measured value in comparison with the central measuring position.

4.1 Influence of the Sensor Placement

The measured ultrasonic data provided the amplitude heights of the bonding and back surface echoes, whose ratio (Eq. 1) was used to assess the bonding quality and processed into distribution mappings of the joining surfaces.

$$ratio = \frac{\hat{A}_{BSE}}{\hat{A}_{BE}} \quad (1)$$

The colouration, ranging from green for better bonding to red for worse bonding, corresponds to the joining ratio. As no correlation study has been carried out between the bonding ratio and tensile tests on the present sample type, it is not possible to draw an exact line between good and bad bonding. Therefore, for the time being, an approximate estimate must be that the BSE should be significantly higher than the BE (acc. to [13]). The data collected from specimens with a planar joining zone resulted in rotationally non-symmetric figures with a similar orientation across all specimens, suggesting the significance of even small tilt angles, which will be examined closer in Chap. 4.2. For this reason, the mean values of all equivalent measuring points across the five samples with flat joint geometry were used to visualize the data as shown in Fig. 5a. It appears that there is an alteration of the joining ratio along the radius, which is presumably due to the temperature distribution during the rotary friction welding process [6].

Based on the data collected from specimens with non-flat joining surfaces, the travelling distance of the ultrasonic waves was used to extend the mapping into a 3D representation, shown in Fig. 5b-c. In order to determine the influence of the sensor

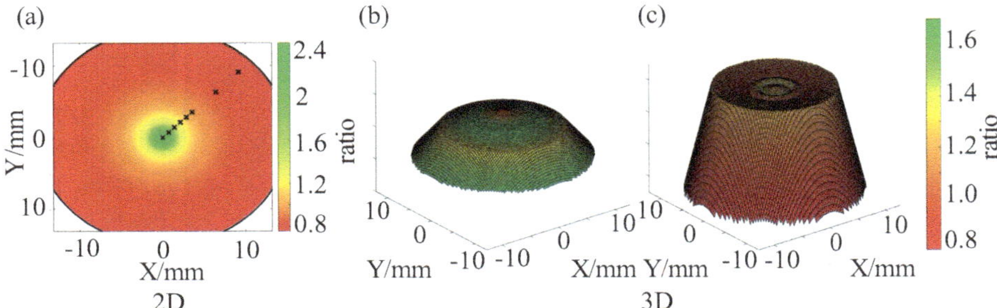

Fig. 5 Joining ratio distribution of the planar joining surfaces (top view) (**a**) and 3D-mapping of the joining surfaces with a spherical (**b**) and a steel-pin geometry (**c**)

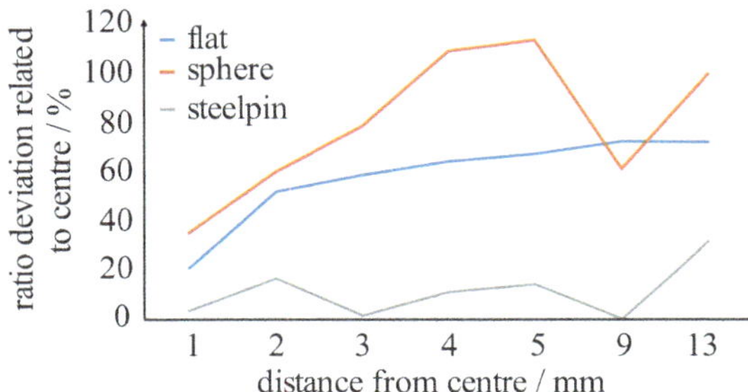

Fig. 6 Percentage difference of the measured ratio values compared to the centre

position on the measured ratio value, the deviations from the corresponding central point values were calculated and normalised (Fig. 6).

Whilst the changing joining ratio of the specimen with a flat joining zone could be based on the temperature distribution during manufacturing process, the measurements of the spherical sample might be affected by the wave refraction at the curved joining surface. The specimen featuring a steel pin geometry exhibits the least deviation towards the centre, yet also display the lowest measured ratio values.

4.2 Influence of the Sensor Tilt Angle

The data obtained from the measurements taken with a tilt angle were processed similarly to the previous evaluation. The mapping in Fig. 7a shows distinct differences to the one measured without tilt angle. Not only is the value of the ratios much smaller, but also its distribution over the joining zone presents a different picture, which is likely due to edge effects. The variation of the tilt angle illustrates its severe influence on the ratio value (Fig. 7b).

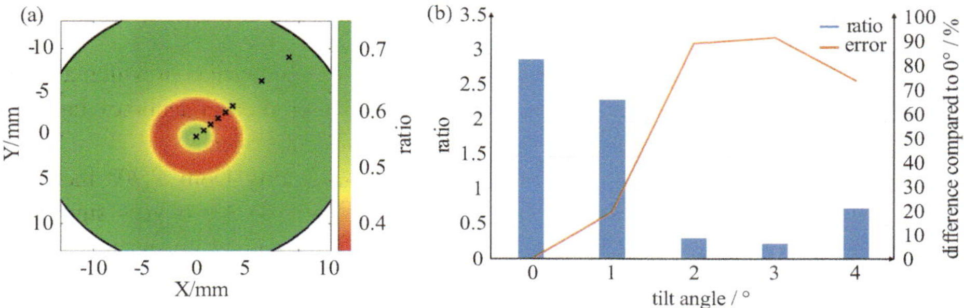

Fig. 7 Mapping of a planar joining surface with 1° tilt angle (**a**) and sensor tilt angle variation at the centre of the sample affecting the measured ratio value (**b**)

5 Discussion

The results indicate that the measured value strongly depends on the tilt angle between the sensor and the sample surface. Even slight deviations in angle, which can be caused by variations in surface quality resulting from prior mechanical preparation such as cutting, can have a significant impact on the results (Fig. 7b). In this setup, the influence of robot positioning error and the inaccuracies of manual sample clamping are not considered. Therefore, an automation concept is required that adapts the sensor tilt to the sample surface. Otherwise, the collected data will not be reliable.

Furthermore, the deviation in signals for the same joining zone geometry needs to be considered while automating the bonding quality assessment process (Fig. 6). With the given data, the automated measuring concept has to take several measuring points on each sample, which is unfeasible to be realised during the brief handling cycle times. Here, more data are required to investigate the possibility of utilizing just one measurement at the central point as a testing value for a known joining ratio distribution, which requires precise positioning of the sensor on the sample.

The measured ultrasonic wave travelling distance enables the determination of the final joining zone geometry and the location of possible defects or bad bonding spots. With this information, the heat treatment and forging can be adjusted for the next workpiece to eliminate the undesired conditions. As depicted in Fig. 5a, the spherical joining zone shows a weaker spot at its half height. Here, heat treatment can provide additional thermal energy, while forging enhances the quality of the joint. The validation of this procedure requires the measurement of workpieces in both unforged and forged states, as well as a correlation with mechanical testing results, which are currently unavailable.

6 Summary

This paper aims to determine the requirements for an automated NDT of Tailored-Forming-Components. Tailored Forming is a new manufacturing process for hybrid bulk components, where the joining occurs at the beginning of the process, resulting in higher bonding quality compared to other hybrid manufacturing processes. The integration of non-destructive testing and monitoring of the joining zone into the manufacturing process in an automated way is crucial for the optimization of the overall process robustness. Therefore, samples with different joining zone geometry were tested to investigate the measuring point location and sensor tilt angle influence on the results. The findings are that both, the tilt angle and the sensor positioning, are essential disruptive factors for the measurement results. Even small angular variations, caused by the surface quality of the sample result in a high deviation. This information will be taken into consideration when designing fully automated handling and testing systems. Additionally, more data are needed to determine the ideal number of measuring points on each sample for assessing bonding quality, which is a key factor for future work.

Acknowledgements The results presented in this paper were obtained within the Collaborative Research Centre 1153 'Process chain to produce hybrid high-performance components by Tailored Forming' -252662854 in sub project A01 and C07. The authors would like to thank the German Research Foundation (DFG) for the financial and organizational support of this project.

References

1. Azamatjon, K., Cho, Y., Kim, Y.H., Kim, J., Park, J., Yi, J.-H., Malikov: Ultrasonic assessment of thickness and bonding quality of coating layer based on short-time Fourier transform and convolutional neural networks. Coatings **11**(8), 909 (2021)
2. Bastianini, F., Di Tommaso, A., Pascale, G.: Ultrasonic nondestructive assessment of bonding defects in composite structural strengthenings. Compos. Struct. **53**, 463–467 (2001)
3. Behrens, B.-A., Uhe, J.: Introduction to tailored forming. In: Production Engineering, vol. 15, no. 2. Springer Science and Business Media LLC, pp. 133–136 (2021)
4. Behrens, B.A., Uhe, J., Petersen, T., Klose, C., Thürer, S., Diefenbach, J., Chugreeva, A.: Challenges in the forging of steel-aluminum bearing bushings. Materials **14**, 803 (2021)
5. Behrens, B.-A., Uhe, J., Ross, I., Peddinghaus, J., Ursinus, J., Matthias, T., Bährisch, S.: Tailored forming of hybrid bulk metal components. Int. J. Mater. Form. **15**(3). Springer Science and Business Media LLC (2022)
6. Bouarroudj, E.-O.: Chikh, Salah, Abdi, Said, Miroud, Djamel: Thermal analysis during a rotational friction welding. Appl. Therm. Eng. **110**, 1543–1553 (2017)
7. Fortunato, J., Anand, C., Braga, D.F.O., Groves, R.M., Moreira, P.M.G.P., Infante, V.: Friction stir weld-bonding defect inspection using phased array ultrasonic testing. Int. J. Adv. Manuf. Technol. **93**, 3125–3134 (2017)

8. Fricke, L.V., Thürer, S.E., Kahra, C., Bährisch, S., Herbst, S., Nürnberger, F., et al.: Non-destructive evaluation of workpiece properties along the hybrid bearing bushing process chain. J. Mater. Eng. Perform. **32**(15), 7004–7015 (2023)
9. Fukumoto, S., Inuki, T., Tsubakino, H., Okita, K., Aritoshi, M., Tomita, T.: Evaluation of friction weld interface of aluminium to austenitic stainless steel joint. Mater. Sci. Technol. **13**(8), 679–686 (1997)
10. Guo, C., Xu, C., Xiao, D., Hao, J., Zhang, H.: Trajectory planning method for improving alignment accuracy of probes for dual-robot air-coupled ultrasonic testing system. Int. J. Adv. Robot. Syst. **16**(2). SAGE Publications (2019)
11. Ince, C.-V., Blümel, R., Raatz, A.: Concept for a resource-efficient process chain for hybrid bulk components with optimized energy utilization. Procedia CIRP **116**, 732–737. Elsevier BV (2023)
12. Ji, C., Na, J. K., Lee, Y.-S., Park, Y.-D., Kimchi, M.: Robot-assisted non-destructive testing of automotive resistance spot welds. Weld. World **65**(1), 119–126. Springer Science and Business Media LLC (2020)
13. Krautkrämer, J., Krautkrämer, H.: Ultrasonic testing of materials. Unter Mitarbeit von W. Grabendörfer. 4th fully revised edition, translation of the 5th revised German edition. Berlin, Heidelberg: Springer-Verlag (1990)
14. Liu, F., Liu, S., Li, L.: Ultrasonic evaluation of friction stir welding. In: 17th World Conference on Nondestructive Testing. Shanghai, China. In: e-J. Nondestruct. Test. **13**(11) (2008)
15. Thürer, S.E., Peters, K.R., Heidenblut, T., Heimes, N., Peddinghaus, J., Nürnberger, F., Behrens, B.-A., Maier, H. J., Klose, C.: Characterization of the interface between aluminum and iron in co-extruded semi-finished products. Materials **15** (2022)
16. Wang, J., Zhang, H., Xu, H., Wang, C., Cao, J., Liu, T., Li, Z., Yang, J.: Robot-assisted ultrasonic testing technology for complex revolved workpiece. J. Phys.: Conf. Ser. **2679**(1). IOP Publishing (2024)
17. Yılmaz, M., Çöl, M., Acet, M.: Interface properties of aluminum/steel friction-welded components. Mater Charact **49**(5), 421–429 (2002)
18. Yılmaz, B., Jasiuniene, E.: Advanced ultrasonic NDT for weak bond detection in composite-adhesive bonded structures. Int. J. Adhes. Adhes. **102**, 102675 (2020)
19. Zhang, J., Cho, Y., Kim, J., Malikov, A.K., Kim, Y.H., Yi, J.-H., Li, W.: Non-destructive evaluation of coating thickness using water immersion ultrasonic testing. Coatings **11**, 1421 (2021)
20. Zhang, W., Xu, J.: Advanced lightweight materials for Automobiles: A review. In: Materials & Design, vol. 221. Elsevier BV (2022)

Force-Based Object Dimension Estimation in Practical Bin Picking Scenarios

Tristan Fogt, Louis Hardinghaus, Georg Siegemund, Timon Adler, Holger Kunz, Arne Glodde, Sina Rahlfs, and Franz Dietrich

Abstract

Bin-picking processes have traditionally relied on camera-based methods, presenting challenges such as high costs and the need for expert training. Recent advancements have introduced camera-less strategies leveraging force and moment data from sensors. However, these strategies lack knowledge of object geometry, limiting precise placement and stacking. This paper proposes an approach for autonomously deducing object dimensions to enhance camera-less bin-picking systems. Evaluations with real-world objects demonstrate accurate predictions of object dimensions within a margin of 8–18%.

Keywords

Random bin picking • Object handling • Force feedback • Soft-gripper • Dimension estimation • Force-torque sensor • Force based

T. Fogt · L. Hardinghaus · G. Siegemund · A. Glodde (✉) · S. Rahlfs · F. Dietrich
Technical University of Berlin, Berlin, Germany
e-mail: arne.glodde@tu-berlin.de

G. Siegemund
e-mail: g.siegemund@tu-berlin.de

T. Adler · H. Kunz
FORMHAND Automation GmbH, Braunschweig, Germany

M.-C. Wanner et al. (eds.), *Annals of Scientific Society for Assembly, Handling and Industrial Robotics 2024*, https://doi.org/10.1007/978-3-031-91463-8_25

1 Introduction

Bin-picking is dominated by camera-based approaches [1], which present several challenges, including the high costs associated with both hardware and software, as well as the necessity for professional expertise to conduct training and install the software [2]. A recent study demonstrated the feasibility of a bin picking strategy that does not use any camera, accomplishing all aspects of the bin-picking process by leveraging force and moment data obtained from a force-moment sensor [3]. The aim of this strategy is to circumvent problems associated with conventional camera-based bin-picking approaches. The functionality of the camera-less strategy is currently limited by a lack of knowledge of the handled object's geometry. The strategy cannot place or stack the objects with precision, as it neither knows the objects dimensions generally nor the orientation of the object it is handling. Knowing the object geometry would additionally be helpful if an object is grasped off-center and a moment is detected: if the system knows the objects dimensions it can more accurately determine how its grasp should be adjusted to reach the objects center of mass.

Although it is possible to manually input the object dimensions, this requires user interaction and escalates configuration efforts. Moreover, this would not help the system identify the orientation of the object it is currently handling. Thus, this paper investigates an approach whereby a camera-less bin picking system can autonomously deduce object dimensions, allowing for more nuanced system behavior.

2 State of the Art

Tactile gripping has been implemented in a number of robotic grasping tasks. A system developed by Pia et al. focuses on object recognition and grasping in visually inaccessible environments with unknown object shapes and free object movement. This system utilizes sparse tactile feedback from fingertip sensors on a dexterous hand to localize, identify, and grasp novel objects without relying on visual feedback. This approach represents an innovative solution to address challenges in demanding scenarios such as extracting items from drawers [4]. Another system by Leão et al. addresses entanglement of complex geometries. This system employs force/torque sensors to ascertain the number of items picked and the necessary rotation direction to untangle them [5]. Outside of tactile object recognition and handling applications, there is limited research focusing on camera-less manipulation or object dimensioning [6]. Most work regarding bin picking focuses on image-based deep learning or machine learning techniques. A study reviewing 18 bin picking approaches from 2014 to 2021 determined that most use RGB information as their data modality, and none use tactile or force-based procedures [1].

Object dimensioning for bin picking tasks is focused chiefly on camera-based methods [7]. Object dimensions are commonly described with a bounding box (BB), which refers to a rectangle (2D) or cuboid (3D) which encloses the object. A 2D bounding box extracted from a 2D RGB image can be extrapolated using known information of the objects geometry to reliably determine a 3D bounding box [6]. With a rotatable bounding box (RBox), information such as center position, object width, object height, object class and orientation can be extracted from images of the bin. To determine the RBox, first a coarse 2D bounding box with axes parallel to those of the captured image (axis aligned bounding box) is identified. An algorithm then transforms the BB to be flush with the edges of the object [8]. This modified BB can be used to adjust the robot end effector's grasping position [9]. Some studies employ 3D sensing technology such as 3D camera systems, stereo cameras or LIDAR sensors to directly obtain a 3D bounding box without the need for prior dimensional knowledge [6, 14, 15]. However, 3D point cloud segmentation still has considerable room for development compared to the progress made with 2D segmentation [10]. Gupta et al. demonstrate a low-cost Bin-Picking solution for logistic applications that uses 3D point clouds for box segmentation, with varying success for objects of different texture characteristics [11].

Further dimensioning methods can be found in other domains. One method used in production metrology utilizes a CMM (coordinate measuring machine). The CMM probes the surface of an object with a touch probe and creates a numerical image of the object from the probed points [12]. Moreover, Paul James Scott describes a *roundness measuring machine* which determines deviations in the roundness of a cylindrical object by rotating the object and systematically probing [13].

3 Approach

The aim of this approach is to determine an objects dimensions by identifying its bounding box. The simplest approach is to determine the axis-aligned bounding box (AABB) by measuring the objects dimensions strictly along the robot's axes. To enhance precision, the granularity can be increased at the cost of additional time by examining various rotations of the bounding box for a tighter fit. Exploring more rotations increases granularity but also lengthens the process time.

3.1 Axis Aligned

To obtain the axis-aligned dimensions of a grasped object, the robot moves the object out of the bin, and lowers it until it contacts the work surface. At this point, the system records the distance from the gripper to the pre-recorded worksurface position, establishing the first axis-aligned dimension (Fig. 1).

The subsequent axis aligned dimensions are determined by probing the object against the outside wall of the bin and measuring the distance from the gripper to the pre-recorded

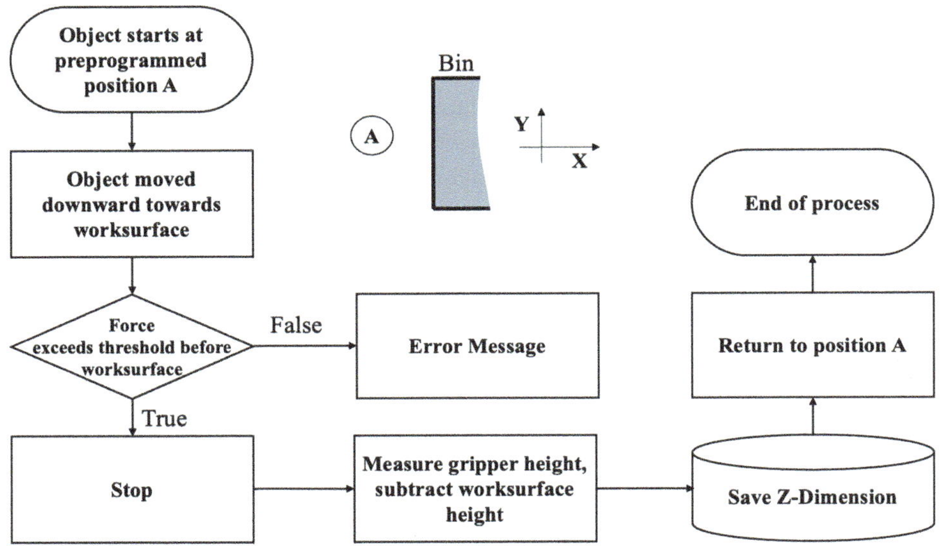

Fig. 1 Flowchart for measurement of axis aligned z dimension

bin edge position. The object is probed at 90-degree intervals, and opposite measurements are added to calculate the remaining dimensions and thus determine the complete AABB (Fig. 2).

3.2 Semi Axis Aligned

To enhance the precision of the second and third dimensions, the object is probed at smaller angle intervals, in this study 45, 30, and 15 degrees. The expected result is a more precise measurement at the cost of increased process time.

Note that even with the higher precision approach, the first dimension remains axis-aligned. This means that the derived dimensions may not match the true minimum bounding dimensions. However, given the tendency for objects to rest on a stable face, the axis-aligned dimension may indeed tend to match the true dimension.

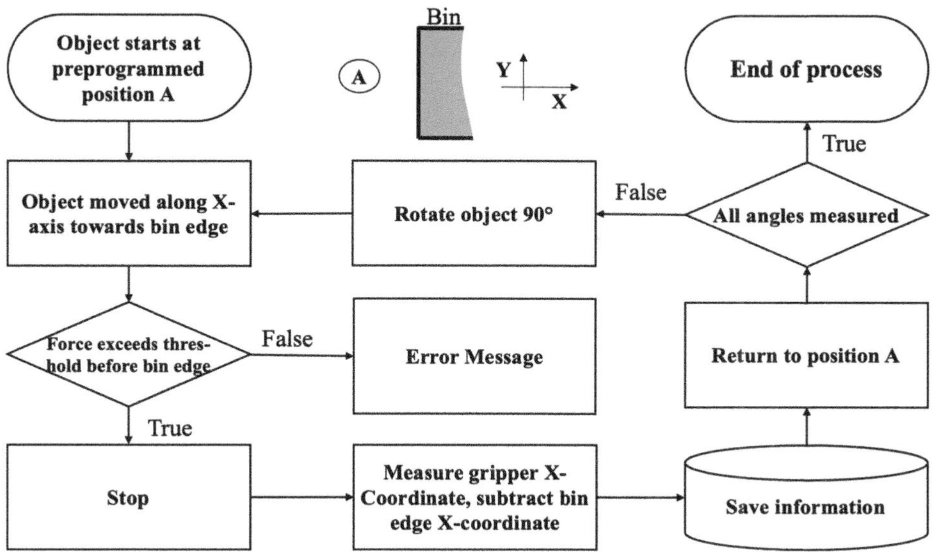

Fig. 2 Flowchart for measurement of axis aligned x and y dimension

4 Experiment

4.1 Hardware

For consistency, the robotic hardware in this study matches the camera-less system described in [3], consisting of a UR10e collaborative robot from Universal Robots with integrated force-torque sensor. The bin is a plastic container with dimensions of 30 cm × 40 cm × 20 cm. Throughout the experiment, the UR10e robot executes commands sent over RTDE via Python, while the force-torque sensor provides continuous feedback to monitor information on the gripper's interaction with the objects and surroundings (Fig. 3).

4.2 Procedure

In order to ensure a comprehensive evaluation of the dimensioning process, real-world objects of various geometries were selected. These include cardboard boxes, 330 ml cans of mushrooms and standard tennis balls (Fig. 4).

To replicate a typical bin-picking task, each group of objects is placed freely in the bin, with unconstrained position and orientation. The bin itself is constrained on two sides

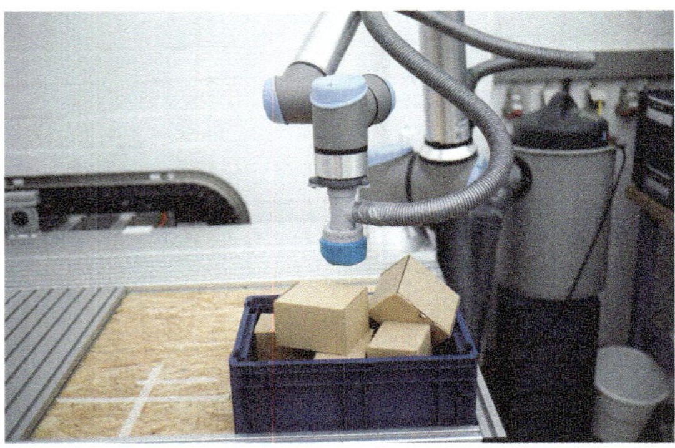

Fig. 3 Hardware used in this study

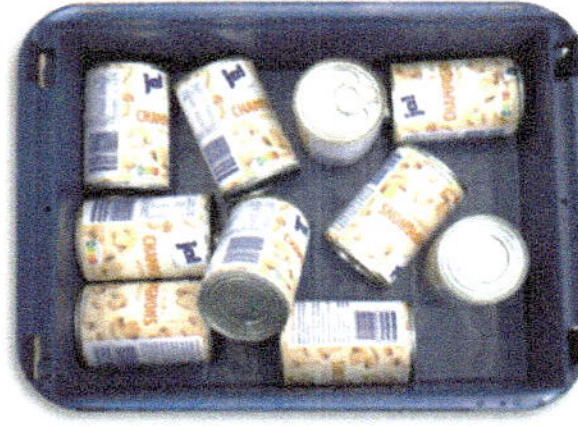

Fig. 4 Objects tested in this study

and on the bottom, ensuring identical placement in all trials. One side of the bin is left free from obstruction to allow space for the object to be probed on that side (Fig. 5).

At first, the robot begins the bin-picking task by searching for objects through tactile recognition of collisions inside the box. Once an object is detected, the robot positions

Fig. 5 Object positioned outside of the bin before measurement

itself above the object and attempts to grip it from the top. After successfully gripping an object, the robot initiates the dimensioning process.

The dimensioning process begins with the axis-aligned strategy using 90-degree rotations, followed by the semi axis-aligned strategy with rotations of 45, 30 and 15 degrees. Each rotation interval is tested with twenty objects from each of the three categories: boxes, cans, and tennis balls.

5 Results

Table 1 shows the average relative deviation (Δ) of the measured dimensions from the actual dimensions of the objects, as well as the mean process duration. For this comparison the manually measured volume of the object ($V_{calculated}$) was compared with the calculated volume of the algorithm ($V_{calculated}$). The deviation hereby was calculated as:

$$\Delta = \frac{V_{calculated}}{V_{measured}} \cdot 100$$

Increment step refers to the rotation angle between measurements, with 90° representing the axis-aligned strategy.

For all three object classes, the relative deviation was decreased with a smaller increment step, which was also led to a longer measurement duration. For the boxes and cans, the deviation was significantly improved when the step size decreased from 30 to 15 degrees. However, this change was not significant for the tennis ball. The time required for the measurement increased proportionally to the reduction in increment step size.

For a more comprehensive insight into the behavior of the data, the variance of the deltas is presented below in a box plot (Fig. 6).

As will be discussed, the deltas for the sphere are higher than expected, despite its complete symmetry Thus, Table 2 presents the data for the dimensioning of the spheres with dimensions viewed separately.

Table 1 Mean trial results

	Boxes		Cans		Tennis balls	
Increment step (°)	Delta (%)	Time (s)	Delta (%)	Time (s)	Delta (%)	Time (s)
90	16.19	18.77	16.25	17.70	13.19	18.22
45	17.56	25.29	13.22	25.75	15.44	26.49
30	16.06	33.02	13.07	33.42	10.69	32.90
15	10.16	55.51	8.53	56.11	10.31	55.29

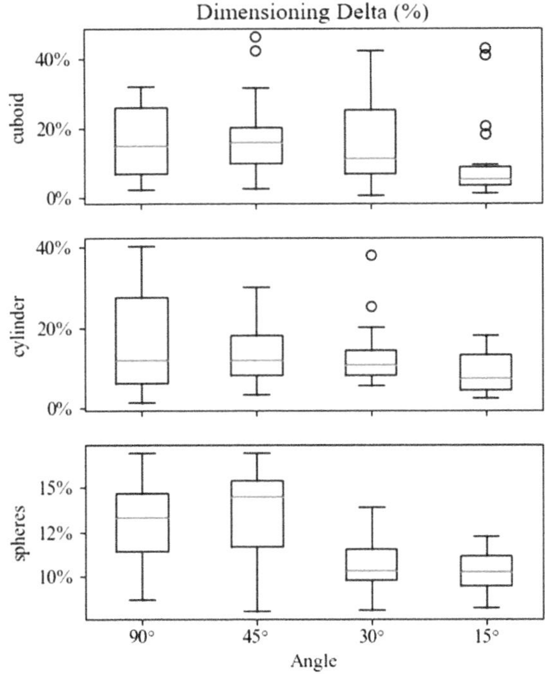

Fig. 6 Box plot of dimensioning delta for different object types

Table 2 Deltas of dimensions viewed separately

Increment step (°)	X Delta (%)	Y Delta (%)	Z Delta (%)
90	5.26	5.77	28.55
45	3.53	6.67	30.12
30	3.91	2.41	25.75
15	2.56	2.34	26.03

Additionally, Fig. 7 shows the delta of sphere dimensions as a percentage. Each circle on the figure represents a single measurement. This figure aggregates the individual dimension data for each sphere and for each increment step.

While smaller increment steps consistently improve the measurement along the x- and y-axes, this is not the case for the z-axis. Additionally, the overall measurement deviation for the z-axis is significantly higher compared to the other two axes.

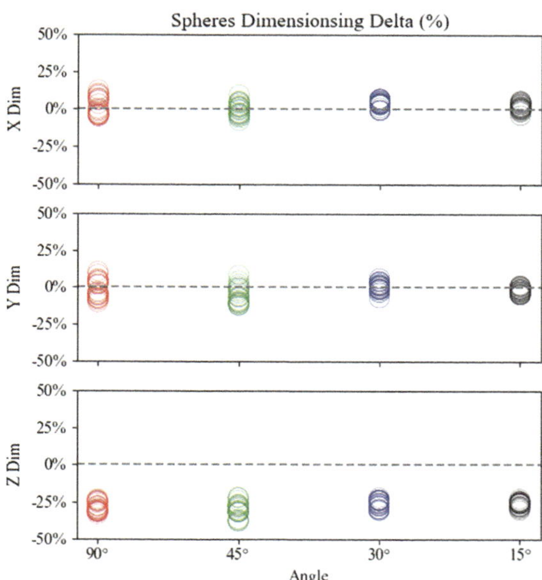

Fig. 7 Spheres dimensioning delta separated into x, y and z dimensions

6 Discussion

The behavior of the data for cuboids and cylinders are consistent with each other, insofar as the mean delta is largely consistent between 90°, 45° and 30° trials, and only drops significantly for the 15° trials. This is likely due to the fact that cuboids and cylinders tend to rest at least partially axis aligned: The cuboids typically rested on a face, while the cylinders usually rested on a face or on their side, and only rarely were either leaning at an angle. Additionally, cylinders on their sides often rolled to an edge, causing the cylinders to be further axis aligned. However, while the mean delta does not decrease appreciably along the first three trials, the variance of the delta does indeed decrease with a finer step size for cylinders, suggesting that the finer step increments better account for outlier cases.

The data for the spheres is unexpected, because the delta should be quite small for such a symmetrical object. While a cuboid can be measured at the hypotenuse, this is not applicable to a sphere. Closer inspection reveals that the large delta in the Z dimension (the up-down dimension relative to the work surface) is the main factor. The other two dimensions show comparatively small deltas. This large Z delta is due to the tennis balls being cupped by the vacuum gripper, causing them to sink approximately 2 cm into the gripper's surface. In contrast, a cuboid gripped on its face does not sink into the gripper.

If the X and Y deltas for the sphere data are observed in isolation, one sees a similar behavior to the other geometries, namely a general reduction in mean delta with finer step increments, with a most pronounced decrease at 15°. Moreover, a decrease in variance with decreased step increment is also observed.

The durations for the different step increments vary considerably, ranging from approximately 18 s for the 90° increment to approximately 55 s for the 15° increment. For a potential automated bin picking process, this duration, along with the use of the bin and handling robot, would allow for the measurement of the objects size at the beginning, providing additional information without requiring additional equipment.

Comparing the developed method with existing methods that use camera depth information [14] or RGB-D cameras [15], it is evident that the tactile approach is less precise. However, integrating this method into the process is more straightforward, as it does not require additional equipment. Moreover, no extra manual work is necessary to carry out the measurement. In contrast, camera-based approaches require additional information, such as the distance from the object [14], or need extensive camera calibrations [15]. Additionally, the computational effort is significantly lower for the tactile approach, which uses basic calculations, whereas camera-based algorithms demand higher computational resources.

7 Conclusion

The proposed approach accurately predicts object dimensions within a margin of 8–18%, depending on increment step angle and geometry.

The data show a mostly consistent trend across the three geometries, in which a reduction in increment step angle does not significantly improve prediction accuracy until it is reduced to 15°, at which point a significant improvement is observed.

Given only a very modest improvement in prediction quality from a 90° increment to a 30° increment, the increase in duration of approximately 83.3% is very significant, and likely outweighs the slight improvement in prediction accuracy. In instances where a higher accuracy is required and duration is not a significant factor, the 15° increment could be employed. This might be advantageous if, for example, only a single object is dimensioned at the start of a bin picking task, and not each object individually.

These results highlight the potential of force-based object dimension estimation to enhance the capabilities of bin-picking systems, especially in scenarios where camera-based methods may be impractical or cost-prohibitive. Future research could focus on refining this approach to improve prediction accuracy and reduce processing time. Additionally, the object dimension data derived with this strategy could be integrated into a camera-less bin picking strategy to further validate its practical utility.

References

1. Cordeiro, A., Rocha, L.F., Costa, C., Costa, P., Silva, M.F.: Bin picking approaches based on deep learning techniques: a state-of-the-art survey. In: IEEE ICARSC (2022)
2. Bogue, R.: Bin picking: a review of recent developments. IR **50**(6), 873–877 (2023)
3. Fogt, T., Müller, A., Adler, T., Kunz, H., Dietrich, F.: Benchmarking of a camera-less random bin picking strategy. In: CIRP CATS (2024)
4. Pai, S., Chen, T., Tippur, M., Adelson, E., Gupta, A., Agrawal, P.: TactoFind: a tactile only system for object retrieval. In: IEEE ICRA (2023)
5. Leão, G., Costa, C.M., Sousa, A., Veiga, G.: Detecting and solving tube entanglement in bin picking operations. Appl. Sci. **10**(7), 2264 (2020)
6. Sahin, C., Garcia-Hernando, G., Sock, J., Kim, T.-K.: A review on object pose recovery: from 3D bounding box detectors to full 6D pose estimators. Image Vis. Comput. **96**, 103898 (2020)
7. Kyprianou, G., tsidis, L., Kapoutsis, A.C., Zinonos, Z., Chatzichristofis, S.A.: Bin-picking in the industry 4.0 era. In: IEEE ICCE (2023)
8. Jamzuri, E., Pinandita, A., Analia, R., Susanto, S.: Object detection and pose estimation using rotatable object detector DRBox-v2 for bin-picking robot. In: Wikanta, P., Uperiati, A., Dwijotomo, A. (eds.) Proceedings of the 5th ICAE, Batam, Indonesia (2022)
9. Olesen, A.S., Gergaly, B.B., Ryberg, E.A., Thomsen, M.R., Chrysostomou, D.: A collaborative robot cell for random bin-picking based on deep learning policies and a multi-gripper switching strategy. Procedia Manuf. **51**, 3–10 (2020)
10. Xu, Y., Arai, S., Liu, D., Lin, F., Kosuge, K.: FPCC: fast point cloud clustering-based instance segmentation for industrial bin-picking. Neurocomputing **494**, 255–268 (2022)
11. Gupta, A., Jadhav, A., Korupolu, P.V.N.: Low cost bin picking solution for e-commerce warehouse fulfillment centers (2021)
12. Dutschke, W. (ed.): Fertigungsmeßtechnik, 3rd edn. Vieweg+Teubner Verlag, Wiesbaden (1996)
13. Scott, P.J.: Roundness measuring. U.S. Patent No. 5926781 (1999)
14. Pu, L., Rui, T., Wu, H.-C., Yan, K.: Novel object-size measurement using the digital camera. In: 2016 IEEE Advanced Information Management, Communicates, Electronic and Automation Control Conference (IMCEC), Xi'an, China, pp. 543–548 (2016). https://doi.org/10.1109/IMCEC.2016.7867270.
15. Long, Y., Wang, Y., Zhai, Z., Wu, L., Li, M., Sun, H., Su, Q.: Potato volume measurement based on RGB-D camera. IFAC-PapersOnLine **51**(17), 515–520 (2018). ISSN 2405-8963

Optimization of Robot Grasping Using Reinforcement Learning in Simulation

Manuel Belke, Mrunmai Vivek Phatak, Oliver Petrovic, and Christian Brecher

Abstract

This paper presents a novel approach to optimizing robot grasping through reinforcement learning in simulation, emphasizing the integration of tactile feedback alongside visual cues to enhance object grasping efficiency. Traditional methods primarily depend on visual feedback, often falling short in achieving optimal grasping. Our methodology proposes a cost-effective solution that significantly improves the software components responsible for object grasping. We introduce a two-phase process: a pre-grasp phase focusing on positioning and a grasping phase that adjusts based on tactile inputs. This approach demonstrates improved grasping outcomes, particularly in environments with high noise levels, underscoring the potential of tactile information in robotic grasping optimization.

Keywords

Robotics • Reinforcement learning • Grasping

1 Introduction

Optimal object grasping is one of the important challenges in the field of robotic operation, especially when taking into consideration the inaccuracies in determining the object's position. The whole grasping process consists of the phases object pose estimation, grasp pose

M. Belke (✉) · M. V. Phatak · O. Petrovic · C. Brecher
Laboratory for Machine Tools and Production Engineering (WZL) of RWTH Aachen University, Aachen, Germany
e-mail: M.Belke@wzl.rwth-aachen.de

M.-C. Wanner et al. (eds.), *Annals of Scientific Society for Assembly, Handling and Industrial Robotics 2024*, https://doi.org/10.1007/978-3-031-91463-8_26

estimation and trajectory generation and grasp routine execution. The object pose estimation task is typically carried out with a vision-based system consisting of one or more vision sensors. Vision sensors output quality can affect the overall efficiency of pick and place or object grasping tasks. Small and medium-sized enterprises (SMEs) often face difficulties to invest in high-end vision sensors which limits their ability to use advanced techniques to increase grasping accuracy. There are various influencing factors affecting the accuracy of object pose estimation, which includes calibration errors, environmental disturbances, or variations in object shapes and sizes.

Robotic arms with grippers often have difficulty aligning precisely with the target pick up locations, which might result in dropped objects, failed attempts or unsuitable grasping configurations. These challenges are reducing the efficiency and reliability of robotic systems, particularly in the applications where precision is critical, such as manufacturing, medical and aerospace industry.

Therefore, from a SME perspective, it becomes essential to develop a solution for optimal object grasping without relying on high-end visual sensors. Instead, the object grasping process should be optimized by incorporating tactile information as well, rather than relying solely on visual feedback as it may not lead to optimal object grasping. Hence, there is a need for an alternative approach which is cost-effective while enhancing the software components responsible for optimizing the object grasping process.

2 Related Work

Object manipulation is a very common industrial process where the grasping strategies are broadly categorized into analytical and empirical (data-driven) methods. Analytical methods focus on constructing force-closure stable grasps, formulated as a constrained optimization problem, while empirical strategies are less computationally demanding and are based on classification and learning methods such as deep learning and reinforcement learning [1–3].

Vision-based methods are open-loop and do not consider material properties like object mass, friction coefficient and contacts with other objects. Grasping objects with uncertainties in surface textures and object pose are still critical in robotic systems. [21] focuses on identifying stable grasp pose which minimize the wrist torque by desiging an analytical method for a center-of-mass (COM) based object grasping task using exteroceptive as well as proprioceptive information obtained from 3D perception unit and force/torque sensor respectively. Feng et al. [20] implemented a COM based object grasping by employing a support vector machine (SVM) and a long-short-term-memory (LSTM) for the feature extraction from tactile sensors. The algorithms are used to implement a slip detection and re-grasp planner, where the grasp stability is identified through an object's COM estimation.

To learn more robust behaviors compared to vision based and analytical appraoches, reinforcement learning (RL) has shown promising results in learning robot manipulation tasks over the past few years. Breyer et al. [9] developed an approach for solving the integrated task of object localization, reaching, grasping and lifting various objects through a RL setup by learning a mapping between inputs and outputs. The input is a depth image and the outputs are end-effector position and gripper control. Reward shaping and curriculum learning is used for effcient learning by gradully increasing the robot workspace. In [12], a policy gradient method with a hindsight experience replay (HER) was implemented to learn a RL policy for the task of object pushing for various dynamics models by parameterizing the dynamics, while employing dynamics randomization to carry out successful Sim2Real transfer. Shahid et al. [14] implemented a model-free RL approach to learn a continuous robot control in an object grasping task by using a proximal policy optimization (PPO) RL algorithm in combinatination with a dense reward to learn joint velocities and gripper's joint positions in a joint space by mapping desired joint velocities to joint torques. Orsula et al. [15] used a vision-based approach for solving the task of object grasping in unstructured scenes as well as on uneven terrains through a deep reinforcement learning (DRL) by learning a policy with a truncated quantile critics (TQC) [16] RL algorithm, to map 3D octree observations into continuous actions in cartesian space. Similarly, numerous vision-based object grasping approaches are implemented using RL in [17]. Chen et al. [19] implemented a vision-based object grasping approach with RL, which involved object detection and object pose estimation, which is later used to determine the grasping pose. Allshire et al. [18] learned the task of object grasping and in-hand manipulation using a single policy trained with a PPO algorithm, where keypoint-based representations are used to describe the object pose which are used in observations and calculation of the reward. Ceola et al. [24] used vision, tactile as well as proprioceptive information in learning multi-fingered grasping with DRL through a grasp planner to collect set of demonstrations and started policy training from these collected demonstrations. Zhang et al. [23] focused on optimizing the robotic grasping process using DDPG, which focuses on using torque information from the force / torque sensor to deal with the positional uncertainties in visual sensory output.

3 Approach

To remove the described effect of uncertainty during the vision-based object pose estimation, force and torque sensors are proposed to be used for optimal object grasping. These force and torque sensors provide the force and torque profile of an external or internal force or torque experienced by the robot during its manipulation in its surrounding environment or interaction with humans or objects. Our approach uses the force and torque feedback from the joint torque sensors which will be tested to adjust the gripper position and orientation in response to small pose errors. Instead of solving the task of optimal robotic grasping analytically, a method using reinforcement learning (RL) will be developed. The task of optimal object grasping is intended to be learned without requiring real world samples or human demonstrations and with no prior system dynamics knowledge.

Table 1 Variations in object physical properties

Property	Min	Max
Mass [kg]	0.4	1
Friction coefficient []	0.4	1
Dimension—width/diameter [mm]	20	60
Dimension—height [mm]	30	75

For implementing the concept of optimal grasping using RL, fundamental shaped objects with even mass distribution like cubes, cylinders and cuboids are selected. The motivation behind choosing these shapes come from the fact that, most industrial engineering objects have parallel and round faces. The use of these shapes allows to address real-world scenarios from a manufacturing industry perspective.

To enforce the adaptability while learning of an optimal grasping RL agent, the physical properties like mass and friction coefficient of the above mentioned objects are varied. This enables to train a robust policy capable of handling diverse settings effectively. Furthermore, the dimensions of these objects are also varied in an iterative manner to learn size-invariant grasping policy. The use of diverse object shapes, masses, friction coefficient and dimensions for learning optimal grasping behavior using RL, addresses the challenge encountered in the industrial world. The maximum values were chosen based on the maximum grasping force and opening width of the gripper. The minimum values were chosen based on the dimensions and masses usually encountered for machining processes in our institute. The variations applied to object mass, friction coefficient and dimensions are listed in the Table 1.

3.1 Grasping Process

The adaptive process starts with reaching a pre-grasp pose after which a grasping attempt is made. If this grasping attempt is not optimal, the RL agent will make another grasping attempt until the grasp is optimal. Between the grasps, the gripper fingers are opened and closed again after the new grasp attempt was calculated. This ensures that the object pose is not changed unintentionally during the reorientation of the gripper. If the grasp pose is optimal, the object is grasped and lifted.

Objects with uniform mass distribution have a COM coinciding with the object's centroid. Hsiao et al. [25] also state that, using COM based grasping pose is preferable and can reduce the effect of torque around the gripper axis due to the object's weight. Feng et al. [20] also employed a COM based approach to optimally grasp an object. Inspired from these methods, an approach based on an object's COM is designed to achieve a pre-grasp pose. The selected approach provides an advantage over traditional analytical methods by also considering the

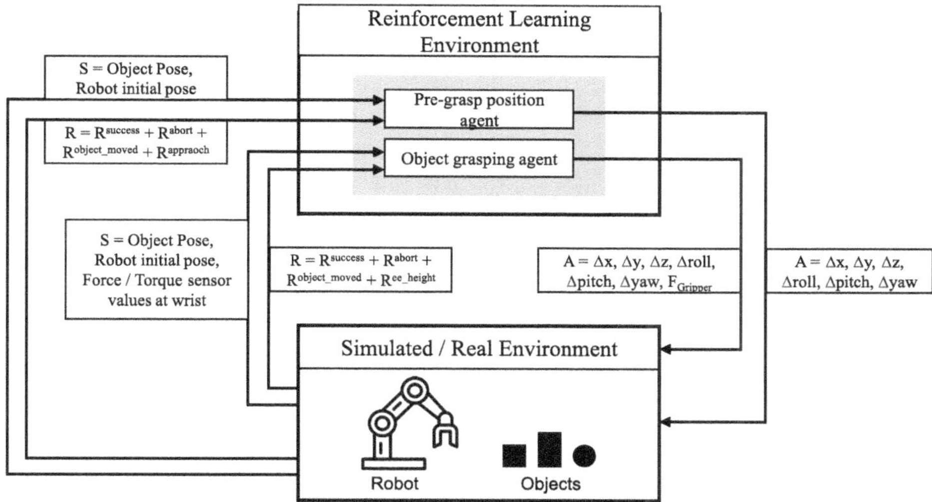

Fig. 1 Optimal object grasping concept using a two-stage RL structure

effect of object's mass distribution. If the gripper is not properly aligned over the object's COM, excessive torque might be experienced by the gripper.

During the grasping attempt, the external force and torque values (wrench) experienced at the robot's end-effector are measured, after the gripper fingers have been closed for a time period of 2–3 seconds. If the gripper orientation is not properly aligned with the object's orientation along the z-axis, peaks are observed for the torques around the axes.

The optimal object grasping concept is shown in Fig. 1. On the upper side, the two RL agents are displayed, whereas the bottom shows the simulated and real environment which executes the actions A and outputs the states S and the rewards R.

3.2 Pre-Grasp Pose Agent

The pre-grasp pose agent has a task of aligning itself over the object. The pose of the end-effector is described with respect to the robot's base coordinate system whose origin coincides with the origin of the world-coordinate system. The object is placed on the ground, positioned with respect to the world coordinate system. The robot control is learned for planning a trajectory from robot's initial pose to the object's pose. To learn a behavior relative to the object pose, randomizations in robot position are also required. Similar to the object pose randomization, for each episode, the initial joint configuration of the robot is also varied by keeping it within a specified workspace.

The observation space for pre-grasp pose can be described as Eq. 1:

$$s_t = [x_{ee},\ y_{ee},\ z_{ee},\ \phi_{ee},\ \theta_{ee},\ \psi_{ee},\ x_{obj},\ y_{obj},\ z_{obj},\ \phi_{obj},\ \theta_{obj},\ \psi_{obj},\ dist] \quad (1)$$

$[x_{ee}, y_{ee}, z_{ee}]$ and $[\phi_{ee}, \theta_{ee}, \psi_{ee}]$ represents the position and the orientation of the end-effector defined with respect to robot's base coordinate system (which coincides with world coordinate system). Similarly, $[x_{obj}, y_{obj}, z_{obj}]$ and $[\phi_{obj}, \theta_{obj}, \psi_{obj}]$ describe the object's position and orientation in robot's base coordinate system. In addition to this, the euclidean distance between object and the end-effector is also considered in order to offer a feedback to the robot about its current position relative to the object.

$$a_t = [\Delta x,\ \Delta y,\ \Delta z,\ \Delta\phi,\ \Delta\theta,\ \Delta\psi] \quad (2)$$

The action space for the pre-grasp pose agent is given as Eq. 2, which includes translation displacement in the x, y, z direction and relative orientation around all three axes. All the action variables are normalized between [–1, 1] and are re-scaled by a factor such that the translational displacement stays in between [–0.02, 0.02]m for x, y, z axes and the rotational displacement is restricted in between $[-0.1(\pi/4), 0.1(\pi/4)]$rad for x and y axis, $[-0.1(\pi), 0.1(\pi)]$rad for z axis. All of these translational and angular displacements are applied relative to the current robot's position in robot's base coordinate system.

$$R_t = R_t^{success} + R_t^{abort} + R_t^{object\_moved} + R_t^{approach} \quad (3)$$

The episode constitutes to a successful episode, when the agent centers the robot's end-effector over the center of an object and aligns itself along the z-axis with the object in between the gripper fingers. For promoting the robot's motion in the direction of object, the euclidean distance based reward structure is used. Moreover, orientation alignment along the axes is also important, which is also crafted into a dense reward. The overall reward function Eq. 3 and its decomposition into individual components $R_t^{success}$, $R_t^{approach}$, R_t^{abort}, $R_t^{object\_moved}$ is described in Fig. 2. The reward of 500 is assigned when *I(cond)* is satisfied, following tests to determine the value based on learning

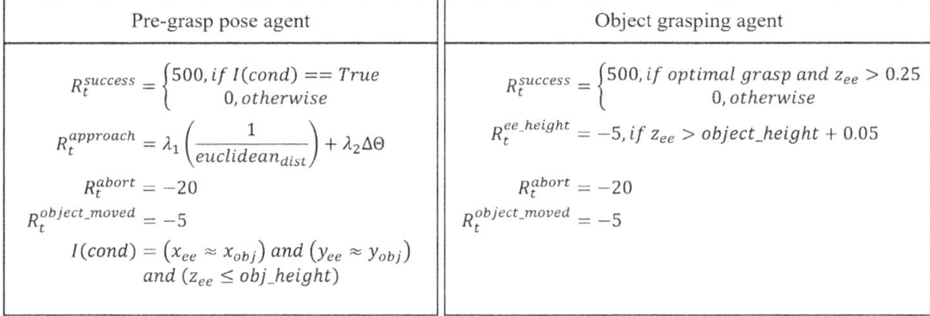

Fig. 2 RL agent rewards

success. The agent receives the rewards $R_t^{success}$, R_t^{abort}, $R_t^{object\_moved}$ only at the end of the episode (sparse reward) and $R_t^{approach}$ is used as a dense reward to accelerate the training.

3.3 Object Grasping Agent

The grasping agent is mainly responsible for small position and orientation re-adjustments in response to external torques experienced by the robot's end-effector. The state-space and action-space are defined in Eqs. 4 and 5.

$$s_t = [F_x, F_y, F_z, M_x, M_y, M_z, F_{finger_1}, F_{finger_2}, grasping\_status] \tag{4}$$

$$a_t = [\Delta x, \Delta y, \Delta z, \Delta\phi, \Delta\theta, \Delta\psi, F_{gripper}] \tag{5}$$

State space stores the observed forces and torques along x, y, z axes at the instance when gripper fingers are closed around the object as well as the sampled force ($F_{gripper}$) is applied by the gripper finger joints. $grasping\_status$ describes whether the grasp action was successful or not, which takes value as -1 (unsuccessful grasp) or 1 (successful grasp). $grasping\_status$ is set to 1, if the object is successfully grasped and lifted to a height of 0.02 m without loosing it, otherwise it is set to -1, when the optimal grasp pose is not achieved or the object is slipped out of the gripper fingers. Slipping is detected when the gripper fingers close because of the force $F_{gripper}$ applied without an object between the fingers. Similar to the pre-grasp pose agent, actions are given as relative translation and rotational displacements. $F_{gripper}$ defines the combined force applied by the gripper finger joints.

The episode is considered to be successfully completed if the object is optimally grasped and lifted above the ground without slipping out of the gripper fingers. In addition to this, position as well as the orientation alignment checks similar to the pre-grasp pose agent are also carried out to ensure that the robot does not move farther away from the object. The reward function structure for grasping agent is described as in Eq. 6 which is again given as a combination of $R_t^{success}$, $R_t^{ee\_height}$, R_t^{abort}, $R_t^{object\_moved}$.

$$R_t = R_t^{success} + R_t^{abort} + R_t^{object\_moved} + R_t^{ee\_height} \tag{6}$$

In $R_t^{success}$, *optimal_grasp* flag refers to a condition where minimal torques (M_x, M_y, M_z) are experienced in all three axes. These torques are compared against a reference value to identify the presence of peaks at the instant of object grasping. If no peaks are found, an attempt to grasp followed by lifting of an object is carried out, where a successful attempt accounts for the reward of 500.

3.4 RL Algorithm

Since the state space as well as the action space of the RL problem are continuous, only algorithms which are able to inherently support this are considered. From the considered algorithms Proximal Policy Optimization (PPO), Twin-Delayed Deep Deterministic (TD3) and Soft Actor-Critic (SAC), SAC and TD3 were considered in detail since these off-policy algorithms are generally more sample efficient which is important in industrial environments to learn the grasping skill with less training. The TD3 algorithm does not perform well with minimal state spaces and takes long to learn new complex behavior [22]. Therefore, SAC was used for the experiments.

4 Implementation and Training

A franka emika panda robot is used for the implementation of the approach. The robot was chosen due to the support for integration in the robot operating system (ROS) and the accuracy of the force and torque estimation which enables the robot to perform closed-loop force control with regulated forces as low as $< 0.05N$ [26]. A Franka Hand is used as gripper which provides a force based grasping process.

The toolchain is implemented in ROS with a node for each the pre-grasp and grasping execution. These nodes call a ROS action for moving the robot with the MoveIt framework and closing the gripper with a gripper control node. MoveIt plans a trajectory to reach the object. The planned trajectory and gripper finger positions are executed in gazebo. The observations and sensor values in gazebo are fed back to the pre-grasp and grasping node.

OpenAI Gym is used as training evironment with the standardized attributes observation_space and action_space and the methods reset(), step() and render(). The stable-baselines3 library implemented in PyTorch is employed as the RL implementation. The hyperparameters used are listed in Table 2.

5 Evaluation

If the robot lifts the object and maintains it in the lifted state for a specified time duration, the task of optimal object grasping is considered to be accomplished. The validation process for optimal object grasping is carried out in a predefined 3D workspace around the robot's end-effector. With the trained agents, the validation of an optimal object grasping agent is carried out by sequentially arranging the pre-grasp pose and the grasping agent.

The training process of the grasping agent is shown in Fig. 3. It can be seen that the grasping agent does not converge. Even though the learning curve is noisy and non-converging, this trained agent learned a behavior for aligning the robot's end-effector with the object's orientation. During the training with multiple objects, the performance of the agent showed

Table 2 Hyperparameters for training

Symbol	Name	Value
γ	Discount factor	0.99
τ	Target update coefficient	0.005
α	Entropy regularization term	0.2
batch_size	Batch size	256
target_update_interval	Target update interval	1
automatic_entropy_tuning	Automatic entropy tuning	True
gradient_steps_per_update	Gradient steps per update	1
replay_buffer	Replay buffer size	100000
initial_random_steps	Initial random steps	100000

Table 3 Validation of grasping agents (values are given in [%])

Cube		Cuboid		Cylinder		Random	
Pre-grasp	Grasp	Pre-grasp	Grasp	Pre-grasp	Grasp	Pre-grasp	Grasp
88	70	100	54	88	38	88	44

observable improvement, however, could not converge fully. The validation performance for the two agents for different objects are shown in Table 3. It decreases with more objects and randomizations introduced. The most likely cause for the non-converging training is the low number of input values to the neural network. The data basis for the decision for the best action is probably not large enough to accurately estimate the necessary actions for object-gripper alignment.

6 Conclusion and Future Work

In the scope of this paper, applicability of RL for the task of optimal object grasping through the use of force-torque sensor was evaluated. In the realization of this concept, robot motion control is learned using a Soft Actor Critic RL agent, which allowed the robot to plan its path as well as to readjust the gripper pose for autonomous optimal object grasping without requiring human intervention. The selected approach for optimal object grasping through force-torque feedback using RL is observed to grasp an object optimally with a moderate success rate. This shows a possibility of using RL in learning a robot control in contact-based tasks. However, further improvements are essential to allow more reliable grasping strategy. The joint torque sensor does not provide a fine-grained force-torque information about contact points observed between the gripper and the object during the grasping operation,

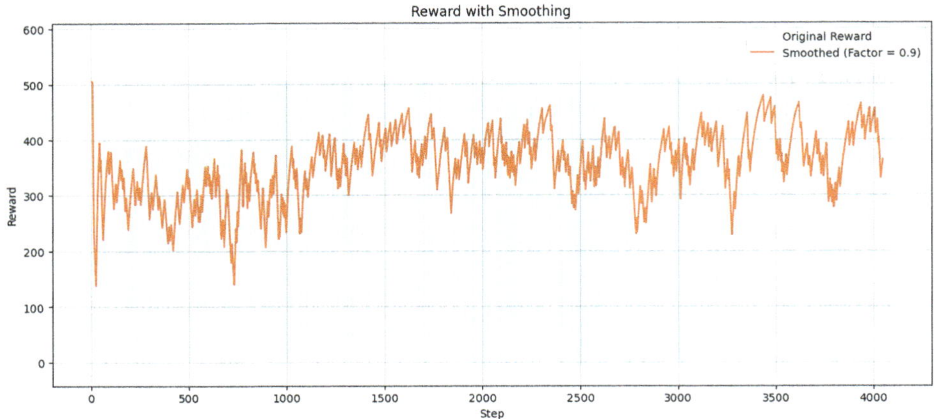

Fig. 3 Training of object grasping agent in simulation

which can prove to be beneficial to ensure optimal object grasping. Therefore, instead of joint torque sensor, tactile sensors or wrist force-torque sensors can be used to obtain exact estimates of accurate contact points and external force-torque respectively. In the next step, we will integrate tactile sensors on the gripper fingers to have more input data to the neural network. We will validate if this increase in input data leads to a converging behavior of the training of the agent.

Acknowledgements The IGF-projekt 22715 N (SPrinteR) of the research association FVP was supported via the AiF within the funding program "Industrielle Gemeinschaftsforschung und –entwicklung (IGF)" by the Federal Ministry for Economic Affairs and Climate Action (BMWK) due to a decision of the German Parliament.

References

1. Sahbani, A., El-Khoury, S., Bidaud, P.: An overview of 3D object grasp synthesis algorithms. Robot. Auton. Syst. **60**(3), 326–336 (2012). https://doi.org/10.1016/j.robot.2011.07.016
2. Shimoga, K.B.: Robot grasp synthesis algorithms: a survey. Int. J. Robot. Res. **15**(3), 230–266 (1996). https://doi.org/10.1177/027836499601500302
3. Bohg, J., Morales, A., Asfour, T., Kragic, D.: Data-driven grasp synthesis-a survey. IEEE Trans. Rob. **30**(2), 289–309 (2013). https://doi.org/10.1109/TRO.2013.2289018
4. Lenz, I., Lee, H., Saxena, A.: Deep learning for detecting robotic grasps. Int. J. Robot. Res. **34**(4–5), 705–724 (2015). https://doi.org/10.1177/0278364914549607

5. Mahler, J., Liang, J., Niyaz, S., Laskey, M., Doan, R., Liu, X., Ojea, J.A., Goldberg, K.: Dex-net 2.0: deep learning to plan robust grasps with synthetic point clouds and analytic grasp metrics (2017). arXiv:1703.09312. https://doi.org/10.48550/arXiv.1703.09312
6. Pinto, L., Gupta, A.: Supersizing self-supervision: learning to grasp from 50k tries and 700 robot hours. In: 2016 IEEE International Conference on Robotics and Automation (ICRA), pp. 3406–3413. IEEE (2016). https://doi.org/10.1109/ICRA.2016.7487517
7. Johns, E., Leutenegger, S., Davison, A.J.: Deep learning a grasp function for grasping under gripper pose uncertainty. In: 2016 IEEE/RSJ International Conference on Intelligent Robots and Systems (IROS), pp. 4461–4468. IEEE (2016). https://doi.org/10.1109/IROS.2016.7759657
8. Liu, K.: Dynamic animation and robotics toolkit (2014)
9. Breyer, M., Furrer, F., Novkovic, T., Siegwart, R., Nieto, J.: Flexible robotic grasping with sim-to-real transfer based reinforcement learning (2018). arXiv:1803.04996
10. Ng, A.Y., Harada, D., Russell, S.: Policy invariance under reward transformations: theory and application to reward shaping. In: Icml, vol. 99, pp. 278–287 (1999)
11. Bengio, Y., Louradour, J., Collobert, R., Weston, J.: Curriculum learning. In: Proceedings of the 26th Annual International Conference on Machine Learning, pp. 41–48 (2009). https://doi.org/10.1145/1553374.1553380
12. Peng, X.B., Andrychowicz, M., Zaremba, W., Abbeel, P.: Sim-to-real transfer of robotic control with dynamics randomization. In: 2018 IEEE International Conference on Robotics and Automation (ICRA), pp. 3803–3810. IEEE (2018). https://doi.org/10.1109/ICRA.2018.8460528
13. Chebotar, Y., Handa, A., Makoviychuk, V., Macklin, M., Issac, J., Ratliff, N., Fox, D.: Closing the sim-to-real loop: adapting simulation randomization with real world experience. In: 2019 International Conference on Robotics and Automation (ICRA), pp. 8973–8979. IEEE (2019)
14. Shahid, A.A., Roveda, L., Piga, D., Braghin, F.: Learning continuous control actions for robotic grasping with reinforcement learning. In: 2020 IEEE International Conference on Systems, Man, and Cybernetics (SMC), pp. 4066–4072. IEEE (2020). https://doi.org/10.1109/SMC42975.2020.9282951
15. Orsula, A., Bøgh, S., Olivares-Mendez, M., Martinez, C.: Learning to grasp on the moon from 3D octree observations with deep reinforcement learning. In: 2022 IEEE/RSJ International Conference on Intelligent Robots and Systems (IROS), pp. 4112–4119. IEEE (2022). https://doi.org/10.1109/IROS47612.2022.9981661
16. Kuznetsov, A., Shvechikov, P., Grishin, A., Vetrov, D.: Controlling overestimation bias with truncated mixture of continuous distributional quantile critics. In: International Conference on Machine Learning, pp. 5556–5566. PMLR (2020)
17. Breyer, M., Furrer, F., Novkovic, T., Siegwart, R., Nieto, J.: Comparing task simplifications to learn closed-loop object picking using deep reinforcement learning. IEEE Robot. Autom. Lett. **4**(2), 1549–1556 (2019)
18. Allshire, A., MittaI, M., Lodaya, V., Makoviychuk, V., Makoviichuk, D., Widmaier, F., Wüthrich, M., Bauer, S., Handa, A., Garg, A.: Transferring dexterous manipulation from gpu simulation to a remote real-world trifinger. In: 2022 IEEE/RSJ International Conference on Intelligent Robots and Systems (IROS), pp. 11802–11809. IEEE (2022). https://doi.org/10.1109/IROS47612.2022.9981458
19. Chen, Y.L., Cai, Y.R., Cheng, M.Y.: Vision-based robotic object grasping–a deep reinforcement learning approach. Machines **11**(2), 275 (2023). https://doi.org/10.3390/machines11020275
20. Feng, Q., Chen, Z., Deng, J., Gao, C., Zhang, J., Knoll, A.: Center-of-mass-based robust grasp planning for unknown objects using tactile-visual sensors. In: 2020 IEEE International Conference on Robotics and Automation (ICRA), pp. 610–617. IEEE (2020). https://doi.org/10.1109/ICRA40945.2020.9196815

21. Kanoulas, D., Lee, J., Caldwell, D.G., Tsagarakis, N.G.: Center-of-mass-based grasp pose adaptation using 3d range and force/torque sensing. Int. J. Hum. Robot. **15**(04), 1850013 (2018). https://doi.org/10.1142/S0219843618500135
22. Mock, J.W., Muknahallipatna, S.S.: A comparison of PPO, TD3 and SAC reinforcement algorithms for quadruped walking gait generation. https://doi.org/10.4236/jilsa.2023.151003
23. Zhang, H., Wang, F., Wang, J., Cui, B.: Robot grasping method optimization using improved deep deterministic policy gradient algorithm of deep reinforcement learning. Rev. Sci. Instrum. **92**(2), (2021). https://doi.org/10.1063/5.0034101
24. Ceola, F., Maiettini, E., Rosasco, L., Natale, L.: A grasp pose is all you need: learning multi-fingered grasping with deep reinforcement learning from vision and touch. In: 2023 IEEE/RSJ International Conference on Intelligent Robots and Systems (IROS), pp. 2985–2992 (2023)
25. Hsiao, K., Chitta, S., Ciocarlie, M., Jones, E.G.: Contact-reactive grasping of objects with partial shape information, pp. 1228–1235. IEEE (2010)
26. Haddadin, S., Parusel, S., Johannsmeier, L., Golz, S., Gabl, S., Walch, F., Sabaghian, M., Jähne, C., Hausperger, L., Haddadin, S.: The franka emika robot: a reference platform for robotics research and education. IEEE Robot. Autom. Mag. **29**(2), 46–64 (2022)
27. Haarnoja, T., Zhou, A., Abbeel, P., Levine, S.: Soft actor-critic: off-policy maximum entropy deep reinforcement learning with a stochastic actor. In: International Conference on Machine Learning, pp. 1861–1870. PMLR (2018)

Drum Gripping Concept for Secured Orientation-Variable Handling in Solid-State Battery Cell Assembly

Do Minh Nguyen, Matthias Strauß, Timon Scharmann, and Klaus Dröder

Abstract

As Lithium-ion batteries (LIBs) are forecasted to reach their technological optimization limits in the coming decade, next generation battery technologies, such as solid-state batteries (SSBs), are investigated to enable even higher energy and power densities. However, depending on the utilized solid electrolytes (SEs), the cell components exhibit low mechanical stability or an adhesive behavior with serious implications on handling operations during cell assembly. Additionally, mainly SSB cell components with one-sided coating are currently available, which limit the automated cell assembly to the production of bicells with challenges in securely changing the component's orientation during stacking with established gripping technologies. In the present research, a drum gripping concept for a secured change of the cell components' orientation along its pitch axis during handling is introduced. Especially relevant is the avoidance of undesirable folds in the materials which occur in dependence of the adjustable pressure difference and the gripper's rotation direction. To validate the applicability of the presented concept, an experimental evaluation of the deposition accuracy in relation to the supplied pressure difference with polymer-based SEs and lithium metal anodes is carried out. As key results, adequate deposition accuracies are achieved with deposition with the original orientation and for pressure differences of 0.5 bar. Overall, the proposed gripping concept can be viewed as a valid solution for a flexible automated cell

D. M. Nguyen (✉) · M. Strauß · T. Scharmann · K. Dröder
Institute of Machine Tools and Production Technology, Technische Universität Braunschweig, Braunschweig, Germany
e-mail: do-minh.nguyen@tu-braunschweig.de

Battery LabFactory Braunschweig, Technische Universität Braunschweig, Braunschweig, Germany

M.-C. Wanner et al. (eds.), *Annals of Scientific Society for Assembly, Handling and Industrial Robotics 2024*, https://doi.org/10.1007/978-3-031-91463-8_27

assembly of early-stage test cells which are required for further developments towards the establishment of industry-scale SSB production.

Keywords

Solid-state battery • Battery production • Cell stacking • Drum gripper • Pneumatic gripper • Solid electrolyte

1 Introduction

The lithium-ion battery (LIB) is a central energy storage technology and currently especially relevant for the electrification of the transportation sector, aiming to reduce greenhouse gas emissions of internal combustion engine vehicles. Due to limitations in energy and power densities in current LIB systems, a shift to next-generation battery technologies, such as solid-state batteries (SSBs), is required. This is mainly defined through the application of solid electrolytes (SEs). Depending on the choice of SE classes, either being oxides, sulfides, or polymers, different characteristics in ionic conductivity, safety, and cycle stability on cell level can be achieved [1, 2]. Along with the highly adhesive behavior of certain SE classes, all currently investigated SE classes are very susceptible to mechanical stresses, thereby posing serious challenges in realizing a material-sensitive and precise handling in automated cell assembly as comparable to the established LIB cell production [1–4]. Lithium metal anodes (LMAs) as an integral part of certain SSB designs have been reported to display a similar adhesive behavior as well as high mechanical sensitivity as polymer solid electrolytes [4–7]. Research on the handling of SSB cell components is at an early stage with a few investigations focusing on the handling of LMAs as well as polymer-based and oxide-based SEs and battery cells [4–8]. In addition, continuously changing material formulations and challenges in throughput-orientated synthesis currently limit the access to SSB cell components with one-sided coating for building test cells, such as bicells [9]. When realizing automated bicell stacking with state-of-the-art gripping technology, additional storage machinery for the provision of cell components with the active material facing upwards or downwards is necessary as these grippers are usually unable to perform an intentional change of the cell component's orientation along its pitch axis during handling. To address this, the present publication presents a drum gripping concept designed to securely and effectively change the handled cell components' orientation during stacking via flipping around its pitch axis. In addition to the presentation of the changing mechanism, an experimental validation of the achievable deposition accuracy in relation to the applied pressure difference is performed and evaluated.

2 Single Sheet Stacking and Handling in Battery Cell Assembly

Handling operations on battery cell components are mostly conducted in cell stacking. Here, cell components are assembled into an alternating compound stack of anodes, separators/electrolytes, and cathodes (see Fig. 1) [10]. Single sheet stacking, where singular electrode and separator/electrolyte sheets are stacked alternatingly onto each other, is the most favored stacking technology for future production systems of SSBs due to the components' high sensitivity towards mechanical stress [3, 11]. The handling of cell components is facilitated by grippers, for which different types can be classified by their working principles. In this paper, pneumatically actuated grippers are investigated which are the preferred handling technology in battery cell assembly, conventionally as vacuum suction grippers (Fig. 1 - (1)) [12, 13]. They generate an area-distributed holding force via a vacuum between gripper and electrode and have been implemented for both LIB and SSB component handling investigations [5, 6, 12–14]. However, potential material adhesion due to the mechanical contact between gripper and gripping object can cause electrostatic charge and surface damage. Contactless pneumatic gripping principles for cell components, such as the Bernoulli (Fig. 1 - (2)) or cyclone gripper (Fig. 1 - (3)), circumvent this issue and utilize overpressure through fast flowing or circularly rotating air flows for generating the holding force. Challenges for this type of grippers are occurring vibrations during handling which hinder a secure electrode fixation [6, 14, 15]. These described gripping principles do not inherently allow for a secured change of the cell component's orientation along its pitch axis and are therefore more suitable for cell assembly with double-sided coating (Fig. 1 - (a)). By including a mechanism for intentionally and effectively flipping cell components with one-sided coating during handling, a novel gripping solution for assembling bicells (Fig. 1 - (b)) would remove the necessity for additional preparation steps in providing alternatingly oriented cell components to the stacking process.

3 Drum Gripping Concept for Cell Stacking

Vacuum drums for handling have previously found applications within the foil, paper, or textile industries [16–18]. In battery production, it has also been investigated as a handling concept for high-throughput-orientated singulation of electrodes for a subsequent z-folding by Mooy et al. [12, 19, 20]. Differing from these previous applications, the proposed drum gripper within the present use case focuses on bidirectional actuation of the drum's rotation for an orientation change of the handled object via flipping around its pitch axis. Figure 2 shows the iteratively designed drum gripper prototype. The vacuum roll (1) (3D printed, Acrylonitrile butadiene styrene (ABS)) functions as the main

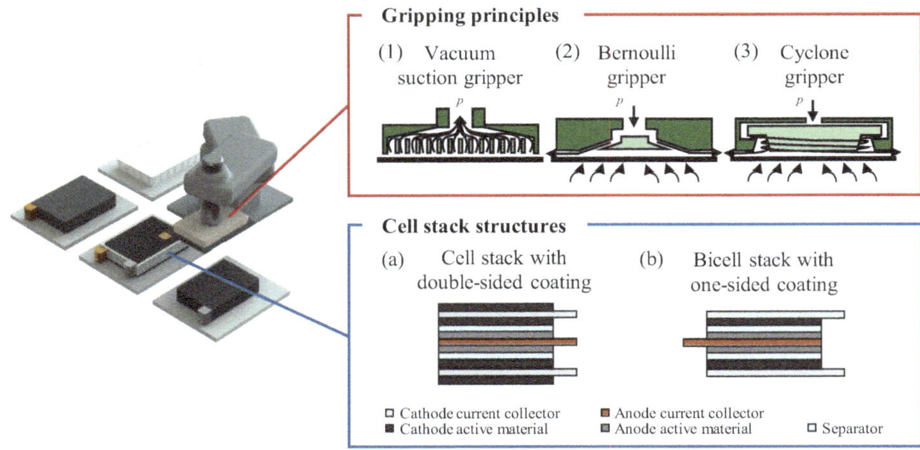

Fig. 1 Single sheet stacking with different gripping principles ((1) Vacuum suction gripper, (2) Bernoulli gripper, (3) Cyclone gripper [6]) and cell stack structures (**a** Cell stack with double-sided coating, **b** Bicell stack with one-sided coating [9])

Fig. 2 Drum gripper prototype, including (1) a vacuum roll, (2) servo drives with drive belts and gear wheels, (3) air supply connectors, and (4) a microcontroller

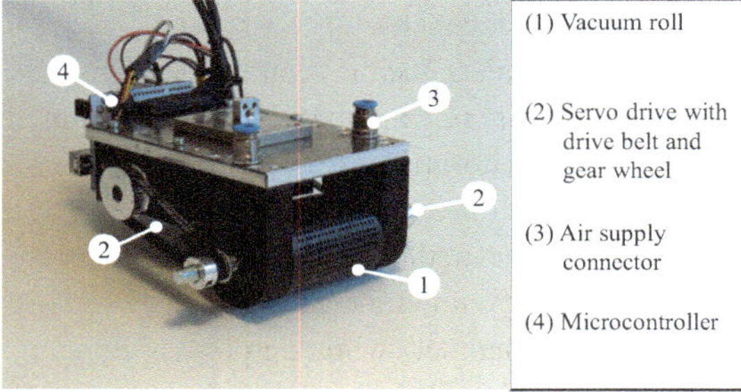

handling component and is post-processed subtractively with a turning process for guaranteeing sufficient concentricity to reduce potential cell component damage during handling. The design consists of an outer roll with 420 evenly spread suction holes (hole diameter: 1.5 mm), divided onto 20 rows of 21 holes, on half of the roll's circumference in which each row is connected with a chamber for air pressure supply (3). A rotary dial inside of the vacuum roll is used for consecutive shutting of the chambers, and therefore pressure supply, depending on its relative rotation with regard the vacuum roll. Thereby, a successive separation of the handled cell component from the vacuum roll is achieved through independent control of both units through a microcontroller (Arduino UNO Rev3) (4) and separated drive belts on servo drives (2).

With this setup, handled cell components can be deposited in two alternative ways onto the forming cell stack: 1) either deposition with original orientation (DOO) where it is

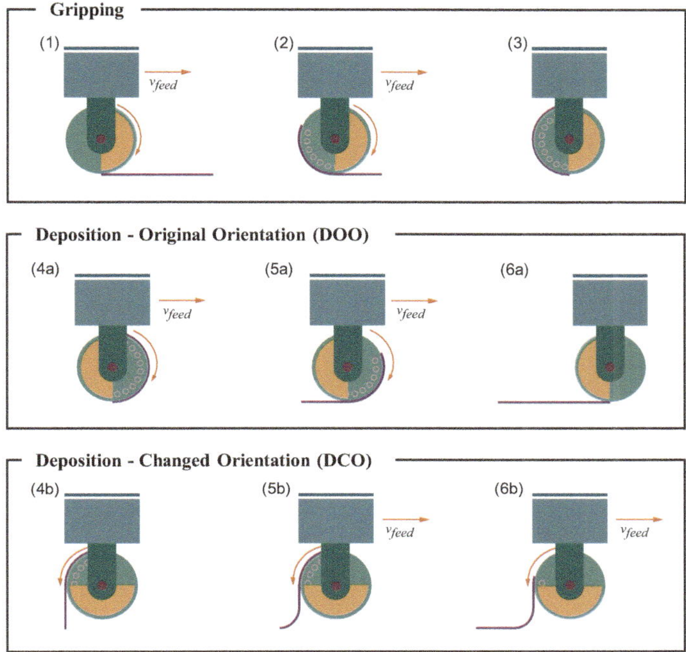

Fig. 3 Handling procedure of the drum gripper with a sheet (purple), (1–3) gripping, (4a–6a) deposition with original orientation (DOO), (4b–6b) deposition with changed orientation (DCO)

rolled by clockwise rotation in feed speed direction (Fig. 3, 4a, 5a, 6a), or 2) deposition with changed orientation (DCO) where it is draped by counterclockwise rotation in feed speed direction (Fig. 2, 4b, 5b, 6b).

Therefore, a low bending resistance of the handled material is a key requirement for a damage-free material condition after adapting to the roll's shape during handling.

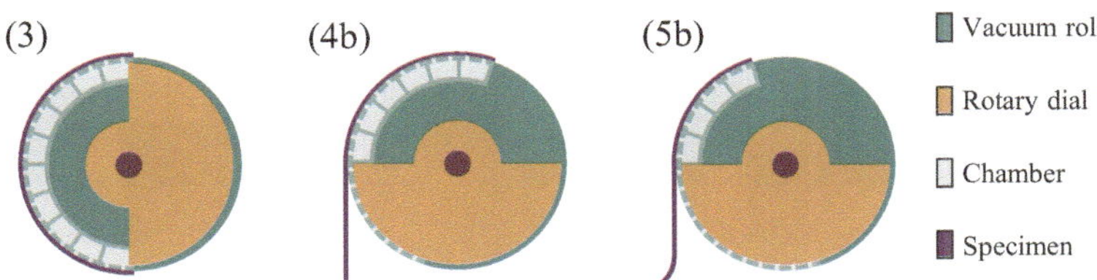

Fig. 4 Sectional view of the vacuum (turquoise) with moving rotary dial (orange) subsequently covering air chambers (grey) for the deposition of the specimen (purple) according to Fig. 3 for the stages 3–5b

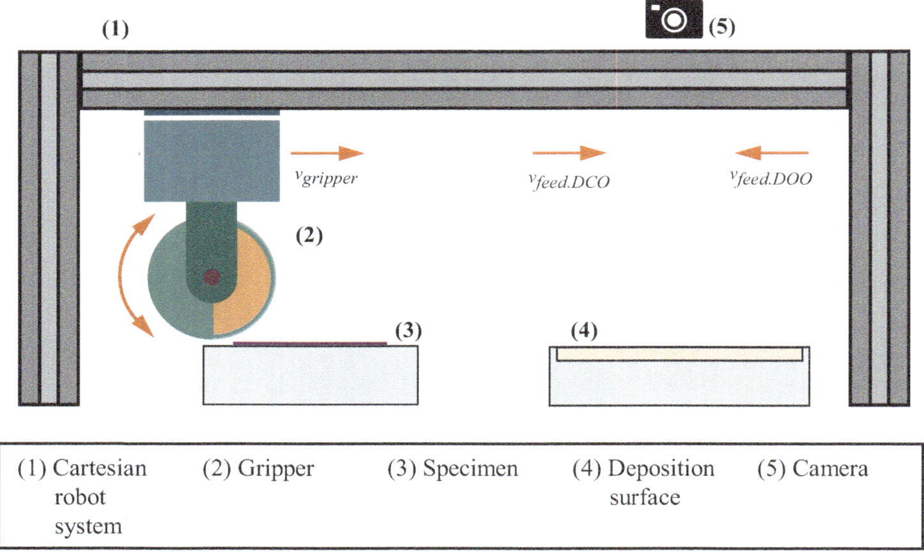

Fig. 5 Schematic test setup, including component details and movement sequences (orange)

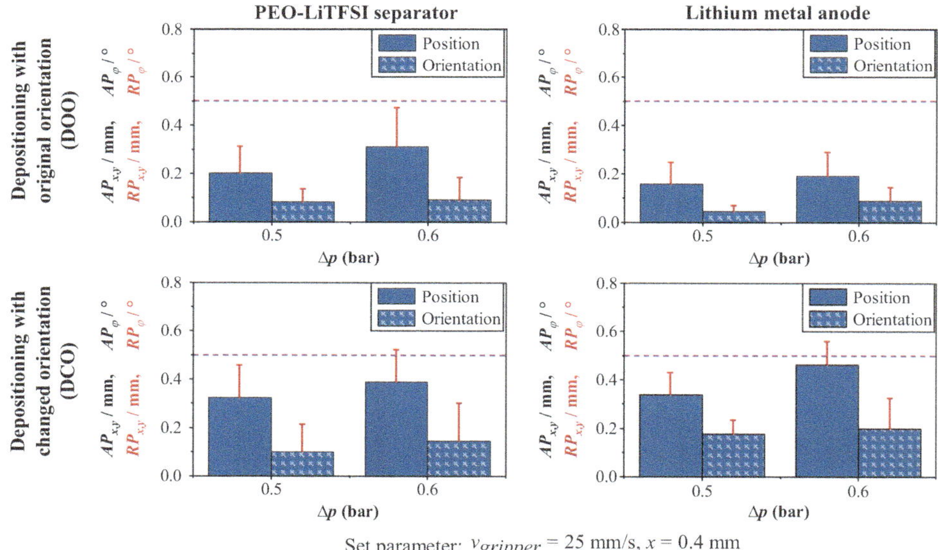

Fig. 6 Result overview of total deposition accuracy achieved with drum gripper for PEO-based separators and LMAs with original orientation (DOO) and changed orientation (DCO)

Figure 4 depicts a sectional view of the vacuum roll (turquoise, with vacuum chambers) with the rotary dial (orange) rotating clockwise for subsequently covering the air chambers (grey). Relative to the counterclockwise rotation of the vacuum roll, a DCO is performed by the vacuum roll with the handled specimen (purple), as visualized in Fig. 3 from stages 3 to 5b.

4 Materials and Experimental Setup

Experiments are conducted on polymer-based solid electrolytes on separators and LMAs. The separator, being synthesized by the Fraunhofer Institute for Manufacturing and Advanced Materials (IFAM), features Polyethylene oxide (PEO) as the polymer matrix and lithium bis(trifluoromethanesulfonyl)imide (LiTFSI) as the conductive salt and is extruded to a material thickness of 48 μm. The LMAs are commercially acquired from GoodFellow with a lithium coating thickness of 20 μm on a copper substrate thickness of 10 μm. Cantilever bending tests on both samples according to DIN 53,362 in our research laboratory confirm the material's low bending resistances, amounting to values of around 0.03 and 6.11 Nmm^2 respectively [21]. Both cell components are cut into sheets with an overall dimension of 50×70 mm2, corresponding to the Battery LabFactory Braunschweig's BLB1 format.

Figure 5 depicts the experimental setup in the dry room of the research laboratory featuring a room temperature of 20 °C and a dew point of −40 °C. The introduced gripper is mounted on a cartesian robot system with a degree of freedom of 4, consisting of a Festo EXCM-30 planar surface gantry with a Festo EGSC-32 lifting axis and a Festo ERMO-12 rotary axis as used in [5]. After gripping the PEO-based separator through counterclockwise rotation of the vacuum roll, the gripper is moved towards the deposition surface. While DCO through clockwise rotation of the vacuum roll is performed in the same feed speed direction when moving towards the deposition nest, DOO of the PEO-based separator sheets requires the gripper to move backwards compared to the initial feed speed direction for the simplification of the deposition. Potential specimen adhesion to the gripper is prevented by using an ion blower. A camera with an optional ring lighting along with a Python-based image processing software are used to visually analyze the deviation in x–y axis as well as the rotation around the z-axis in comparison to a reference position and orientation for calculating the deposition accuracy.

The image processing software uses a Gaussian filter and Canny algorithm for noise reduction and subsequent edge detection to calculate these deviations. The total deposition accuracy as the superposition of position accuracy and repeatability as well as orientation accuracy and repeatability is calculated with the formulas for pose accuracy AP_i as well as the pose repeatability RP_i from the ISO 9283 [22]. Considering the impact of poor deposition accuracies on undesired discharge capacity losses [23], values of 0.5 mm and

0.5° define the acceptable limits for a sufficiently adequate stacking process during this investigation [20].

While the gripper speed $v_{gripper}$ is set at a constant level of a 25 mm/s for synchronization with the drum rotation speed through the servo drives, the gripping distance x is set to 0.4 mm throughout all experiments with regards to the preliminary tests as greater or lower values facilitated no gripping or visible folds in the specimens. A similar behavior for variations of the applied pressure difference on the drum gripper is noticeable during the preliminary tests, therefore the pressure difference Δp is set to 0.5 and 0.6 bar. This results in four experimental variations with 30 iterations per orientation direction.

5 Results and Discussion

Figure 6 shows the achieved total deposition accuracies as the superposition of position and orientation accuracy (columns) as well as the position and orientation repeatability (red error bars) for both tested rotation directions and tested materials.

In general, experiments with DOO generally exhibit more adequate deposition accuracies compared to DCO. The characteristic overhang of the handled cell component during DCO (Fig. 3, 4b) in combination with the specimens' adhesive behavior towards the deposition surface are suggested to be the main cause for this observation, as they occasionally result in the formation of unwanted folds in the deposited specimens. These folds are not completely avoided through the integration of the ion blower, especially when handling PEO-based separators. A similar trend in elevated deposition accuracy values is also visible for increasing pressure differences of $\Delta p = 0.6$ bar, exceeding the set accuracy limit in these experiments for both tested cell components. This observation can be attributed to the cell components' adhesive behavior and their partial adhesion tendency to the gripper due to electrostatic charge upon mechanical contact between both surfaces which tends to increase with higher pressure difference and therefore holding force. Additionally, the higher air pressure difference combined with the round shape of the vacuum roll induces high irreversible bending of the specimens as observed during several iterations, so that the handled cell components experience plastic deformation. In all experiments, the orientation accuracy and repeatability meet the set requirement and account for smaller accuracy values with $AP_{\varphi} \approx RP_{\varphi} \approx 0.09°$ compared to the position accuracy and repeatability.

For both cell components, the most adequate deposition accuracy is observed during DOO with the application of a lower pressure difference of $\Delta p = 0.5$ bar, ranging from $AP_{x,y} = 0.24$ to 0.31 mm in position and $AP_{\varphi} = 0.07$ to 0.13° in orientation accuracy. Here, the LMA can be deposited more accurately than the PEO-based separator. In these calculations, the position repeatability of the planar surface gantry accounts for ±0.05 mm on the deviation in accuracy.

Despite these results, partial specimen adhesion hindering accurate deposition during these experiments should be addressed in future testing setups. Considering the gripper's design, this could include the application of coatings onto the vacuum roll which reduce the adhesive effect or the electrostatic charge upon contact with tested cell components. Other approaches could focus on the application of overpressure for blowing off of the specimens. As the used ejector facilitates a sufficient gripping vacuum only with high leakage and noise emissions, a constructive redesign of the drum air chambers for efficient utilization of vacuum, aided by computational fluid dynamics simulations, along with improvements in sealings in the vacuum roll could be beneficial for minimizing occurring leakage.

6 Conclusion and Outlook

The main goal of this research was the development of a drum gripping concept for secured orientation-variable handling of mechanically sensitive SSB cell components with one-sided coating for sufficiently accurate cell stacking. The presented concept was experimentally validated for the achievable deposition accuracy in relation to both rotation directions and the applied pressure difference. Especially for pressure differences of 0.5 bar, sufficient accuracies could be demonstrated. Nevertheless, the formation of folds and wrinkles as well as the remaining specimen adhesion necessitate the investigation of further adaptions and improvements to the concept. With visible surface damages at higher air pressure differences as well as the negative influence of particulate contamination during electrode handling on the resulting cell performance [14], a thorough investigation of the specimens' surface damage after gripping via e. g. microscopic imaging should also be performed. The applicability for further SE types, such as sulfide-based specimen with a sufficient flexibility should be analyzed. Additionally, future challenges lie in the transfer of this gripping concepts to industry-scale production considering miniaturization and necessary kinematic acceleration while guaranteeing process stability. Overall, the presented gripper concept contributes towards the establishment of a scalable SSB production which plays an important role in the further and more efficient implementation of electromobility.

Acknowledgements The authors acknowledge the support of the German Federal Ministry of Education and Research in funding the research project "FB2-Prod" (03XP0432A). The authors thank Frederieke Lange, Andrea Wiegandt, and Ruben Schmonsees from the Fraunhofer Institute for Manufacturing and Advanced Materials IFAM for providing the separator material.

References

1. Schmaltz, T., Hartmann, F., Wicke, T., Weymann, L., Neef, C., Janek, J.: A roadmap for solid-state batteries. Adv. Energy Mater. (2023). https://doi.org/10.1002/aenm.202301886
2. Janek, J., Zeier, W.G.: Challenges in speeding up solid-state battery development. Nat. Energy (2023). https://doi.org/10.1038/s41560-023-01208-9
3. Duffner, F., Kronemeyer, N., Tübke, J., Leker, J., Winter, M., Schmuch, R.: Post-lithium-ion battery cell production and its compatibility with lithium-ion cell production infrastructure. Nat. Energy (2021). https://doi.org/10.1038/s41560-020-00748-8
4. Nguyen, D.M., Scharmann, T., Dröder, K.: Material-adapted Gripping and Handling of PEO-based Cell Components for All-Solid State Battery Cell Stacking (2023)
5. Fröhlich, A., Masuch, S., Dröder, K.: Design of an automated assembly station for process development of all-solid-state battery cell assembly. In: Schüppstuhl, T., Tracht, K., Raatz, A. (eds.) Annals of Scientific Society for Assembly, Handling and Industrial Robotics 2021, pp. 51–62. Springer International Publishing, Cham (2022)
6. Fröhlich, A., Gresens, D., Vervoort, B., Dröder, K.: Design and evaluation of a material-adapted handling system for all-solid-state lithium-ion battery production. Procedia CIRP (2020). https://doi.org/10.1016/j.procir.2020.03.040
7. Konwitschny, F., Schnell, J., Reinhart, G.: Handling cell components in the production of multi-layered large format all-solid-state batteries with lithium anode. Procedia CIRP (2019). https://doi.org/10.1016/j.procir.2019.03.300
8. Plocher, L., Plumeyer, J.F., Wennemar, S.: Automatisierter Stapelprozess zur Herstellung von Festkörperbatterien. Zeitschrift für wirtschaftlichen Fabrikbetrieb (2022). https://doi.org/10.1515/zwf-2022-1116
9. Pettinger, K.-H.: Fertigungsprozesse von Lithium-Ionen-Zellen. In: Korthauer, R. (ed.) Handbuch Lithium-Ionen-Batterien, pp. 221–235. Springer, Berlin Heidelberg, Berlin, Heidelberg (2013)
10. Kampker, A., Hohenthanner, C.-R., Deutskens, C., Heimes, H.H., Sesterheim, C.: Fertigungsverfahren von Lithium-Ionen-Zellen und -Batterien. In: Korthauer, R. (ed.) Handbuch Lithium-Ionen-Batterien, pp. 237–247. Springer, Berlin Heidelberg, Berlin, Heidelberg (2013)
11. Schnell, J., Günther, T., Knoche, T., Vieider, C., Köhler, L., Just, A., Keller, M., Passerini, S., Reinhart, G.: All-solid-state lithium-ion and lithium metal batteries—paving the way to large-scale production. J. Power. Sources (2018). https://doi.org/10.1016/j.jpowsour.2018.02.062
12. Schröder, R., Glodde, A., Aydemir, M., Seliger, G.: Increasing productivity in grasping electrodes in lithium-ion battery manufacturing. Procedia CIRP (2016). https://doi.org/10.1016/j.procir.2016.11.134
13. Fleischer, J., Ruprecht, E., Baumeister, M., Haag, S.: Automated handling of limp foils in lithium-ion-cell manufacturing. In: Dornfeld, D.A., Linke, B.S. (eds.) Leveraging Technology for a Sustainable World, pp. 353–356. Springer, Berlin Heidelberg, Berlin, Heidelberg (2012)
14. Fröhlich, A., Leithoff, R., von Boeselager, C., Dröder, K., Dietrich, F.: Investigation of particulate emissions during handling of electrodes in lithium-ion battery assembly. Procedia CIRP (2018). https://doi.org/10.1016/j.procir.2018.08.322
15. Li, X., Li, N., Tao, G., Liu, H., Kagawa, T.: Experimental comparison of Bernoulli gripper and vortex gripper. Int. J. Precis. Eng. Manuf. (2015). https://doi.org/10.1007/s12541-015-0270-3
16. Angerer, A., Ehinger, C., Hoffmann, A., Reif, W., Reinhart, G.: Design of an automation system for preforming processes in aerospace industries. In: 2011 IEEE International Conference on Automation Science and Engineering. 2011 (CASE 2011), Trieste, Italy, 24.08.2011–27.08.2011, pp. 557–562. IEEE (2011). https://doi.org/10.1109/CASE.2011.6042411

17. Divoux, M., Dorez, M.: Apparatus for stacking thin flexible objects. USA Patent US4440388A
18. Götz, R.: Strukturierte Planung flexibel automatisierter Montagesysteme für flächige Bauteile. Zugl.: München, Techn. Univ., Diss.: 1991. IWB-Forschungsberichte/Institut für Werkzeugmaschinen und Betriebswissenschaften, Technische Universität München, vol. 39. Springer, Berlin, Heidelberg, New York, London, Paris, Tokyo, Hong Kong, Barcelona, Budapest (1991)
19. Mooy, R.J.M., Aydemir, M., Glodde, A.: Increasing productivity in assembly automation through the utilization of continuous processes. J. Ind. Intell. Inf. (2018). https://doi.org/10.18178/jiii.6.2.38-44
20. Mooy, R.J.M.: Beitrag zur Produktivitätssteigerung in der Vereinzelung, Positionierung und Orientierung von Elektrodenfolien durch eine kontinuierliche Materialbewegung (2019)
21. Deutsches Institut für Normung: DIN 53362:2003–10, Prüfung von Kunststoff-Folien und von textilen Flächengebilden (außer Vliesstoffe), mit oder ohne Deckschicht aus Kunststoff_- Bestimmung der Biegesteifigkeit_- Verfahren nach Cantilever. Beuth Verlag GmbH, Berlin (2003)
22. Internationale Organisation für Normung: Manipulating industrial robots—performance criteria and related test methods. International standard ISO, ISO 9283 (1998)
23. Leithoff, R., Fröhlich, A., Dröder, K.: Investigation of the influence of deposition accuracy of electrodes on the electrochemical properties of lithium-ion batteries. Energy Technol. (2020). https://doi.org/10.1002/ente.201900129

Robotics Simulation

Sim2Rob—A Sim2Real Framework for Smart Robotics in Assembly

Oliver Petrovic, Petar Tesic, and Christian Brecher

Abstract

Robot-based assembly automation struggles with flexibility and adaptability in today's dynamic production landscape. Intelligent, AI-based robotic systems bring great potential for addressing those challenges as well as raising the level of automation. However, the implementation of powerful deep learning (DL) algorithms poses the so-called data problem. Acquiring real, annotated training data sets of sufficient quality and quantity usually requires a significant manual effort. Realistic simulation environments open up the possibility of generating synthetic training data in order to train models for real-world application. Combined with techniques for stable Sim2Real transfer to address the domain gap, this enables rapid adaptation to new products and processes. This work proposes a novel holistic framework for autonomous assembly based on synthetic data generation that leverages the synergy of lightweight robots, AI and advanced simulation capabilities while incorporating domain knowledge.

Keywords

Simulation • Synthetic data • Machine learning • Assembly • Computer vision • Sim2Real transfer

Oliver Petrovic and Petar Tesic are contributed equally

O. Petrovic (✉) · P. Tesic · C. Brecher
Laboratory for Machine Tools WZL, RWTH Aachen University, Aachen, Germany
e-mail: o.pertrovic@wzl.rwth-aachen.de

M.-C. Wanner et al. (eds.), *Annals of Scientific Society for Assembly, Handling and Industrial Robotics 2024*, https://doi.org/10.1007/978-3-031-91463-8_28

1 Introduction and Motivation

In times of high quality standards and shortage of skilled labour, the automation of industrial assembly processes is a crucial aspect of todays production. Furthermore, with the ever-increasing demand for product variety and customization in a volatile manufacturing landscape, there is a growing need for flexible and adaptable automation solutions, that can be easily deployed and reconfigured. In recent years, there have been significant advances in the field of AI-based robotics that offer promising solutions for the flexible automation of assembly. Modern DL enhances the capabilities of robots by equipping them with cognitive abilities like perception, decision-making and learning, thus allowing assembly systems to adapt to changes in the production environment.

However, AI applications are still rarely utilized in German companies [1]. One of the reasons for this is the data problem, that describes the need for large amount of often manually annotated data for training robust DL models [2]. The training data acquisition often accounts for over 80% of the costs of data science projects, which is particularly significant when assembly processes need to be reconfigured frequently [3]. Learning assembly steps using deep reinforcement learning (DRL) on the other hand requires many explorative repetitions for the algorithm to achieve optimal results and the robot cannot perform any value-adding work in the meantime. In addition, there is a risk of damaging production resources due to exploratory behavior during training.

The use of synthetically generated data and training in simulation environments is a promising approach to address the data problem. Based on simulated environments, diverse data sets are automatically generated and annotated, which takes only a fraction of the time of manual generation, is less prone to human error and highly accurate [4]. RL algorithms can also learn new tasks significantly faster and in a safe, experimental context by using a large number of parallel simulated environments. However, the transfer of these synthetically trained AI models to real-world applications poses a new challenge, the so called Sim2Real gap, that can lead to underperforming solutions [5]. This problem can be tackled through the methods of Sim2Real transfer. Here, the integration of domain-specific knowledge is crucial to success [6]. This, and the fact that there are still comparatively few approaches in industrial Sim2Real research [7], leads to a great need for research in the manufacturing domain.

Therefore, to the best of our knowledge, this work proposes Sim2Rob, the first fully comprehensive and modular Sim2Real robotics framework for industrial applications, that is based on industrial domain knowledge and corresponding information models. It combines pipelines for computer vision applications based on synthetic data and the training of DRL policies in the simulation and finally executes an adaptive assembly process via control modules and interfaces to the real robot. To successfully achieve the Sim2Real transfer, a realistic simulation environment is used as a domain adaptation technique on the one hand and methods of domain randomization are applied on the other hand.

2 Background and Related Work

2.1 AI-Based Robotics in Assembly Based on Synthetic Data

Addressing the limitations of the data problem and benefiting from the latest developments in graphics computing, training based on synthetic data and in simulators offers an opportunity to bring AI-based robotics applications into widespread use [2]. In recent years, numerous synthetic datasets that are relevant for robotics have been made publicly available and DL models have been developed based on them. However, these often originate from domains such as urban space, autonomous driving or general household objects and have limited to no industrial relevance [7].

Nevertheless, there are some approaches relevant to the manufacturing domain, which can be categorized into vision tasks and control tasks according to [8]. For the training of vision tasks, synthetic photorealistic RGB images or depth data are usually rendered and annotated in gaming engines or robot simulators [2]. Based on the data, models can for example be trained for object recognition and segmentation [9–11], 6D pose estimation [12–14], grasping [15, 16] or optical inspection [17]. To solve control tasks on the other hand, a common approach is to utilize DRL policies that are trained in (physical) simulation environments and then applied in reality. This approach allows the autonomous, goal-oriented learning of robot behavior, for example for complex object manipulation [18, 19], force-based processes such as the contact-rich joining of components [20–22] or the handling of deformable objects [23].

2.2 The Sim2Real Transfer

Sim2Real Transfer describes a set of methods that aim to bridge the gap between simulation and reality, which inevitably results from the modeling nature of simulation environments. Without the application of such approaches, simulatively trained models tend to fail in physical reality [5]. One obvious approach is to model the simulation as accurately as possible. Kaspar et al. [20] use system identification methods for precise modeling of robot dynamics for the training of force-based joining processes. Numerous other approaches use photorealistic rendering of image or depth data for computer vision applications in robotics [4, 11, 24]. Local optimization of the image in the regions of interest can also be considered to save computing costs, as proposed in [13].

Another common approach is domain randomization (DR), where relevant parameters in the simulated environment are randomized, such that reality is contained within the spectrum. This can lead to robust policies and models for object recognition [9, 25], object localization [14, 15], dexterous handling tasks [18], or force-based operations [26], especially with the appropriate utilization of specific domain knowledge [10].

Finally, domain adaptation (DA) is frequently utilized in robotics, optionally in combination with DR. This involves either using hybrid datasets consisting of synthetic and real data [17] or alternatively performing subsequent fine-tuning in reality [16, 27] to increase the similarity between the domains and their features. Alternatively, generative adversarial networks (GAN) can also be utilized for this purpose [22]. Even though these and other methods have proven their effectiveness, there is still little research in the field of industrial production [7]. Yet it is precisely here that Sim2Real transfer represents a major challenge, because of the complexity of the domain with high robustness requirements and objects of high geometric but low optical variance [6].

2.3 Knowledge-Based Synthetic Data Generation for Industrial Robotics

One issue with the generation of synthetic data in simulation for training robust DL models for industrial robotic applications is its complex and time-consuming nature [6]. This is addressed by integrating the process of generating synthetic data into an automated pipeline to make it scalable. Widespread examples of such synthetic data generators (SDG) are NViSII [28] and BlenderProc [29]. While widespread SDGs incorporate domain-specific knowledge in a very limited manner, there are some approaches that build on domain knowledge from the industrial sector.

In CAD2Render, Moonen et al. combine CAD models with textures, modeled surface defects and environmental images to create synthetic data sets for manufacturing use cases [30]. In [6], Rawal et al. present a comparable pipeline that automatically generates data from the components of entire assemblies and Yang et al. cascades the data generation to save computing power [12]. In [31], synthetic data generation is based on assembly sequences to automatically generate data from each intermediate assembly step and subassemblies. In addition, semantic technologies, such as ontologies, can be used to model domain knowledge for the use in cyber-physical systems. In [32, 33], an approach was proposed for the knowledge-based generation of synthetic data for industrial applications, including an ontology developed for this purpose, on which, in combination with further manufacturing and assembly ontologies, the framework presented here is built upon.

3 Sim2Rob Framework Design

In this paper, we present a novel framework called Sim2Rob for autonomous assembly that combines state-of-the-art technologies from robotics, simulation and artificial intelligence. This enables the generation of Sim2Real-capable synthetic data, accelerates the learning of RL-based skills in simulation environments and creates a virtual representation of the assembly process as a planning basis. The system acquires semantic knowledge

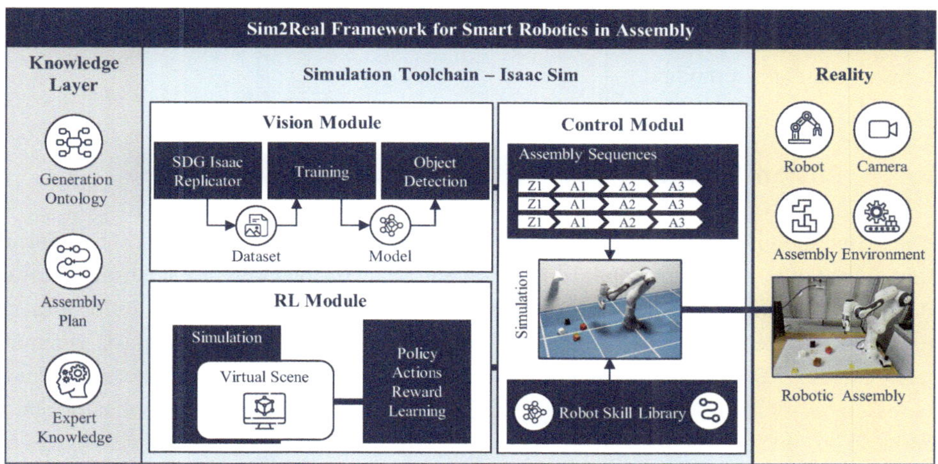

Fig. 1 Sim2Rob–Sim2Real framework for smart robotics in assembly

models for meaningful parameterization of these virtual environments. Structurally, the framework can be divided into the three main parts control module, vision module and RL module, as shown in Fig. 1. Thanks to the modularity, the system components and models can be selected according to the application, which leads to a high transferability and applicability of the framework in industrial applications.

The skill-based control module, models the assembly system including robot, derives robot actions and control commands based on the virtual scene and deploys them to the real robot system. The coordination of the assembly steps and the selection of the performed skills is carried out by a Decider-Network, which was specially developed for the execution of joining-based assembly processes.

Skills based on DRL policies can be learned in the RL module. For this purpose, a training environment was developed exemplarily, which enables the learning of sensitive joining processes. The focus was specifically on the Sim2Real capability of the resulting policies by using highly realistic physical simulations and robot models as well as domain randomization methods.

The vision module for 6D pose estimation of the parts, based on object and key-point detection, serves as an interface between the virtual scene of the Control Module and the reality. The model is trained on purely synthetic data, which is generated application-specifically. Here, again, domain randomization methods are used in addition to photorealistic rendering to close the Sim2Real gap.

Each of the three modules requires specific knowledge about the workpieces and the assembly process. The previously mentioned ontology-based knowledge models were developed in advance among others in [32, 33]. The modules were designed so that these knowledge models can be used to parameterize the respective simulation environment by serving as input for the automatic generation of the synthetic data set, for defining the training and policy hyperparameters in the RL module, and for planning the assembly steps in the control module.

3.1 Vision Module

The vision module is used to identify and localize the assembly components. The information collected by the vision module serves as the basis for planning the next robot actions in the control module. The module consists of two sections, which are used consecutively. In the preparatory Synthetic Data Generation section, synthetic data is generated from existing 3D models in a photorealistic simulation environment parameterized based on assembly domain knowledge. It is implemented in NVIDIA Isaac Replicator and in addition to the high realism of the physically rendered images, domain randomization (DR) is used to achieve a robust Sim2Real transfer. The randomizable parameters are derived from the entities present in the simulation environment and are shown in Fig. 2. Setting these parameters meaningfully requires expert knowledge about the process, which is acquired through the ontology developed in [32, 33].

The synthetic training data set contains rendering of the scenes and the corresponding annotation incorporating precise ground truth information including the object classification, bounding boxes and keypoint positions. The DL model, in this case YOLOv8, can be trained on the basis of this data. The trained model can then be deployed in the inference part and thus applied to reality. Based on the camera data, collected by an

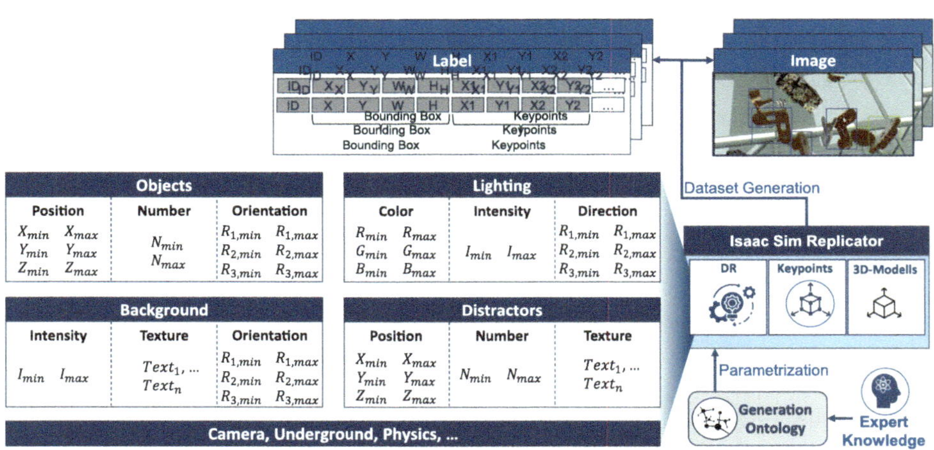

Fig. 2 DR-Parameters and Structure of the annotated synthetic dataset for YOLOv8

Intel Realsense and published via a ROS node, the model can be used for object identification and keypoint localization. By combining YOLOv8 for keypoint detection and Perspective-n-Point (PnP) algorithms, we compute the transformation between the camera's and object's coordinate systems, given knowledge of intrinsic camera parameters and the keypoint positions in the object coordinate system. From these transformations the 6D-Pose can be calculated.

3.2 RL-Module

The RL module is designed to enable the robot system to learn various skills for complex assembly processes like contact-rich joining operations. These skills are initially trained completely synthetically in a simulative environment in order to subsequently integrate the resulting policy into the control module, which makes the skills available for industrial assembly processes. The module contains the robotic simulation environment with the physics engine, the models of the robot and the components as well as the virtual sensors and the robot controller. To ensure Sim2Real capabilities, the latter must correspond to the one that controls the real robot. Within the simulation, the policies for the assembly skills are trained through the actions performed by the agent and the resulting observation. For a robust Sim2Real transfer, parameters like the initial spawn position of components and the end effector as well as physical parameters such as friction and damping constants are randomized. Due to the high compatibility with the tools used in the Vision module and the possibility to simulate many instances in parallel in vectorized environments (VE), we have chosen Isaac Sim but because of the modularity, other simulators such as PyBullet, Gazebo or MuJoCo can also be used.

Another advantage of Isaac Sim is the Isaac Gym extension, which can be used as the DRL Submodule and offers end-to-end GPU-accelerated simulation capabilities, that can significantly speed up the training process in parallel VE. However, other RL environments such as OpenAI Gym can also be used modularly. While the simulation can be generalized across many applications, the DRL submodule that interacts with the individual virtual scene must be selected based on the application. This is where the RL algorithm, the policy, the model and learning parameters as well as the agent with its action and observation space together with the reward function are implemented.

3.3 Control-Module

The control module forms the nucleus of the framework and connects the vision and RL modules with the real assembly system as a central hub. Here, the assembly sequence, available resources and DL-models are orchestrated and coordinated in a central decision-making system, which is implemented in an acyclic graph as a decider network organizing

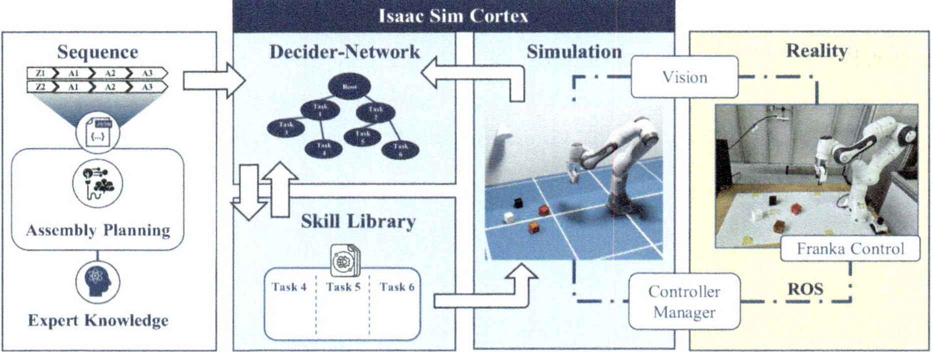

Fig. 3 Structure of the control module and its interfaces to assembly planning and reality

various AI-based and classic robot skills. A digital twin of the assembly System in the simulation environment, which gets updated in real-time based on the Input from the Vision-Module, is used to fine-tune the executing skills by performing dynamic-reactive path planning. Using a bidirectional ROS interface to control the real robot, the control module also handles the execution of real-world operations and reads the robot's joints and sensors in order to synchronize the digital twin. The control module is also implemented in Isaac Sim and uses the Isaac Cortex extension for the decider networks and the robot interface (Fig. 3).

3.4 Validation Use Case

To test and validate the framework, an assembly use case was selected, encompassing all the described modules. The use case involves a demo assembly from the Siemens Robo Learning Challenge, consisting of a five-stage joining process (Fig. 4). The synthetic generated Dataset based on the 3D models was utilized for training the keypoint and object detection using a YOLOv8-m model. First, not yet fine-tuned experiments of the methodology show a good performance with a MAP50 of 0.856 at the keypoints and 0.975 in the bounding box. This leads to a positioning accuracy of <1 mm over large parts of the workspace despite the low-cost camera for 6D pose estimation and handling with a Franka Emika Panda. The assembly process is characterized by contact-rich joining operations, which were learned within the RL module. Here, a PPO model combined with an impedance controller was chosen, achieving a success rate of 93% in initial simulation experiments and 85% after the Sim2Real transfer. Further fine-tuning of the model and controller parameters can potentially yield even greater stability. The Decision Network in the Control Module was manually parameterized, which enabled the Control Module

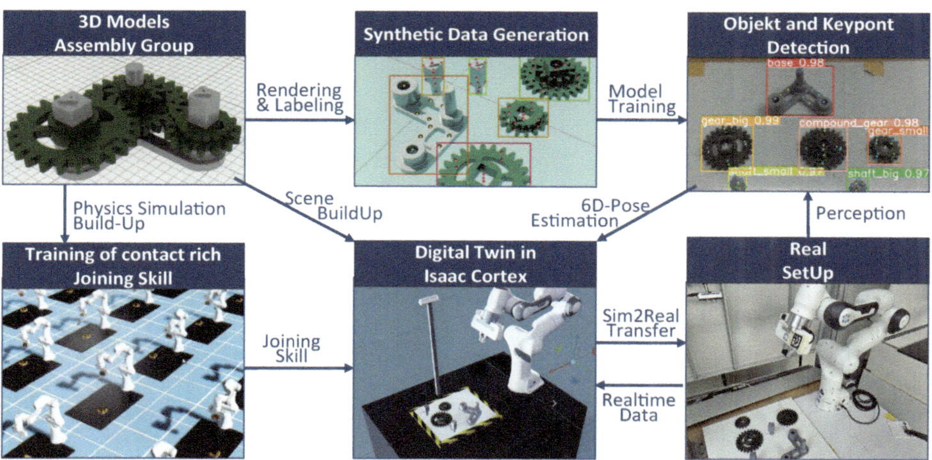

Fig. 4 Validation use case set-up

to sequentially assemble the correct workpieces by selecting the appropriate skills. The setup and intermediate steps are shown in Fig. 4.

4 Summary and Outlook

In this article, we have presented the novel Sim2Real framework for industrial, robot-based assembly called Sim2Rob. Thanks to its modular structure and combining synthetic data generation for computer vision applications with the DRL of policies for robot operations via a central control system, the framework offers broad applicability for smart robot systems in assembly. Extending to other industrial Sim2Real applications, such as quality inspection or logistics, is also easily possible due to the high transferability. Therefore, the framework has great potential for addressing the data problem and implementing DL-based robot systems in the production domain. The functionality of the modules has already been demonstrated in initial experiments with promising results. In ongoing work, these experiments are being expanded in order to fine-tune and extend the framework as well as benchmark it against other research approaches. In addition, the framework provides a good basis for research to increase the robustness of the Sim2Real transfer by allowing various methods to be tested at little effort. In the current expansion stage, experts in the fields of assembly, simulation and Sim2Real transfer are still required to use the framework. In further work, the pipelines will be successively automated to implement applications with a lower competence hurdle.

Acknowledgements The IGF-project 22648 N/2 (ROOKIE) of the research association FVP was supported via the AiF within the funding program "Industrielle Gemeinschaftsforschung und-entwicklung (IGF)" by the Federal Ministry for Economic Affairs and Climate Action (BMWK) due to a decision of the German Parliament.

References

1. Bitkom Research: Künstliche Intelligenz—Wo steht die deutsche Wirtschaft? https://www.bitkom.org/sites/main/files/2022-09/Charts_Kuenstliche_Intelligenz_130922.pdf. Last accessed 04 Apr 2024 (2022)
2. Melo, C.M. de, et al.: Next-generation deep learning based on simulators and synthetic data. Trends Cogn. Sci. (2022)
3. Nikolenko, S.I.: Synthetic Data for Deep Learning, vol. 174. Springer International Publishing, Cham (2021)
4. Eversberg, L., Lambrecht, J.: Generating images with physics-based rendering for an industrial object detection task: Realism versus domain randomization. Sensors (2021)
5. Sudhakar, S., et al.: Exploring the Sim2Real gap using digital twins. In: 2023 IEEE/CVF International Conference on Computer Vision (ICCV), pp. 20361–20370. IEEE (2023)
6. Rawal, P., Sompura, M., Hintze, W.: Synthetic Data Generation for Bridging Sim2Real Gap in a Production Environment (2023)
7. Akar, C.A., et al.: Synthetic object recognition dataset for industries. In: 35th SIBGRAPI Conference on Graphics, Patterns and Images, pp. 150–155. IEEE (2022)
8. Zhao, W., Queralta, J.P., Westerlund, T.: Sim-to-real transfer in deep reinforcement learning for robotics: a survey. In: 2020 IEEE Symposium Series on Computational Intelligence (SSCI), pp. 737–744. IEEE (2020)
9. Mayershofer, C., et al.: LOCO: Logistics objects in context. In: 19th IEEE International Conference on Machine Learning and Applications (ICMLA), pp. 612–617. IEEE (2020)
10. Mayershofer, C., Ge, T., Fottner, J.: Towards fully-synthetic training for industrial applications. In: Liu, S., Bohács, G., Shi, X., Shang, X., Huang, A. (eds.) LISS 2020, pp. 765–782. Springer Singapore, Singapore (2021)
11. Ivanovic, A., et al.: Render-in-the-Loop Aerial Robotics Simulator: Case Study on Yield Estimation in Indoor Agriculture (2022)
12. Yang, X., et al.: Image translation based synthetic data generation for industrial object detection and pose estimation. IEEE Robot. Autom. Lett. (2022)
13. Zheng, T., et al.: Deep learning-based 6-DoF object pose estimation considering synthetic dataset. Sensors (2023)
14. Tobin, J., et al.: Domain randomization for transferring deep neural networks from simulation to the real world. In: 2017 IEEE/RSJ International Conference on Intelligent Robots and Systems (IROS), pp. 23–30. IEEE (2017)
15. Horváth, D., et al.: Object detection using Sim2Real domain randomization for robotic applications. IEEE Trans. Robot. (2023)
16. Bousmalis, K., et al.: Using simulation and domain adaptation to improve efficiency of deep robotic grasping. In: IEEE International Conference on Robotics and Automation (ICRA), pp. 4243–4250. IEEE (2018)
17. Nguyen, H.G., Habiboglu, R., Franke, J.: Enabling deep learning using synthetic data: a case study for the automotive wiring harness manufacturing. Procedia CIRP (2022)

18. Andrychowicz, O.M., et al.: Learning dexterous in-hand manipulation. Int. J. Robot. Res. (2020)
19. Moosmann, M., et al.: Separating entangled workpieces in random bin picking using deep reinforcement learning. Procedia CIRP (2021)
20. Kaspar, M., Munoz Osorio, J.D., Bock, J.: Sim2Real transfer for reinforcement learning without dynamics randomization. In: 2020 IEEE/RSJ International Conference on Intelligent Robots and Systems (IROS), pp. 4383–4388. IEEE (2020)
21. Petrovic, O., et al.: Sim2Real Deep reinforcement learning of compliance-based robotic assembly operations. In: 2022 26th International Conference on Methods and Models in Automation and Robotics (MMAR), pp. 300–305. IEEE (2022)
22. Yuan, C., et al.: Sim-to-real transfer of robotic assembly with visual inputs using CycleGAN and force control. In: 2022 IEEE International Conference on Robotics and Biomimetics (ROBIO), pp. 1426–1432. IEEE (2022)
23. Tong, D., et al.: Sim2Real neural controllers for physics-based robotic deployment of deformable linear objects. Int. J. Robot. Res. (2023)
24. Zhang, X., et al.: Close the optical sensing domain gap by physics-grounded active stereo sensor simulation. IEEE Trans. Robot. (2023)
25. Prakash, A., et al.: Structured domain randomization: bridging the reality gap by context-aware synthetic data. In: 2019 International Conference on Robotics and Automation (ICRA), pp. 7249–7255. IEEE (2019)
26. Valassakis, E., Ding, Z., Johns, E.: Crossing the gap: a deep dive into zero-shot sim-to-real transfer for dynamics. In: 2020 IEEE/RSJ International Conference on Intelligent Robots and Systems (IROS), pp. 5372–5379. IEEE (2020)
27. Tang, H., Jia, K.: A new benchmark: on the utility of synthetic data with blender for bare supervised learning and downstream domain adaptation. In: IEEE/CVF Conference on Computer Vision and Pattern Recognition, pp. 15954–15964. IEEE (2023)
28. Morrical, N., et al.: NViSII: A Scriptable Tool for Photorealistic Image Generation (2021)
29. Denninger, M., et al.: BlenderProc: reducing the reality gap with photorealistic rendering. In: Toussaint, M., Bicchi, A., Hermans, T. (eds.) Robotics: Science and Systems XVI. Robotics Science and Systems Foundation (2020)
30. Moonen, S., et al.: CAD2Render: A modular toolkit for GPU-accelerated photorealistic synthetic data generation for the manufacturing industry. In: 2023 IEEE/CVF Winter Conference on Applications of Computer Vision Workshops (WACVW), pp. 583–592. IEEE (2023)
31. Dümmel, J., Kostik, V., Oellerich, J.: Generating Synthetic Training Data for Assembly Processes. In: Dolgui, A., Bernard, A., Lemoine, D., Cieminski, G. von, Romero, D. (eds.) Advances in production management systems. In: Artificial Intelligence for Sustainable and Resilient Production Systems, vol. 633. IFIP Advances in Information and Communication Technology, pp. 119–128. Springer International Publishing, Cham (2021)
32. Petrovic, O., et al.: Towards knowledge-based generation of synthetic data by taxonomizing expert knowledge in production. In: Proceedings of the 6th International Conference on Intelligent Human Systems Integration (IHSI 2023). AHFE International (2023)
33. Petrovic, O., Duarte, D.L.D., Herfs, W.: Generating synthetic data using a knowledge-based framework for autonomous productions. In: 2023 IEEE/ASME International Conference on Advanced Intelligent Mechatronics (AIM), pp. 1086–1093. IEEE (2023)

Experimental Semi-Automatic 3D Model Generation of a Robot in Brownfield for Simulation Models Using Camera-Based Methods

Nico Brandt, Erik-Felix Tinsel, Alexander Verl, Oliver Riedel and Michael Neubauer

Abstract

In response to the increasing complexity of production systems, automated methods for creating simulation models are being explored. Accurate simulation models are necessary for virtual commissioning to effectively model the real operation of a factory. However, manual model creation is labor-intensive and requires expertise. To address this challenge, this work investigates automated approaches by reviewing current methods and proposing the automated creation of simulation models from video data, especially for brownfield machines. Experimental results confirm the feasibility of using photogrammetry and Neural Radiance Field for automated mesh generation.

Keywords

Simulation • Modeling • 3-D reconstruction

1 Introduction

Due to the increasing complexity of production systems, simulation technology is becoming increasingly important for planning and improving the real factory [7]. Tools of the *digital factory* [27] are used to develop a digital model of the planned and existing factory systems. Virtual commissioning is part of the *digital factory* [23]. It involves preparing and testing systems using simulations. In order to make sufficiently accurate predictions about the real

N. Brandt (✉) · E.-F. Tinsel · A. Verl · O. Riedel · M. Neubauer
Institute for Control Engineering of Machine Tools and Manufacturing Units, University of Stuttgart, Stuttgart, Germany
e-mail: nico.brandt@isw.uni-stuttgart.de

M.-C. Wanner et al. (eds.), *Annals of Scientific Society for Assembly, Handling and Industrial Robotics 2024*, https://doi.org/10.1007/978-3-031-91463-8_29

system, it is important to adequately represent reality in a simulation model with regard to all relevant aspects. The quality of the predictions based on the simulation depends directly on the quality of the model. The modeling process required to obtain high quality models that are verified and validated is time consuming and usually requires a simulation expert. According to [25], physics-based models for virtual commissioning can consist of 3D CAD, triangulation, collision, physics and simulation models. Accordingly, the model can be built in this order. In the first step, the 3D CAD model is converted into a triangulation model by pre-processing (e.g. tessellation). This is then encapsulated within a simplified collision model and augmented by physical parameters such as mass and inertia. Finally, kinematic chains (e.g. gears, joints) are added to the simulation model.

In addition, simulation technology is very interdisciplinary and requires expertise in various fields of knowledge, including computer science, economics, mechanical engineering, industrial engineering and statistics. The latter is often a challenge for the use of simulation in production and logistics, especially in small and medium-sized companies. The benefits that can be derived from simulation depend heavily on the skills of the simulation expert [3].

As the digitalization of industrial processes progresses, automatic generation of the required simulation models is increasingly being discussed as a way of reducing the effort involved in creating models. Various efforts have been made to generate simulation models directly from the digital factory design data. In general, these approaches are suitable for automating model generation for specific application scenarios. However, for broader application and reusability of such approaches, standardized interfaces and file formats are required, which are insufficiently supported by today's digital factory tools. The automation of model generation also promises positive effects in the areas of model quality, reproducibility, and model standardization [18].

In order to reduce the development costs of creating and adapting simulation models for the planning and production processes of industrial plants, it is necessary to at least partially automate the generation of initial model parts for simulation. In this regard, this work aims to demonstrate that it is possible to generate meshes that align with a CAD model and to quantify the precision of these meshes. This will reduce the workload for companies, allowing them to focus more on optimising and integrating existing models instead of simply creating and adapting them. This provides a further basis for the economical use of digital models.

2 Current Methods for the Automated Creation of Simulations

An important application for a physics–based simulation model is virtual commissioning, where the behavior of a factory plant can be tested by coupling it to a (virtual) control system [29]. Subsequent modification of the plant components necessitates the execution of a new test, even after the initial commissioning phase. In this case, the simulation model goes through several development phases and must be kept up to date, leading to incremental model development [3]. In order to reduce the high effort [6] involved in creating and adapting physics-based simulation models, a data-driven approach, i.e. the transformation of existing information into the simulation model, can be used to automatically create the models. The following section shows some examples using different data.

2.1 Automatic Generation of Simulation Models

The AutomationML standard is a widely used data model for simulation generation. For example, in [19], AutomationML and the eCl@ss[1] standard are used to assign a role class to abstract model objects. The purpose of this is to enable the semantic mapping of user-defined classes (from vendor catalogues or from simulation libraries). A corresponding object of the class is then instantiated for each abstract object in AutomationML. This enables role-based mapping of the vendor-specific object to a simulation model from an existing library. Furthermore, object relationships can be transferred to the simulation model. However, the initial definition of these relationships must still be done entirely manually.

The authors of [12] use AutomationML to generate ROS simulation models in Gazebo. A model generator first iteratively reads all components from the AutomationML file and then loads geometry and kinematics from referenced URDF or COLLADA data into a simulation environment. To do this, the model generator exports the configuration of a reconfigurable manufacturing system from AutomationML, defined by its GUID or name. A similar approach is shown in [11], using YAML as an alternative generation base.

In [9], a concept for transferring the AutomationML topology model into a co-simulation in OpenModelica is presented. It includes a sub-concept for automatic compilation and configuration of co-simulations during the commissioning phase. An assistance system identifies modules in the plant and loads the required simulation models from FMU libraries. The FMUs are coupled via interfaces in the topology model and the co-simulation is mapped in various tools such as Simulink, Dymola or PyFMI. The OpenModelica simulator is used for the prototype realization, where a script is generated based on the topology model to configure and start the simulation.

In [1], an approach for the automatic derivation of simulation models from Piping & Instrumentation Diagrams (P&IDs) is described. The approach consists of two steps: Information extraction and formalization, and generation of the simulation model. However, the resulting model is only suitable for simple simulations, as some specific information cannot be extracted directly from the P&ID. In [2], it is shown how code is generated from P&IDs in Modelica and how a graphical representation including an animated model is created.

The approach presented in [13], focuses on the automated creation of machine models for hardware-in-the-loop simulation of assembly machines. This is done by modeling the structure and behavior of components and assemblies. The structure is generated from a bus configuration and a given I/O list, while the behavior models of the components are created

[1] https://eclass.eu/.

separately. The machine model is then imported into a simulation tool, where manual post-processing is required to link the behavior models to the geometry.

An architecture for Automatic Model Generation (AMG) from planning data for Virtual Commissioning was tested in [26]. 3D models were inserted automatically, but changes to the layout or the models had to be made manually. In contrast, [20] generated an extended hardware-in-the-loop simulation with the help of a commissioning list.

The current state of research shows that a (semi-)AMG can be created from a variety of different data sources. Established data exchange formats, structured model description data or 3D CAD models can be used to allow a transfer to an initial simulation model using defined rulesets. These models can also be reused across projects through the use of module libraries. However, the requirement for a valid database as a basis poses a problem for AMG, as the available data may deviate from the physical entity over time [3]. A method is therefore required to keep existing simulation models consistent throughout the lifecycle of a plant with as little effort as possible.

2.2 3D Reconstruction for the Modeling Process

The methods presented for AMG using design data can provide an up-to-date digital image of the system if the models are updated on an ongoing basis. However, the modeling process requires up-to-date 3D CAD models to enable AMG. In the case of brownfield sites, the creation of new models and their integration in simulation models is associated with considerable effort [17] and costs [28]. Although simulation without a 3D model may be sufficient for some applications, some tests, such as collision tests, require 3D models. 3D reconstruction methods such as Gaussian splatting [14], Neural Radiance Field (NeRF) [16], photogrammetry [5], multi-view stereo [22], structured light scanning [30] or time-of-flight sensor [4] technology are available to digitize existing equipment without the need for time-consuming reconstruction using CAD tools. The following section presents some examples of this approach from the current state of research.

The method presented in [21] consists of two main phases: First, 3D digitization is carried out to obtain an image of the existing production structures. This is achieved by capturing and processing 3D scans and color photographs. The result is a colored point cloud that digitally represents the real production. The second phase of the process is to modify the point clouds in order to generate new design states.

In [15] is described how a drone system equipped with various assistance systems and two cameras can be used to generate a simulation model. The image data recorded during the flight is used to create a 3D model of a plant. The point cloud is loaded into Matlab to analyze the data. The first step was to scale the proportions within the model. The objects are segmented into smaller point clouds based on distance and color features. A labeling algorithm is used for recognition, combining points that have no connection to other points in the model into a separate point cloud. The objects are automatically interpreted by a machine learning algorithm. The object databases used for this are previously created by humans. The algorithm learns from these object databases and can assign the learned object. Once a factory object, such as a pallet, has been created in the database, every object in the factory is compared to that object. As errors in the point cloud can also lead to false detections, manual modification of the detection is possible. After segmentation and object recognition, the overall model is reassembled from the individual models. The objects are displayed in the layout as a separate point cloud, similar to the existing objects, and can be subsequently modified.

In [24], a method for the automatic generation of simulation models is presented using accelerated scans for production and object recognition. The process entails the scanning of the object, the creation of a 3D model based on the scanned data, and the generation of a simulation based on the model. The 3D model is created using high-quality point cloud scans, with object recognition using deep learning, in particular VoxNet. The recognized objects are inserted into the simulation model. In addition to the above example, there are cross-domain use cases in architecture for deriving building information models from point clouds [8, 10].

The creation and maintenance of a brownfield simulation models requires an automated process that does not depend on design information. Furthermore, it must not disrupt ongoing operations or even lead to an extended shutdown. The next section therefore presents a concept for the automated creation of a simulation from videos and images.

3 3D Simulation Experiments with Video Images

In order to provide a simple and cost-effective alternative to conventional methods of simulation modeling, this paper proposes the creation of a simulation model from video data. The main benefit lies in the creation and application of simulation models for brownfield machines. For this purpose, a real existing robot is recorded with video cameras. 3D-reconstruction methods such as photogrammetry and NeRF are used to generate 3D models from the video data. These 3D models are then compared with existing premade models to automatically generate a simulation.

The approach presented in this paper is an attempt to automatically create a simulation model of a robotic cell. The goal of this work is to investigate the extent to which the various methods are suitable for generating models from video data that can be visualized in a simulation. For the purpose of this investigation, a single robot is selected as the test object, with the objective of generating its meshes through the utilisation of 3D-reconstruction methods. The generation of meshes serves as an exemplary initial step in the creation of a simulation model of the robotic cell. In the subsequent step, these meshes are compared with the manufacturer's CAD data to ascertain the sufficiency of the meshes generated from video data for utilization. The objective is to achieve a high degree of agreement with the CAD data. This step may be omitted in future applications. An alternative approach would be to conduct an automated comparison with CAD data from simulation libraries, thereby integrating existing simulation components into a simulation instead of the generated meshes.

First, the robot is moved to a position corresponding to the existing CAD. Then a camera is used to record video at 1920×1080 pixels. Both steps are done manually. In the next step the background is automatically removed from the resulting images using CarveKit.[2] The result is shown as an example in Fig. 1a. The processed images are fed into pipelines to generate meshes with photogrammetry and NeRF. To generate meshes from photogrammetry, Meshroom[3] and Colmap[4] are used, see Fig. 1b and c. To generate meshes using NeRF, Nerf2Mesh[5] and Nerfacto[6] are used, see Fig. 1d and e. Nerfstudio and the Sugar extension were used to generate meshes with Gaussian splatting, see Fig. 1f.

Once the various meshes have been created, they are compared with the robot's CAD data. This was done using CloudCompare.[7] The pipelines have generated a set of meshes with varying dimensions and orientations. In order to facilitate a comparative analysis, it is necessary to register the mesh and the CAD data in relation to each other. This process is conducted in multiple stages. Initially, a preliminary registration is performed manually

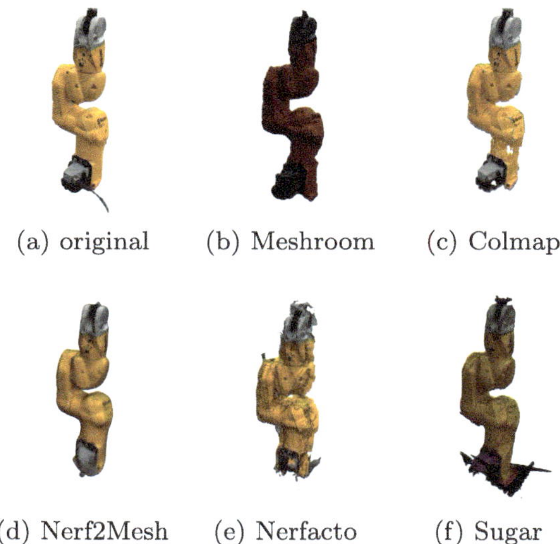

Fig. 1 Comparison of the original image with the differently generated meshes

[2] https://github.com/OPHoperHPO/image-background-remove-tool.

[3] https://github.com/alicevision/Meshroom.

[4] https://colmap.github.io/.

[5] https://github.com/ashawkey/nerf2mesh.

[6] https://docs.nerf.studio/nerfology/methods/nerfacto.html.

[7] https://www.cloudcompare.org/presentation.html.

by marking five points on each object. These points are then automatically aligned with precision. Subsequently, all existing points are incorporated into the registration procedure, enabling an automated fine adjustment. Upon completion of the registration of the objects, the comparison is also conducted automatically. The results, shown graphically in Fig. 2, show the deviation of the reconstruction from the reference model using a color spectrum. Green represents a high level of agreement, blue a negative deviation and red a high positive deviation. The color distribution refers to the maximum deviations in each image.

Figure 3 presents a comparison of the distances between the points of the generated meshes of the methods used for automatic mesh generation and the original CAD data. A comprehensive examination of the data indicates that a considerable number of data points, which are not visible in Fig. 1, are distributed in a cloud-like formation around the generated robot. The majority of these data points can be identified as outliers (indicated in red) in Fig. 3. The identification of outliers is conducted through the application of the interquartile range (IQR) method. The presence of outliers results in a distortion of the comparison result with the CAD data. Therefore, it is not possible to determine which of the methods used produces the most accurate results when creating the meshes. Additionally, a fully automated comparison may be challenging to achieve, as a preliminary positioning using bounding boxes, positioned around all existing data points, could be erroneous. Consequently, an automated removal of the data points based on their Euclidean distance is necessary. However, there is a risk that intentional data points will be incorrectly identified as outliers. Nevertheless, the tests conducted demonstrate that it is generally feasible to utilize photogrammetry, NeRF, and Gaussian splatting to automatically generate meshes that can be recognized by matching algorithms.

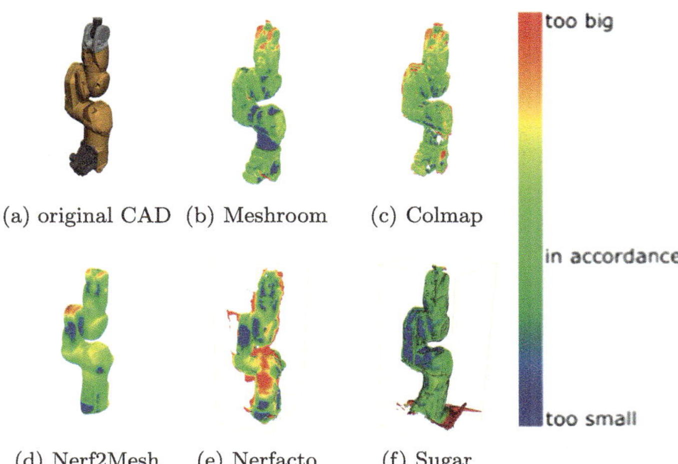

Fig. 2 Comparison of the original CAD with the differently generated meshes using CloudCompare

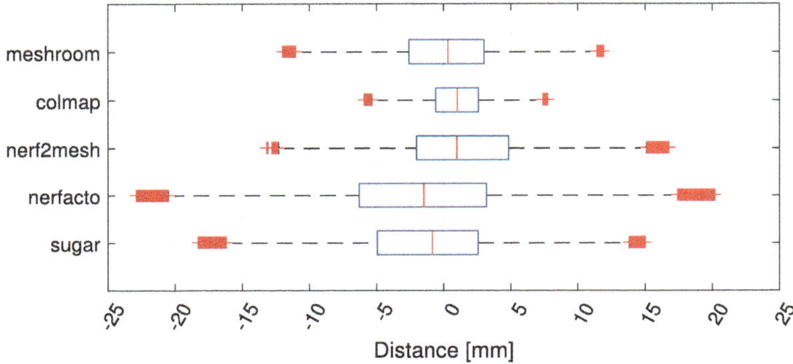

Fig. 3 Comparison of the points of the generated meshes with the points of the CAD model

It is now possible to automatically integrate an object into an existing simulation, be it a generated mesh of a previously unknown object or a corresponding element from a simulation library. Not only is it easier to create simulations models this way, but the deviation of a simulation from the real system can also be reduced more quickly in a semi-automated process.

4 Future Research and Conclusion

Since the robot to be modeled is not necessarily in a position that corresponds to a specific CAD model during the video recording, algorithms must be investigated that allow the generated models to be segmented. In this way, different robot positions can be compared with the CAD data, thereby increasing the probability of a match. It is also necessary to perform further research in order to improve the quality of the generated models. This necessitates a more detailed investigation of the methods used in this work. Further research is also required to assess the efficacy of the methods on various surfaces. In addition to testing the methods on a matte surface, it is essential to evaluate its performance on reflective or transparent surfaces, where notable differences are expected in the methods used for 3D-reconstruction. To this end, more machines will be digitized using these methods, up to and including a complete test bed. This will make it possible to better integrate unknown objects or a large number of different objects into the simulation.

This work shows the suitability of digitizing industrial robots using selected 3D-reconstruction methods based on existing concepts such as Gaussian splatting, NeRF or photogrammetry. This opens up the possibility of simple and fast generation of simulation models using widely available camera technology. The method has the potential to enhance complex and costly methods such as laser scanning or completely manual generation.

Acknowledgements The authors would like to thank the Federal Ministry for Economic Affairs and Climate Action (BMWK) for funding the joint project: SDM4FZI as part of the "Future Investments in the Automotive Industry" funding program.

References

1. Arroyo, E., Hoernicke, M., Rodríguez, P., Fay, A.: Automatic derivation of qualitative plant simulation models from legacy piping and instrumentation diagrams. Comput. & Chem. Eng. **92**, 112–132 (2016)
2. Barth, M., Fay, A.: Automated generation of simulation models for control code tests. Control Eng. Pract. **21**(2), 218–230 (2013). https://doi.org/10.1016/j.conengprac.2012.09.022
3. Bergmann, S., Strassburger, S.: Challenges for the automatic generation of simulation models for production systems. In: Proceedings of the 2010 Summer Computer Simulation Conference, pp. 545–549 (2010)
4. Conti, M., Bendriem, B., Casey, M., Chen, M., Kehren, F., Michel, C., Panin, V.: First experimental results of time-of-flight reconstruction on an LSO PET scanner. Phys. Med. & Biol. **50**(19), 4507 (2005)
5. Forlani, G., Roncella, R., Nardinocchi, C.: Where is photogrammetry heading to? State of the art and trends. Rendiconti Lincei **26**(S1), 85–96 (2015). https://doi.org/10.1007/s12210-015-0381-x
6. Fowler, J.W., Rose, O.: Grand challenges in modeling and simulation of complex manufacturing systems. Simulation **80**(9), 469–476 (2004). https://doi.org/10.1177/0037549704044324
7. Gutenschwager, K., Rabe, M., Spieckermann, S., Wenzel, S. (eds.): Simulation in Produktion und Logistik: Grundlagen und Anwendungen. Springer, Berlin, Heidelberg (2017)
8. Hachisuka, S., Tono, A., Fisher, M.: Harbingers of NeRF-to-BIM: a case study of semantic segmentation on building structure with neural radiance fields. In: Computing in Construction. European Council for Computing in Construction (2023). https://doi.org/10.35490/ec3.2023.284
9. Härle, C.: Automatisierte Generierung, Adaption und Rekonfiguration von Co-Simulationen für modulare Produktionsanlagen. Karlsruher Institut für Technologie (KIT) (2023). https://doi.org/10.5445/IR/1000154191
10. Hellmuth, R., Wehner, F., Giannakidis, A.: Approach for an update method for digital factory models. Procedia CIRP **93**, 280–285 (2020)
11. Kaiser, B., Littfinski, D., Verl, A.: Automatic generation of digital twin models for simulation of reconfigurable robotic fabrication systems for timber prefabrication. In: Proceedings of the 38th International Symposium on Automation and Robotics in Construction (ISARC). IAARC (2021). https://doi.org/10.22260/isarc2021/0097
12. Kaiser, B., Reichle, A., Verl, A.: Model-based automatic generation of digital twin models for the simulation of reconfigurable manufacturing systems for timber construction. Procedia CIRP **107**, 387–392 (2022). https://doi.org/10.1016/j.procir.2022.04.063
13. Kufner, A.: Automatisierte erstellung von maschinenmodellen für die hardware-in-the-loop-simulation von montagemaschinen. Ph.D. thesis, Universität Stuttgart (2012). https://doi.org/10.18419/opus-6786
14. Matsuki, H., Murai, R., Kelly, P.H.J., Davison, A.J.: Gaussian splatting slam. https://doi.org/10.48550/arXiv.2312.06741

15. Melcher, D., Küster, B., Stonis, M., Overmeyer, L.: Optimierung von fabrikplanungsprozessen durch drohneneinsatz und automatisierte layoutdigitalisierung. Logist. J. Proc. **2018**(01), (2018)
16. Mildenhall, B., Srinivasan, P.P., Tancik, M., Barron, J.T., Ramamoorthi, R., Ng, R.: Nerf: representing scenes as neural radiance fields for view synthesis. arXiv:2003.08934v2
17. Mittelstand-Digital: Mehrwerte aus daten: Potenziale und handlungsoptionen für den mittelstand (2019). https://www.mittelstand-digital.de/MD/Redaktion/DE/Publikationen/mehrwerte-aus-daten.pdf
18. Oppelt, M., Wolf, G., Barth, M., Urbas, L.: Simulation im lebenszyklus einer prozessanlage. atp magazin **57**(09), 46 (2015). https://doi.org/10.17560/atp.v57i09.526
19. Puntel Schmidt, P.: Methoden zur simulationsbasierten absicherung von steuerungscode fertigungstechnischer anlagen: Methoden zur simulationsbasierten absicherung von steuerungscode fertigungstechnischer anlagen. Ph.D. thesis, Helmut-Schmidt-Universität Hamburg (2017). https://doi.org/10.24405/4182
20. Scheifele, S., Verl, A.: Automated control system generation out of the virtual machine. Procedia Technol. **26**, 349–356 (2016). https://doi.org/10.1016/j.protcy.2016.08.045
21. Schindler, M.: System und methode zur planung von produktionssystemen auf basis der 3d-digitalisierung bestehender strukturen mit farbinformation. Ph.D. thesis, Universität Stuttgart (2020)
22. Seitz, S.M., Curless, B., Diebel, J., Scharstein, D., Szeliski, R.: A comparison and evaluation of multi-view stereo reconstruction algorithms. In: 2006 IEEE Computer Society Conference on Computer Vision and Pattern Recognition (CVPR'06), pp. 519–528. IEEE
23. Silva, F., Gamarra, C.J., Araujo Jr, A.H., Leonardo, J.: Product lifecycle management, digital factory and virtual commissioning: analysis of these concepts as a new tool of lean thinking. In: Proceedings of the 2015 International Conference on Industrial Engineering and Operations Management, pp. 3–5. Dubai, United Arab Emirates (2015)
24. Sommer, M., Stjepandić, J., Stobrawa, S., von Soden, M.: Automatic generation of digital twin based on scanning and object recognition. In: Transdisciplinary Engineering for Complex Socio-Technical Systems, pp. 645–654. IOS Press (2019). https://doi.org/10.3233/ATDE190174
25. Stich, P.M.A.G.: Interaktiver simulationsgestützter entwurf mechatronischer verarbeitungssysteme. Ph.D. thesis, Technische Universität München (2017). https://mediatum.ub.tum.de/1352936
26. Tinsel, E.F., Riedel, O.: A virtual assistant system for the requirements elicitation and initial simulation model variant generation of modular industrial plants. In: 2022 28th International Conference on Mechatronics and Machine Vision in Practice (M2VIP), pp. 1–4. IEEE, Piscataway, NJ (2022). https://doi.org/10.1109/M2VIP55626.2022.10041099
27. VDI-Gesellschaft Produktion und Logistik: Die digitale fabrik - treiber der digitalen transformation (2023-02). https://www.vdi.de/ueber-uns/presse/publikationen/details/die-digitale-fabrik-treiber-der-digitalen-transformtion
28. Verband Deutscher Maschinen- und Anlagenbau e.V.: Leitfaden virtuelle inbetriebnahme: Handlungsempfehlungen zum wirtschaftlichen einstieg (2020)
29. Wünsch, G.: Methoden für die virtuelle Inbetriebnahme automatisierter Produktionssysteme. Utz (2008)
30. Zhang, S.: High-speed 3d shape measurement with structured light methods: a review. Opt. Lasers Eng. **106**, 119–131 (2018)

Simulation-Based Investigation of the Feasibility of Robotic Machining of On-Wing Engine Components

Vincent Ahrens, Lukas Bath, and Thorsten Schüppstuhl

Abstract

Foreign object damage on the compressor blades of a turbofan engine reduces engine efficiency and compromises safety. This necessitates to repair such defects as soon as possible. As it is time-consuming and costly to disassemble the entire engine to access the compressor blades, it can be advantageous to perform such repairs on-wing, while the engine is still assembled. The main challenge in machining compressor blades on-wing is the limited accessibility within the engine, which precludes the use of traditional machining tools and demands the development of new solutions. To address these obstacles, this work proposes a flexible simulation framework including a shape optimization technique to aid in the development of suitable robotic kinematics for space-constrained machining tasks. Further simulation functionalities such as an interactive mounting simulation foster detailed feasibility studies of kinematic concepts.

Keywords

Simulation • Robotic machining • On-wing • MRO • Foreign object damage

V. Ahrens (✉) · L. Bath · T. Schüppstuhl
Institute of Aircraft Production Technology, Hamburg University of Technology, Hamburg, Germany
e-mail: vincent.ahrens@tuhh.de

L. Bath
e-mail: lukas.bath@tuhh.de

T. Schüppstuhl
e-mail: schueppstuhl@tuhh.de

M.-C. Wanner et al. (eds.), *Annals of Scientific Society for Assembly, Handling and Industrial Robotics 2024*, https://doi.org/10.1007/978-3-031-91463-8_30

1 Introduction

Defects on compressor blades of turbine engines can pose significant safety risks and reduce operational efficiency. Repairing these defects on-wing, while the engine is still assembled, promises to reduce costs while enhancing safety and efficiency. The first kind of prevalent defects addressed in this work includes nicks and dents on the leading edge (Fig. 1 left), caused by the impact of larger foreign object debris such as birds, ice or hail [1]. They can cause instability in the airflow, stress concentration, and potential crack initiation, which may ultimately result in rapid fatigue-induced failure [2]. Another relevant type of defect is erosion caused by the impact of fine particles such as sand and dust (Fig. 1 right), which blunts the leading edge, degrades its surface and results in reduced engine efficiency [3]. These defects are particularly pronounced in the first stage of the high pressure compressor, making it a focal point for repair efforts.

Blending and recontouring are crucial processes for addressing these defects. Blending (Fig. 2a) involves smoothing and rounding nicks and dents to prevent crack initiation and impede crack propagation. This is typically achieved using a rotary grinding tool. Although manual on-wing blending is currently practiced, ongoing research is exploring automated solutions [4]. Recontouring (Fig. 2b), on the other hand, entails grinding and reshaping blunted leading edges to restore their aerodynamic profile, with the aim of improving engine efficiency and longevity. This process also causes a minor loss of chord length, which is neglectable. Recontouring is currently not possible on-wing and only performed on removed blades.

Performing such repair processes on-wing introduces several challenges, primarily due to space constraints within the engine. This applies especially if robots shall be used for the machining operations to improve quality and save time trough automation. These challenges include inserting a grinding tool into the engine, reaching the required poses on the blades with the grinding tool and, if a robot is inserted into the engine, actuating it with size

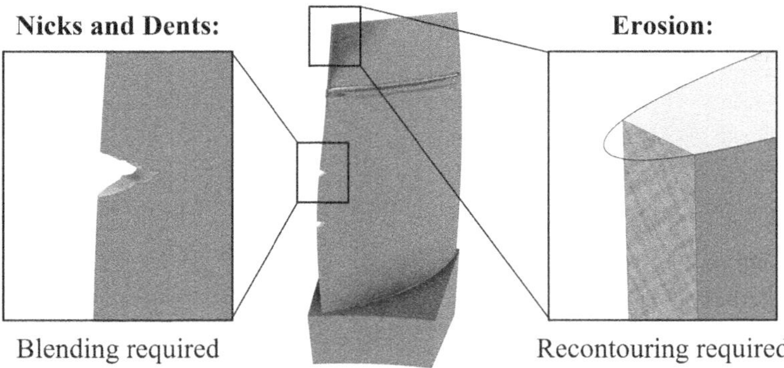

Fig. 1 Blade damage

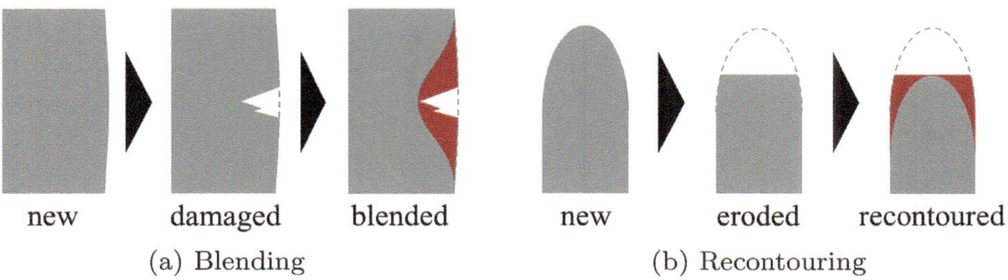

Fig. 2 Blade repair processes

limited actuators. Given the limitations of real-world testing and the resource-intensive nature of prototype development, simulations emerge as promising tools for addressing these challenges. This paper introduces a simulation process tailored to the development and evaluation of space-restricted machining solutions, addressing collision and design constraints. The simulation process is tested using two use-cases of innovative concepts for both on-wing blending and recontouring through bypass valves in order to reach to the blades of the high pressure compressor.

2 Related Work

This section provides an overview of the current state of research concerning on-wing blending, covering both manual and automatic blending methods, as well as the emerging use of continuum robotics for on-wing blending tasks. The complex tasks to be carried out in combination with the highly constrained access possibilities to the inside of the engines result in the very little number of known approaches. Especially for on-wing recontouring there are no known solutions.

2.1 Boreblending

Currently, on-wing blending is carried out manually, involving skilled technicians and is limited to the use of specialized handheld grinding tools developed by Richard Wolf [5] or Schoelly Fiberoptic [6] that are inserted into the engine through narrow inspection ports, to blend and smooth out defects on compressor blades. This operation is called boreblending. While it has been effective to some extent, it is labor-intensive, time-consuming, and may lack consistency in results. In recent years, efforts have been directed towards automating the blending process, leading to the development of automatic boreblending systems [4]. These systems utilize robotic machining techniques to automate the blending process, offering potential advantages in terms of efficiency, accuracy and repeatability compared to manual methods.

2.2 Continuum Robotics

The emerging field of continuum robotics presents a novel approach to on-wing blending tasks. Continuum robots, characterized by their long, slender, snake-like structures and high degrees of freedom, offer unique advantages in accessing confined spaces within the engine. By using the redundant degrees of freedom they can maneuver around complex geometries to reach damaged areas on compressor blades. The current research efforts are mostly focused on performing blending in the intermediate pressure compressor, through the main engine inlet [7]. While these solutions promise to perform automated blending on-wing they are not currently able to recontour the leading edges of compressor blades.

3 Simulation for Concept Development and Evaluation

The aim of the simulation is to aid in the development and evaluation of specialised machining solutions, with a particular focus on addressing the aforementioned challenges. The following chapter discusses the fundamental concept of the simulation process and its objectives. Afterwards, the individual simulation functionalities are described.

3.1 Simulation-Based Design Process

In order to develop new concepts for on-wing machining solutions, the design process must be able to evaluate whether the concepts can overcome the key challenges that hinder the on-wing blending and recontouring of compressor blades in fully assembled aircraft engines. These challenges include inserting the tool into the engine, reaching the required workspace with the tool, and, if a robot is inserted into the engine, actuating it inside of the confined space. In addition to three simulation functionalities implemented to assess these properties, an additional functionality is integrated into the design process, as shown in Fig. 3, to generate collision-optimized shapes for robot components. The simulation environment CoppeliaSim [8] was chosen for its stability and wide range of features based on a quantitative comparison of robot simulation tools [9], to provide supporting functionalities and modeling capabilities.

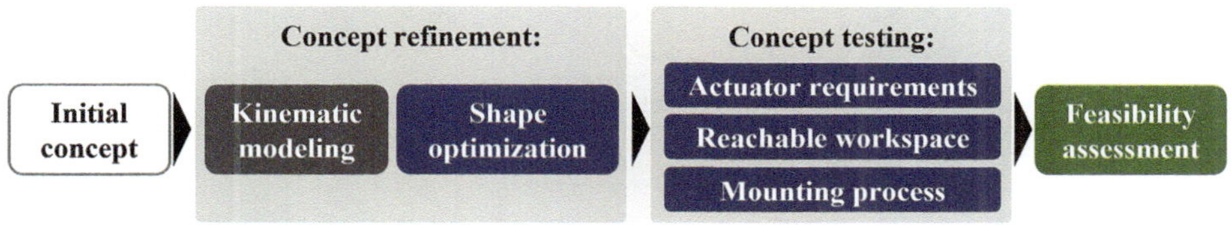

Fig. 3 Process for simulation-based design and testing of kinematic concepts

3.2 Calculation of Actuator Requirements

As the selection of actuators is considerably limited by the restricted space inside the engine it is crucial to ensure that actuators with the required performance are technologically feasible. For this, the required joint speed and torque are determined for each joint on every reachable tool pose in the workspace. The joint speed is calculated using the tool speed and the inverse of the Jacobian provided by CoppeliaSim for each robot configuration. To determine the required joint torque based on the process force and the weight of the robot components the physics engine Bullet [10] is used. It was chosen for its stability and special ability to deal with collisions of concave bodies in the simulation of the mounting process.

3.3 Workspace Evaluation

The different machining operations necessitate reaching various areas of the compressor blades. Recontouring the leading edge requires precise tracking of its shape and orientation, while blending requires reaching both the leading edge and the area immediately behind it, albeit tolerating a less restricted orientation. Given the confined space within the engine and the precision required by the machining process, forward kinematic methods are too computationally intensive due to frequent collisions and are thus not suitable for determining the robot's workspace. An inverse kinematic approach is therefore chosen and presented in Fig. 4. For this, discrete target poses are defined which describe the task space on the blade that needs to be reached in order to perform blending and recontouring. For each target pose, an inverse kinematic search for joint values is performed using the pseudoinverse method. If suitable joint values are found, they are applied to the robot. Collision checks are then performed between the robot and the engine. If collisions occur, the inverse kinematic search is repeated until a collision-free configuration is found or a predefined maximum number of iterations n is reached.

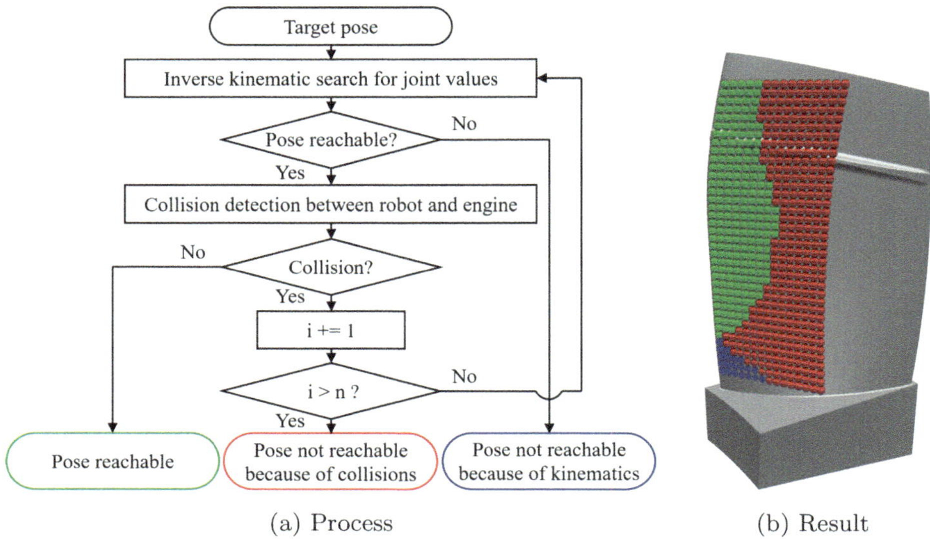

(a) Process (b) Result

Fig. 4 Discrete evaluation of the workspace's reachability

3.4 Shape Optimization of Robot Components

In order to perform the machining tasks, the robot must be able to move inside the confines of the engine. This requires components to be shaped in a way that balances sufficient stiffness with the ability to maneuver without colliding with the engine's structure. Achieving this balance often necessitates highly intricate shapes that exceed the capabilities of conventional design techniques. A novel approach to shape optimization tailored to restricted work environments is proposed to address this challenge. The proposed methodology generates collision-optimized shapes for robot components by considering the required tool poses, an initial search volume representing approximately the eventual moving component of interest and the collision geometry. The search volume is defined as a cuboid that is evenly discretised into voxels. The tool poses required for the various machining operations are then defined as a multitude of target poses on the compressor blade. Based on these, the poses for all robot components can be determined via the inverse kinematics. The cuboid with the voxels is then iteratively transformed in place of the robot component to be designed. For each of the poses every voxel is checked for collision with the engine. Finally, any voxel that has been subject to a collision is removed from the cuboid. The resulting shape (shown in blue in Fig. 5) represents the volume that the final robot component to be designed must not exceed in order to operate free of collisions. This allows the designer to judge the achievable cross-sections of collision free robot components to easily achieve good geometric compromises.

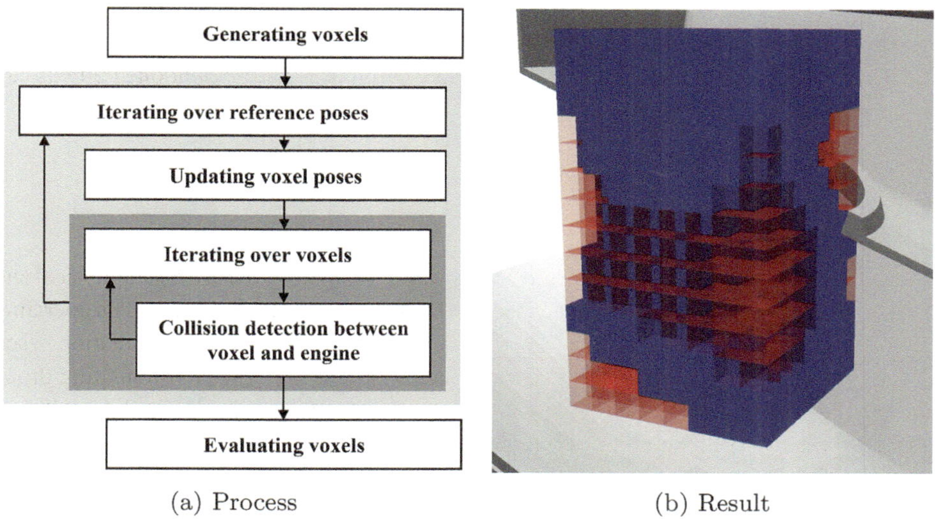

Fig. 5 Voxel-based creation of collision-optimized components

3.5 Simulation of the Mounting Process

Due to the limited space within the engine, it is difficult to determine whether a robot can be inserted. To overcome this challenge, an interactive simulation approach, illustrated in Fig. 6 has been proposed. This simulation allows the user to plan and test the mounting process virtually, without the need for a physical prototype. At the beginning of the simulation, the robot is positioned outside the engine. By using a 6-degree-of-freedom input device, the user can control the robot's movement and steer it into the engine. Collisions between the robot and the engine are continuously monitored throughout the process. The simulation's physics engine enforces collision responses, ensuring that insertion paths that are not possible in the real world are also not possible in the simulation. Through its visualization, intuitive control mechanisms, and collision enforcement, this simulation enables users to explore insertion strategies intuitively, without the need for access to an engine or robot prototypes. An example of such an insertion strategy is shown in the centre of Fig. 6 as an overlay of all robot poses along the insertion path.

4 Trial of the Design Process

The design process is tested using two exemplary use-cases of machining solution concepts. These use-cases address deficits in current solutions for on-wing machining which are limited in their ability to perform recontouring operations on the first stage of the high

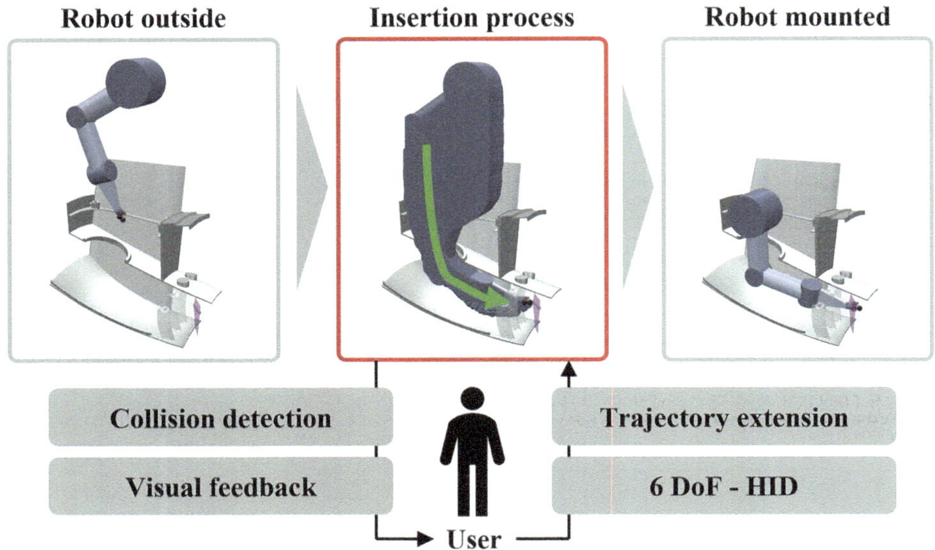

Fig. 6 Simulation of mounting process

pressure compressor. The restricted entryway, through inspection ports or the main air inlet, significantly hinders the machining operation. Therefore, an alternative entryway through the bypass valve is used for the two use-cases presented below. The use-cases were selected to provide contrasting examples and do not represent the full range of potential solutions.

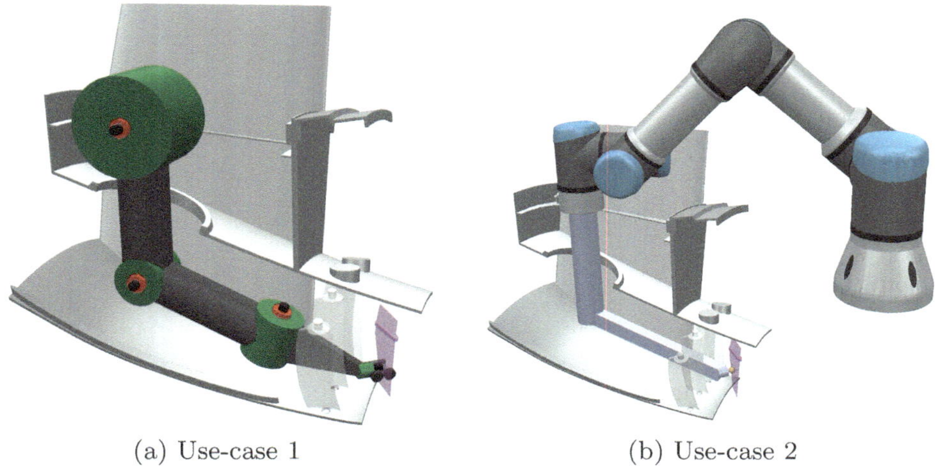

Fig. 7 Use-cases for the simulation-based design process

4.1 Use-Case 1

Use-case 1 places the robot and most of its joints inside the engine. The implementation, as shown in Fig. 7a, aims to perform the machining operation with only three degrees of freedom. This reduces the design's complexity but also limits control over the tool's orientation relative to the blade. In assessing the feasibility of the first use-case, the simulation plays a crucial role, particularly in evaluating whether the actuators are small enough to fit inside of the engine while providing the required speed and torque. The simulation results in Fig. 9 promise the feasibility of use-case 1. Suitable servo drives can be found that fit into the limited space, provide the required maximum torque $\tau_{max} = 1.17\ Nm$ and the required angular speed $\dot{q}_{max} = 0.73\ \frac{rad}{s}$ (Fig. 9), respectively. Most of the target workspace can be reached without collisions and the mounting process is also feasible.

4.2 Use-Case 2

In contrast to use-case 1, the concept of use-case 2 places the robot and its joints outside of the engine. The advantage of this approach is that the size of the robot is not limited by the engine and standard industrial robots can be used to perform the machining. The problem with this approach is that a rigid extension is required to connect the tool inside the engine to the robot on the outside. This extension makes the tool position sensitive to orientation errors from the robot's joints. Furthermore, a specific extension must be designed for every engine type. As the initial design for the extension (shown in Fig. 7b) was prone to collisions and significantly limited the reachable workspace on the blade, the shape optimization was used to improve the shape of the extension. In Fig. 8a the result of the shape optimization is shown and in Fig. 8b the final geometry manually derived from it. It can be seen where material had to be removed from the original shape to avoid collisions with the valve rim and compressor vanes. As in use-case 1, the results in Fig. 9 for use-case 2 also promise feasibility. In this case, a small industrial robot can be used to provide sufficient actuation capabilities and the improved extension can be inserted into the engine and reaches the majority of the target workspace.

5 Conclusion and Future Work

In this work, multiple simulation functionalities were developed to assess the feasibility of on-wing machining solutions and to aid in their development. Even for diverse engine types that lack of geometric synergies between them this can help to develop highly individualized tools without significant additional effort. The proposed design process was applied to two use-cases to evaluate their potential feasibility in regard to real-world applications and to test the simulation-based design process. A notable contribution of this work is a novel

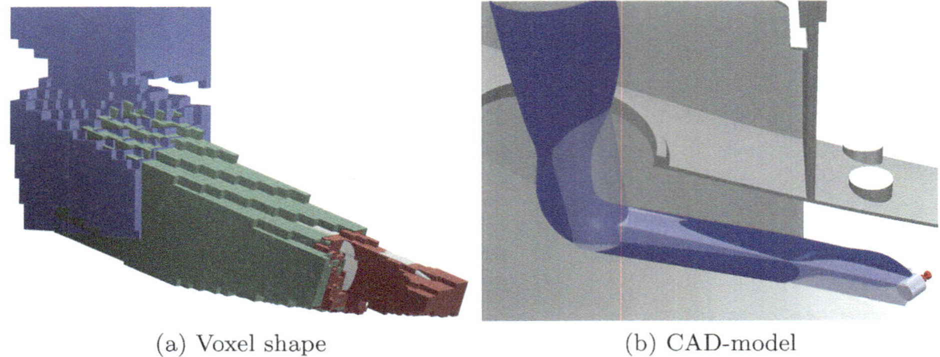

(a) Voxel shape (b) CAD-model

Fig. 8 Design of robot extension

Concept	Actuator requirement	Reachable workspace	Insertion path
1 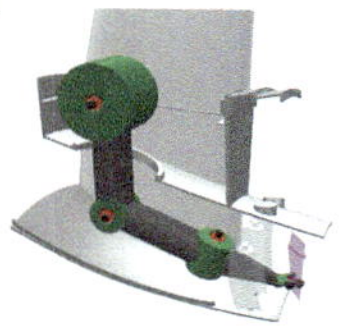	$\dot{q}_{max} = 0.73 \frac{rad}{s}$ $\tau_{max} = 1.17\ Nm$		
2 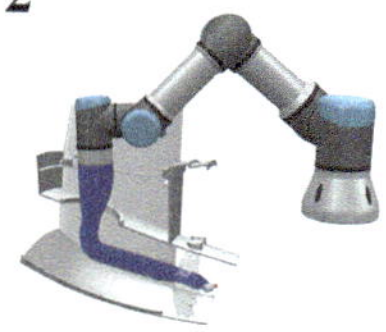	$\dot{q}_{max} = 0.03 \frac{rad}{s}$ $\tau_{max} = 23.21\ Nm$	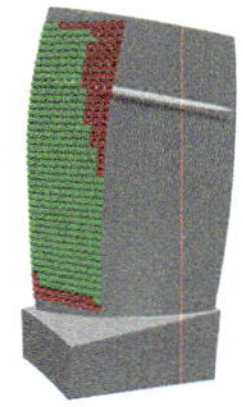	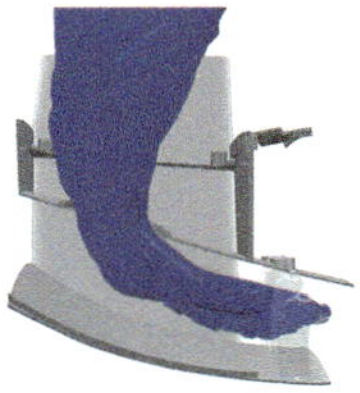

Fig. 9 Simulation results to assess the feasibility of the concepts

shape optimization technique, enabling the design of components that would be challenging to create without this approach. This optimization significantly enhances the performance of the machining solution in use-case 2. Additionally, the interactive mounting simulation was vital in assessing the feasibility without the need for physical prototypes. The results of the simulations conducted so far are promising, indicating that both use-case concepts have potential for addressing the challenges of on-wing machining. However, further investigation is necessary to validate their performance under real-world conditions. In future work, functionalities that not only assess feasibility but also evaluate performance could be implemented. This requires modeling the machining task in greater detail, incorporating accurate tool trajectories, and a realistic representation of process forces. This research lays the groundwork for future advancements in on-wing machining, potentially enhancing aircraft maintenance and improving operational efficiency.

References

1. Aust, J., Pons, D.: Taxonomy of gas turbine blade defects. Aerospace **6**(5), 58 (2019)
2. Carter, T.J.: Common failures in gas turbine blades. Eng. Fail. Anal. **12**(2), 237–247 (2005)
3. Shi, L., Guo, S., Yu, P., Zhang, X., Xiong, J.: A review on leading-edge erosion morphology and performance degradation of aero-engine fan and compressor blades. Energies **16**(7), 3068 (2023)
4. Alatorre, D., Nasser, B., Rabani, A., Nagy-Sochacki, A., Dong, X., Axinte, D., Kell, J.: Robotic boreblending: the future of in-situ gas turbine repair, pp. 1401–1406 (2018). https://nottingham-repository.worktribe.com/output/2550844
5. Richard Wolf GmbH: Aerospace applications (2024). https://www.richard-wolf.com/en/solutions/aerospace-applications/
6. Schoelly Fiberoptic GmbH: Engine maintenance (2024). https://www.schoelly.de/en/visual-inspection/technical-endoscopy/areas-of-applications/aviation-and-aerospace/engine-maintenance
7. Dong, X., Axinte, D., Palmer, D., Cobos, S., Raffles, M., Rabani, A., Kell, J.: Development of a slender continuum robotic system for on-wing inspection/repair of gas turbine engines. Robot. Comput. Integr. Manuf. **44**, 218–229 (2017)
8. Rohmer, E., Singh, S.P.N., Freese, M.: V-rep: a versatile and scalable robot simulation framework. In: 2013 IEEE/RSJ International Conference on Intelligent Robots and Systems, pp. 1321–1326 (2013)
9. Farley, A., Wang, J., Marshall, J.: How to pick a mobile robot simulator: a quantitative comparison of CoppeliaSim, Gazebo, MORSE and Webots with a focus on the accuracy of motion simulations. Simul. Model. Pract. Theory **120**, 102629 (2022)
10. Bullet Real-Time Physics Simulation | Home of Bullet and PyBullet: Physics simulation for games, visual effects, robotics and reinforcement learning (2022)

A Simulation Framework for Reinforcement-Learning-Based In-Flight Optimization of UAVs' Acoustics

Andreas Gründer, Christian Hofmann, Julian Benz, Stefan Becker,
Jörg Franke and Sebastian Reitelshöfer

Abstract

Due to the complexity of the physics of unmanned aerial vehicle's (UAV) acoustics, its optimization is a challenging topic. Considering correlations between noise and UAV's flight operations, its acoustics can be optimized by intelligent navigation. In this paper we show an approach to build a simulation environment for simulating acoustic radiation of multicopters. The simulation can be used for training reinforcement learning approaches using physical effects in drone navigation for acoustic optimization. An extension of a given simulation environment based on a drone autopilot and a multi-body simulation program by using the popular robot operating system (ROS2) is proposed and modeling acoustic signals based on an experimental setup is introduced.

Keywords

UAV • Acoustic optimization • Reinforcement learning • PX4-Autopilot • ROS2 • Gazebo simulator

---

A. Gründer (✉) · C. Hofmann · J. Franke · S. Reitelshöfer
Friedrich-Alexander-University Erlangen-Nürnberg, Institute for Factory Automation and Production Systems (FAPS), Erlangen, Germany
e-mail: andreas.gruender@faps.fau.de
URL: https://www.faps.fau.de/

J. Benz · S. Becker
Friedrich-Alexander-University Erlangen-Nürnberg, Institute of Fluid Mechanics (LSTM), Erlangen, Germany

M.-C. Wanner et al. (eds.), *Annals of Scientific Society for Assembly, Handling and Industrial Robotics 2024*, https://doi.org/10.1007/978-3-031-91463-8_31

1 Introduction

The safe use of unmanned aerial vehicles (UAVs) for operational purposes such as intralogistics, parcel delivery or observation is an important research topic today. In addition to the obvious problems such as a drone crash due to a malfunction, which is addressed by Lieret et al. [1], the acoustic effect of drones on people in their vicinity is an annoying impact that is not to be neglected. Considering the current presumed growth of drone market over the next decade [2], there is a need for optimizing the acoustic radiation of drones towards people. Publications can mainly be divided into mechanical optimization methods and noise optimization through drone control.

König et al. address the problem of mechanical noise reduction of drones [3]. Those approaches can be divided into categories such as blade design, rotor configuration, rotor integration and overall vehicle design. The second approach is noise optimization by drone control, in which the noise generation of UAVs is optimized through flight path planning. For instance, Bian et al. [4] and Gao et al. [5] use concepts for noise-aware flight path planning for UAVs by creating a virtual 3D environment of the operational area and using acoustic ray tracing to take reflections of the environment into account. There are also calculations using a finite element method with Computational Aero Acoustics Simulation (FEM CAA), like shown by Dbouk et al. [6]. However, such simulations are very computationally intensive and slow. Therefore they cannot be performed during the flight in a flexible manner. With the increasing power of lightweight computer hardware, a self-learning approach to reduce acoustic radiation of UAVs by Artificial Intelligence-based path finding becomes a topic of interest. A suitable approach for self-learning algorithms is reinforcement learning (RL), in which an agent selects actions from a predefined action space based on environment observation, in order to maximize rewards associated with the observations. There exist several approaches where RL is used to optimize UAV's navigation behavior, for instance for providing drone commands like position setpoints through RL [7] or camera pictures based path finding and obstacle avoidance [8]. Consequently, the combination of both research fields, RL-based noise optimized path-finding, seems beneficial. For this reason, a method of simulating a drone's noise depending on the rotational speeds of it's motors based on experimental data is introduced within this paper. This simulation architecture can be utilized as a RL training-environment for noise self-optimization of UAVs. Figure 1 gives a conceptual overview of the given problem. It shows a drone as a noise source which, in order to disturb a group of people less, independently takes an acoustically more pleasant path to its destination. The physical influences on the drone noise are shown in the green boxes.

The paper is structured as follows: Sect. 2 introduces the proposed simulation architecture for RL-based drone acoustics optimization, Sect. 3 describes a first step in acoustic mapping for drone audio signal simulation and its uncertainties. Finally, Sect. 4 summarizes the paper and provides an outlook on open research questions.

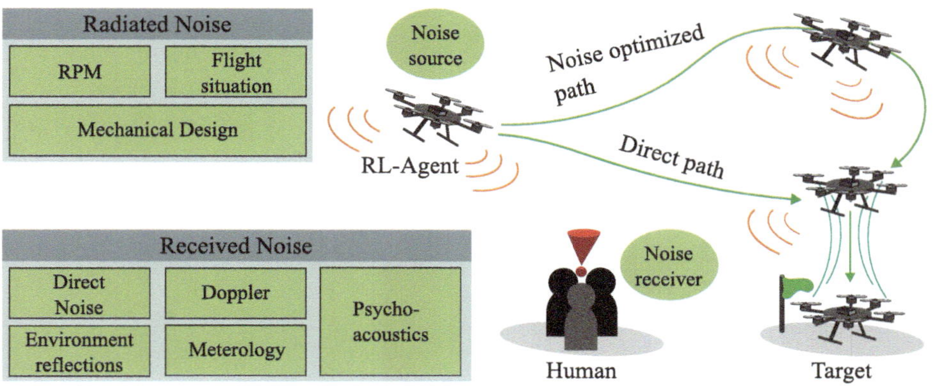

Fig. 1 Scheme of acoustic optimization of UAVs by intelligent path-finding and subdivision of drone noise into radiated noise and human-received noise with its influences

2 Simulation-Architecture

The following section presents the proposed architecture for training RL agents for drone control based acoustic optimization.

We claim that there are six crucial requirements to describe the complexity of drone physics with regard to their in-flight acoustics within a simulation:

1. A Multi-body simulation framework with realistic enough physics of drone kinematics and environmental impacts like wind to take control-based changing of motor rounds per minute (RPM) into account.
2. A Simulation of a complex and realistic drone flight controller unit (FCU) with its sensors and control API to ensure a given transfer-ability to real systems.
3. A Suitable framework with API to the simulation environment to map the complex acoustic behavior in drone operation with frequency characteristics.
4. A quantification of the simulated audio behavior to address psychoacoustic optimization as input for RL-Agent and object of optimization.
5. A RL-Agent with proper chosen reward shaping and simulation-control-program for episodic training to learn acoustic optimized navigation to a target.
6. Experiments as a basis for modeling to ensure behavior close to reality and validation of the simulation environment.

2.1 Simulation Environment with PX4-Autopilot and Gazebo

Today's drones are often controlled by a microcontroller based FCU running a suited control software. In this paper, the PX4-Autopilot [9] is chosen, which is a widely used open-source

Fig. 2 Real Hexcopter system (left) and GUI of gazebo multibody simulation with the simulated model (right)

flight control software for robotic applications and well-supported by the community. The PX4 software comprises a Software in the Loop (SITL) simulation, that can be used with the open-source program Gazebo Simulator [10]. With this simulator, it is also possible to integrate custom models. In this work, the model of a hexcopter is used to provide an RL-Agent redundant motors that can temporarily switch motor operation to lower speed ranges. This gives the agent more flexibility in controlling the UAV's acoustic radiation. The such chosen simulation setup thus comprises a drone model, that matches the real drone system very well according to its kinematic behavior, see Fig. 2.

2.2 Enhancement of the Simulation Environment for Acoustic Optimization

For the automatic computer-controlled flight, signal processing and implementing the RL-Agent, the Robot Operating System 2 (ROS2) [11] Humble is used to set up the robot infrastructure and the communication with the simulated PX4 SITL and Gazebo, cf. Fig. 3. The MAVROS communication bridge is used to send control commands to the drone and to make sensor data from the FCU accessible in ROS2. Additionally a ROS-Gazebo bridge is used for receiving sensor data from the Gazebo Simulator like the RPM of the single motors. All the sensor data is fetched from a data server node and can be accessed by the other parts of the architecture. It contains the kinematic data as observations for the RL-Agent and information about the flight situation (e.g. forward flight, take-off, landing and turn) to map the acoustics of the drone. The simulated audio signal must then be post-processed in order to optimize not only the physical but also the psychoacoustic properties of the drone noise, such as sharpness, loudness, roughness and tonality. The RL-Agent selects actions according to these observations and receives positive rewards for good

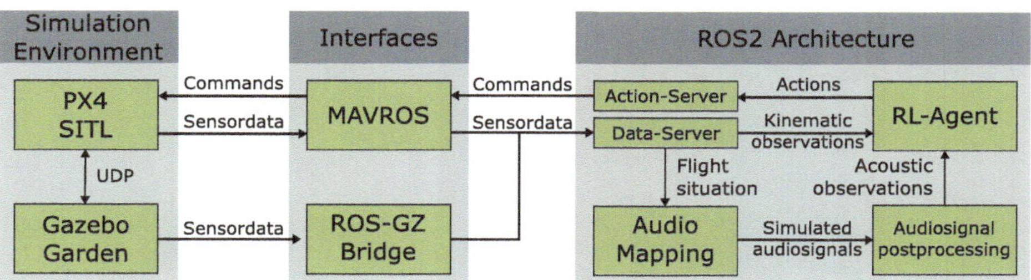

Fig. 3 Simulation framework for Reinforcement-Learning-based inflight acoustic optimization and communication infrastructure between PX4-Autopilot, Gazebo and ROS2

decisions or negative rewards for bad decisions. In this way, it adjusts the probability with which its actions are selected and executed by a action-server. Possible specifications for drone commands can be, for example, an UAV's target position and speed. For RL-training, information about the drone's contact with the environment is also available via the ROS-GZ bridge by a corresponding plugin. For the detection of the collisions that occur during the simulation and to truncate training episodes for the RL-Agent prematurely in order to punish the actions with a strong negative reward. Due to the FCU simulation within the environment, a transfer of the architecture to a real drone system is easily possible by replacing the simulated FCU with a real drone controller and configuring the used interface accordingly.

3 Audio Mapping of Drone Noise

Since, Gazebo does not provide physical audio sources as plugin, an more accurate simulation needs to be investigated to describe the complex audio behavior the UAVs. The acoustic behavior of drones can be seen in two reference systems: the drone as a source of noise radiation and the off-board view of a receiver in a specific distance (cf. Fig. 1). The following chapter describes a method of experimental based modeling of the drone radiation depending on the RPM of the drone motors and possible extensions of this simulation due to uncertainties in the used method to enhance the simulation.

3.1 Experimental Setup for Modeling Rotorblade Acoustic

To model a mapping between the RPM and the sound pressure level (SPL), measurements of a single rotorblade were recorded. As experimental setup, an array of $N = 8$ Microtech Gefell MTG MM210 microphones was build up in an acoustic optimized chamber, with 1 m radius from the microphones to a single BLDC motor (T-MOTOR MN3110, KV780) and a MS1101 11 in. carbon polymer propeller. The microphones were calibrated at 94dB

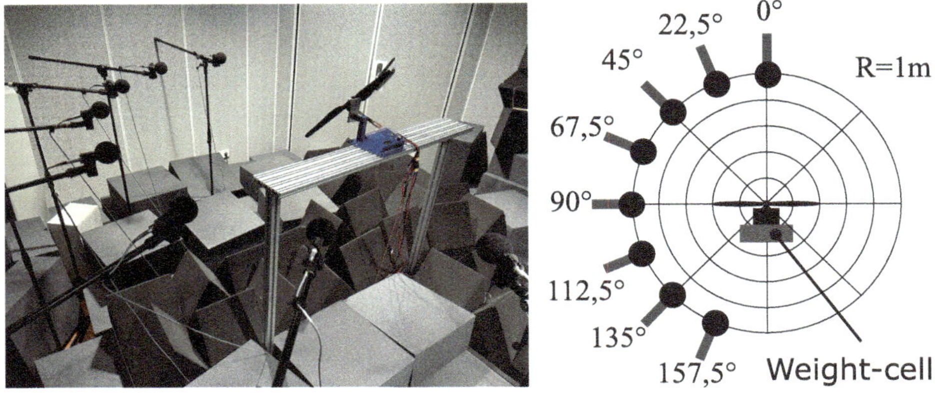

Fig. 4 Experimental setup for measuring the sound pressure depending on the thrust of a motor with rotor-blade in an acoustic optimized chamber (left) and its top-view as a schematic sketch (right)

1000 Hz. For the test-setup a professional microphone protection foam was used to prevent hydrodynamic masking of the acoustic noise signal. The thrust was measured by an weight cell for drones due to rising RPM. The experimental setup is shown in Fig. 4. The microphone positioned at 180o was excluded from the setup due to its direct exposure to the turbulent propeller airflow, which would distort the sound pressure measurements. Since the estimation of the sound pressure level in the simulation is not aimed at the representation of detailed acoustic values, but at the pre-training of RL agents, this simplification is initially sufficient for the application. The sound pressure P_i for each microphone i with $i \in [1, 8]$ has been recorded over 30 s and then averaged in terms of location and time.

The resulting averaged sound pressure P_a can be used to calculate the SPL in decibels with the used reference pressure for air $P_{\text{ref}} = 2 \cdot 10^{-5} Pa$ shown in (1).

$$SPL = 20\log\left(\frac{P_a}{P_{ref}}\right) = 20\log\left(\frac{\sum_{i=1}^{N} P_i}{N \cdot P_{ref}}\right). \tag{1}$$

The relation between RPM and thrust is based on the data sheet for the given configuration of motor, rotor-blade and battery-setup (see Fig. 5 left). The measured average sound pressure level depending on the thrust is based on the described experimental setup (see Fig. 4 right).

To build an easy transition for the RPM into the thrust per rotor, the 2nd order polynomial in (2) describes the real correlation well.

$$T(RPM) = a_2 \cdot RPM^2 + a_1 \cdot RPM + a_0. \tag{2}$$

Similar an exponential function for the description of the thrust-SPL behavior can be written as in (3).

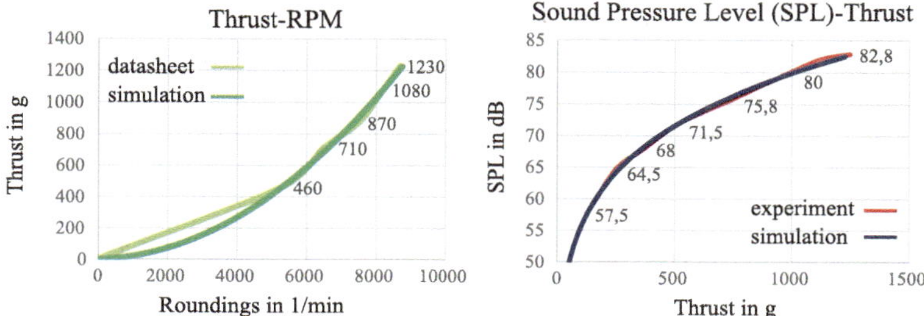

Fig. 5 Mapping diagrams for data sheet based RPM-Thrust (left), experimental based Thrust-Sound pressure level (right) and their calculation

Table 1 Hyperparameter for acoustic mapping

Parameters of (2)	Value	Parameters of (3)	Value
a_2	$1.5529 \cdot 10^{-5}\text{g} \cdot \text{min}^2$	α	$26.741\,\frac{\text{db}}{\text{g}}$
a_1	$0.4416 \cdot 10^{-2}\text{g} \cdot \text{min}$	β	0.1583
a_0	0.7827g		

$$SPL_{\text{simulation}}(T) = \alpha \cdot T^{\beta}. \tag{3}$$

The parameters of both equations listed in Table 1 were determined with the aid of a compensation function using a tabular calculation program.

The functions are implemented in a Audio Mapping node which takes the RPM from the ROS-Gazebo bridge as input and returns the sound pressure level in decibels. Figure 6 shows the simulation of the SPL for the launch of the UAV for a single motor. The minimum SPL is constant at 25 dB from $t \in [0, t_1]$ to represent environmental noise. Withing $t \in [t_1, t_2]$ a

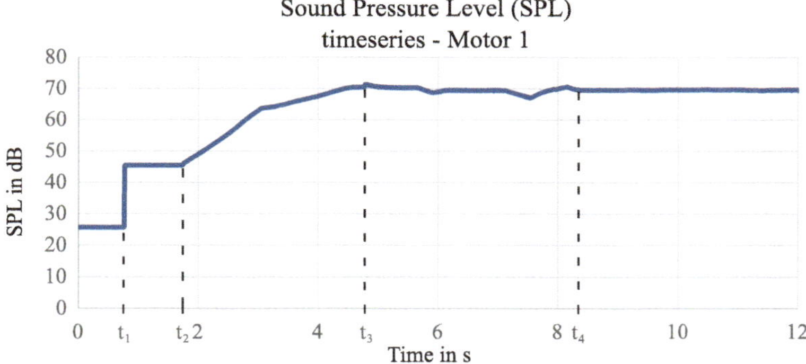

Fig. 6 Simulation of SPL while UAV takeoff for a single motor

step to 46 dB is visible due to the transient to the rotor spinning with desired RPM. From $t \in [t_2, t_3]$ the drone starts lifting from the ground to reach it's supposed height at t_3. After the supposed setpoint is reached, a stabilization due to transient behavior is visible within $t \in [t_3, t_4]$. After this, the drone hovers in its position and there are just small changes in the SPL caused by positional corrections due to simulated noise of the sensors, actuators and environmental conditions.

3.2 Uncertainties in Building the Simulation Model

The previously described mapping as acoustic source by its sound amplitude only depending on the RPM is not sufficient to adequately simulate the complexity of drone acoustics. There are many additional effects in the sound development that must be investigated in the future and included in the simulation environment. One of the most important effects is the flight situation, where the noise development changes significantly. According to [12], there are other effects like the dependence on meteorological data such as air pressure and humidity, air absorption, turbulence effects, background noise, Doppler frequency shifts, ground effect and more. These effects add probabilistic characteristics to the simulation, including a frequency spectrum of the simulated acoustic signals, which can be used for psycho-acoustic optimization in future research. In addition, it will be important not only to simulate the sound radiation from the drone, but individual receiver points in the vicinity of the drone, like in [13], to optimize these points by intelligent control. With the introduced concept and simulation environment, an implementation of the aforementioned effects can be added with low effort if needed.

4 Conclusion

The acoustics of UAVs is a complex field as it is highly dependent on non-linear physical and aerodynamic effects that depend on the mechanical properties of the system and its operating points. Since such a simulation was not previously available, the simulated radiation based on the mapping of the RPM to the SPL was taken as a first step towards simulating drone acoustics in a multi-body simulation framework to optimize the radiated noise by intelligent drone control. The simulation environment was successfully tested on the basis of a flight situation, the take-off. In view of the model uncertainties described in the last chapter, the simulation environment will be investigated and improved in future research work using data from a real drone equipped with onboard microphones and an offboard receiver as a test setup. By applying different RL strategies, the simulation environment will lead to a self-learning algorithm for acoustic optimization with independence from the vehicle mechanics.

References

1. Lieret, M., Fertsch, J., Franke, J.: Fault detection for autonomous multirotors using a redundant flight control architecture. In: 2020 IEEE 16th International Conference on Automation Science and Engineering (CASE). pp. 29–34. IEEE (2020)
2. Ed Alvarado: Drone market analysis 2022-2030 (2022). https://droneii.com/drone-market-analysis-2022-2030
3. Koenig, R., Foell, M., Stumpf, E.: Potentials for acoustic optimization of electric aerial vehicles
4. Bian, H., Tan, Q., Zhong, S., Zhang, X.: Assessment of uam and drone noise impact on the environment based on virtual flights. Aerosp. Sci. Technol. **118**, 106996 (2021)
5. Gao, Z., Porcayo, A., Clarke, J.P.: Developing virtual acoustic terrain for urban air mobility trajectory planning. Transp. Res. Part D: Transp. Environ. **120**, 103794 (2023)
6. Dbouk, T., Drikakis, D.: Computational aeroacoustics of quadcopter drones. Appl. Acoust. **192**, 108738 (2022)
7. Zhou, S., Li, B., Ding, C., Lu, L., Ding, C.: An efficient deep reinforcement learning framework for UAVs. In: 2020 21st International Symposium on Quality Electronic Design (ISQED), pp. 323–328. IEEE (2020)
8. Kalidas, A.P., Joshua, C.J., Md, A.Q., Basheer, S., Mohan, S., Sakri, S.: Deep reinforcement learning for vision-based navigation of uavs in avoiding stationary and mobile obstacles. Drones **7**(4), 245 (2023)
9. Dronecode Foundation: Px4 autopilot, https://px4.io/
10. Open Robotics: Gazebo simulator, https://gazebosim.org/home
11. Macenski, S., Foote, T., Gerkey, B., Lalancette, C., Woodall, W.: Robot operating system 2: Design, architecture, and uses in the wild. Sci. Robot. **7**(66), eabm6074 (2022)
12. Heutschi, K., Ott, B., Nussbaumer, T., Wellig, P.: Synthesis of real world drone signals based on lab recordings. Acta Acustica **4**(6), 24 (2020)
13. Wunderli, J.M., Meister, J., Boolakee, O., Heutschi, K.: A method to measure and model acoustic emissions of multicopters. Int. J. Environ. Res. Public Health **20**(1) (2022)

Mobile Robotics

Mobile Platform for the Inspection of Solar Panels

Arne Blossei, Jan Sender, and Tobias Runge

Abstract

The number of installed solar parks is growing steadily as electricity generation from renewable energies expands. This leads to an increase in the maintenance work to be carried out to maximise yield and lifetime while reducing the risk of failure of photovoltaic panels. In temperate climatic regions, relevant tasks include mowing the grass as well as cleaning and inspecting the photovoltaic panels. Due to the size and number of solar parks, manual maintenance is hardly feasible, making automation and parallelisation of the work necessary. This paper presents the development of an automated platform for the automatic execution of these tasks with a focus on inspection. In the development of the overall system, technologies from the fields of robotics, image processing and data analysis are being used. The inspection is based on passive thermography, which is used to record temperature differences, on the surface of the solar panels, called hotspots. By automatically analysing the solar panels in a software program, potential defects and damages can be detected and rectified at an early stage. The automatic inspection and maintenance system is a cost-effective solution

A. Blossei (✉) · J. Sender
Fraunhofer Institute for Large Structures in Production Engineering IGP, Rostock, Germany
e-mail: arne.blossei@igp.fraunhofer.de

J. Sender
e-mail: jan.sender@uni-rostock.de

J. Sender
Chair for Production Organization and Logistics, University of Rostock, Rostock, Germany

T. Runge
HIT Service and Metal Construction GmbH &Co. KG, Rostock, Germany
e-mail: t.runge@hit-rostock.de

M.-C. Wanner et al. (eds.), *Annals of Scientific Society for Assembly, Handling and Industrial Robotics 2024*, https://doi.org/10.1007/978-3-031-91463-8_32

for improving the efficiency of solar installations and reducing the amount of manual inspections and maintenance work.

Keywords

Inspection • Photovoltaic • Infrared thermography • Mobile robotics

1 Introduction

Climate change and the population's growing environmental awareness are leading to a desire for electricity from renewable energy sources. As a result, global installed photovoltaic capacity increased by an average of 26% per year from 2012 to 1053 GW in 2022 [1]. With the installed power also the need of inspection rises to maximize the outcome of electrical power, profit, durability and preventing the solar plants from catching fire [2]. Due to the design of solar panels, there are various methods for inspection, ranging from measuring the electrical power, measuring the magnetic field on the panel surface and optical damage detection to various thermographic methods [3]. The most common inspection method is passive thermography from the front of the solar panel, as it is quick, does not affect the operation of the solar park and provides a high level of information. Among other things, short circuits, defective cells, defective bypass diodes and broken cells can be detected in the thermographic data and in some cases even rough power estimates are possible [2].

For automated inspection, the thermographic camera is almost entirely used together with a drone to quickly obtain a rough overview of the condition of the solar park [4]. However, these images often have a low resolution per panel, which means that certain defects cannot be identified. With a higher resolution, the condition of the panels can be determined more precisely, although with drones a compromise must always be made between area performance and resolution [5]. In addition, airspace in the EU is strictly regulated, which means that a complex authorisation process is necessary for flying over some solar parks. Also cleaning and lawn care work can only be covered to a limited extent by drones, as they require heavy equipment which tremendously reduces the flight time.

For mowing work some (semi-) automated ground-based systems are on the market [6] and a view also specialized at solar parks, like the Vector-robots [7]. Also, various methods and systems for cleaning photovoltaic panels are already established [8]. They range from coating to active cleaning systems, some of which are permanently installed, through to manual cleaning [9].

However, there is not yet a system that can perform all three tasks: inspecting, cleaning and simultaneous grass mowing automatically. The development of this overall system is the goal of the presented project. The status is primarily concerned with the inspection unit, as cleaning units depend on many parameters such as the structure and accessibility of the solar park and the availability of water, so there is no standardized solution [10].

2 Concept of the Inspection and Cleaning System

2.1 Challenges

The combination of inspection and maintenance tasks in solar parks poses specific requirements for the mobile platform. To perform these different tasks, a modular approach is taken for the system so that it can clean or inspect the solar panels as required and perform a parallel mowing process.

To inspect solar panels, passive thermography is a suitable test method for the non-destructive and non-tactile detection of hotspots. For the inspection process it is necessary to align the thermographic camera at an angle between 5° and 60° to the normal of the panel surface [11, 12]. The reflection effects of the glass on the surface of the solar panel are significant orthogonal to the panel surface and generally reflectance increase with increasing angle. At the same time, the emissivity of glass decreases, and undesirable distortion effects occur as the angle increases. For this reason, a recording angle between 10° and 20° is selected.

The reflection effects of the glass surface and the rack of the solar panels are also relevant for optical sensors for position determination.

Furthermore, for an efficient combination of the mowing and inspection process, it is necessary to be able to record at least one panel at a time. Otherwise the routes would have to be covered several times and the evaluation would become significantly more complex, as the thermographic images of a panel would have to be put together. The different arrangements of the solar panels within the solar parks pose further challenges. These can be arranged lengthways or upwards, individually or on top of each other to optimize the area utilization. There are even solar parks with a comprehensive panel arrangement. The aim here is to cover as many different applications as possible with the choice of vehicle and installed arm.

2.2 System Requirements

Carrier platform and positioning system

The various tasks that are to be covered by the system diversify the requirements widely. For example, the closely interlinked parameters of energy requirements, system performance and range. While a low drive power and only energy for the sensors and computing units is required for inspection drives, the drive power and energy requirement increase significantly when mulching the grass, which greatly reduces the range. In addition, there

Table 1 Arm requirements for inspecting or cleaning a 2.5 m large solar panel (values are rounded)

Arm requirements	Inspection	Cleaning
Horizontal reach (m)	1.7	2.7
Vertical reach (m)	1.3	0.5
Estimated weight of the end effector (kg)	1	5
Static torque at base (without weight of arm) (Nm)	17	132

are various drive concepts, some of which enable the range to be topped up quickly, making it difficult to define specific requirements in this area. Sufficient off-road mobility is necessary for travelling through the unpaved ground in solar parks. Furthermore, most solar parks are densely built-up and have narrow aisles between the rows of panels, which means that the carrier platform should be narrow and manoeuvrable. In addition to the required payload, the platform shape should also provide mounting options for the robot arm and the sensors for determining the position.

To ensure a reliable positioning between the solar panel strings the positioning system should not be based exclusively on optical sensors due to the high amount of highly reflective material in the solar park. For guiding the vehicle safely through narrow solar parks, an absolute positioning accuracy of less than 10 cm should be aimed for.

Arm for inspecting and cleaning solar panels

For the inspection process, experience has shown that camera angles between 10° and 20° from the panel normal to the thermographic camera have proven to be desirable. Between these angles the emissivity is high while the reflection and distortion effects are low. With the desire to capture at least one complete panel, a wide aperture angle of the thermographic camera is required. With a wide-angle lens, aperture viewing angles of around 90° are achieved with acceptable distortion. Taking an angle of 10° to the normal of the panel, a minimum distance of approximately 1.2 m above the panel surface is still required to be able to record the solar panels with a height of up to just under 2.5 m [13]. A horizontal projection of 1.5 m is required to inspect a solar panel at a distance of 0.75 m from the base of the arm. (Table 1). To clean solar panels, a much heavier end effector is required, which also has to cover a greater range. In addition, there are the dynamic loads caused by the vehicle moving, which were not taken into account in Table 1. Everything together leads to a significantly higher torque in the base, which must also be absorbed by the carrier platform.

2.3 Overall System Concept

The aim of the overall system (Fig. 1) is to automate mowing, cleaning and inspection work in solar parks. As a base, an all-terrain platform is required that allows the system to

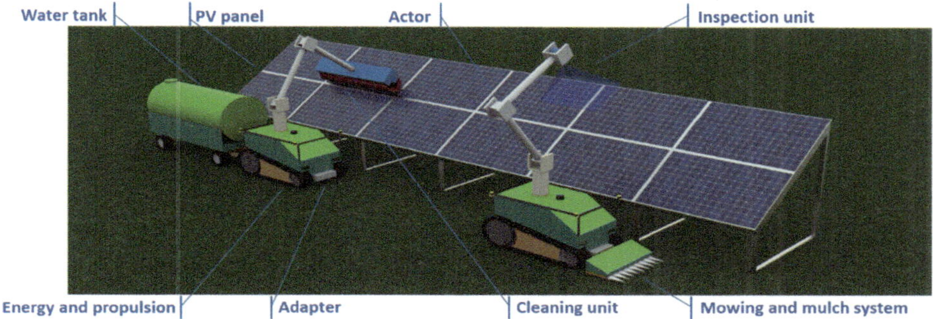

Fig. 1 Schematic image of the system in operation

move freely around the solar park. To perform the tasks automatically, the platform must move independently through the solar park as well as recognise and follow the paths between the solar panels. The cleaning and inspection system is positioned in relation to the solar panel using an arm. This means that a large working area can be covered. The tools required for the work should be attached modularly so that each individual task can be carried out efficiently.

3 Mobil Platform for the Inspection

3.1 Platform Components

Carrier platform

The carrier platform is the ROVO developed by HAWE Mattro [14]. This vehicle has a high degree of off-road mobility and manoeuvrability thanks to the tracks. With its dimensions of around 1.2 m × 1.2 m (L x W), it is compact. Furthermore, the heavy-duty version of the platform has a payload of 500 kg and can travel up to 40 km on a single battery charge. Mounting options and a CAN interface for external control are available, and connections for the power supply of external devices and a hydraulic hoist can be configured.

Positioning system

The RowCropPilot automation kit from RobotMakers [15] is integrated into the controller area network (CAN) interface of the ROVO. This package has both, a Teach&Repeat function including obstacle detection and a function for orientation on the rows. For this purpose, the ROVO is equipped with a global navigation satellite systems (GNSS) antenna, a lidar, an inertial measurement unit (IMU), a computing unit and a communication antenna. While in the Teach&Repeat approach the positioning is based purely on

GNSS and IMU and the lidar is only used for obstacle detection, in the RowCropPilot the lidar sensor is actively involved in the positioning by measuring the distance to the row of panels and this is actively taken into account in the positioning algorithms.

Inspection system

The aim of the inspection system is to clearly identify defects in the solar panels. The PIR uc 605 from InfraTec [16] is used as thermographic camera. With a 5 mm wide-angle lens, the camera has an aperture angle of $95° \times 78°$ and weighs less than 110 g. It uses the typical wavelength range of 8 μm to 14 μm for the examination of solar panels and has a temperature resolution of 0.06 K, which is below the recommended 0.08 K [11]. The camera SDK also enables access to the raw measuring data. This means that the measured infrared radiation values and the temperature values of each pixel can be read out.

The ZED2 stereographic camera from Stereo-labs is used to determine the distance and alignment to the panel and for potential optical defect detection [17]. With a field of view of 120°, it covers a larger area than the thermographic camera, which means that the camera can be used optimally to take over the positioning.

Arm

During the development process, two different arms with different degrees of freedom of movement and automation options were installed and tested on the platform.

For one, the UR10/CB3 cobot from Universal Robots [18] is installed on the mobile carrier platform. This has a range of 1.3 m and can lift a maximum payload of 10 kg. The maximum payload decreases as the centre of gravity shifts. With a centre of gravity shift of 800 mm, the maximum payload is less than 5 kg. However, as shown in chapter 2. 2, an arm of at least 1.5 m is required. When passing the panels, additional dynamic loads occur due to unevenness in the ground surface, which must be considered when dimension the arm.

In a second approach, a separate arm is developed within the project. This has only one prismatic and three swivel joints instead of six in the UR10. Only the swivel joint is motorised, and the other joints have to be adjusted manually before driving. The length of the telescopic arm, its tilt angle and the tilt angle of the camera can be adjusted manually, and the motorised turntable can be used to rotate the arm around the travelling axis. This design increases the covered range and is mechanically much more resilient due to the small number of joints. Furthermore, the movement of the inspection unit on the extension arm is passively damped. This actuator system is also significantly cheaper to procure due to its reduced complexity.

The characteristics of both arms are summarised in Table 2. The developed arm has a reduced number of degrees of freedom that provide enough adjustment options to carry out the inspection process. Furthermore, a potential position control by motorising selected axes of the arm is less time-consuming to develop and more cost-efficient due to the reduced degrees of freedom. The telescopic arm also makes the radius of action

Table 2 Comparison of UR10 and developed arm

Arm	UR10/CB3	Developed arm
Picture		
Type	Revolute arm	Polar arm
Joint types (R-Revolute, P-Prismatic)	R-R-R-R-R-R	R-R-P-R
DOF	6	4
Reach	1.3 m	2.1 m
Motorised at	All joints	Only the yaw joint
Control	Active	Not implemented

significantly more variable for the inspection of different panel arrangements and panel dimensions. Ready-made solutions that do not overload the platform's load capacity and energy system usually have a too low range to payload ratio.

Mowing and Cleaning
At the front of the mobile platform is a hydraulic hoist with an adapter to which standardized mowing and mulching systems can be attached. They usually have a short reach under the solar panels, which leads to the use of an offset mowing system.

For cleaning the photovoltaic panels, an adaptation of already established cleaning systems for automated use is planned. A completely new development of cleaning systems is uneconomical, as different types of soiling and dust as well as environmental conditions must be taken into account when selecting the cleaning system [8].

3.2 Current Status of the Development of the Mobile Platform

The current status of the mobile platform is shown in Fig. 2. At the front of the vehicle a mulcher can be mounted on the hydraulic hoist. Lidar sensor and GNSS antenna are installed on the front mast, on top of the vehicle. This is necessary to ensure a free field of view to the front of the vehicle for the lidar sensor and to minimize shadowing for an optimal GNSS signal.

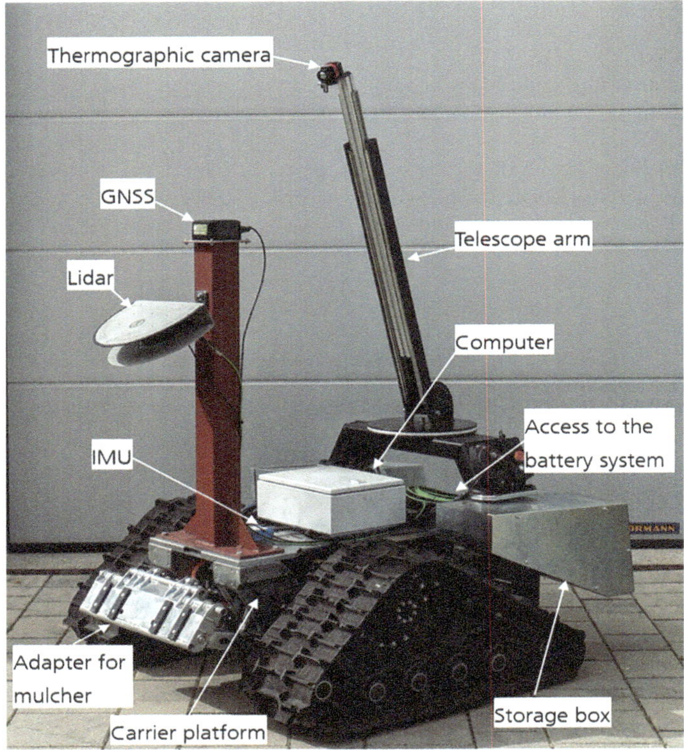

Fig. 2 Current state of the mobile platform for solar panel inspection

The IMU and computing unit of the positioning system are located behind the mast on the ROVO. The silver boxes above the chains contain the hydraulics for the hoist and space for the computing unit of the inspection system. To position the infrared thermography camera the constructed arm with turntable is mounted at the back of the vehicle. The rear edge is kept clear to allow free access to the charging ports and batteries for quick replacement.

On the control side, the inspection system is separated from the positioning system. After the inspection, the images are assigned to the corresponding coordinates via time referencing in the evaluation software.

4 Evaluation of the Thermographic Data

The data analysis consists of two sub-steps. In the first one, the panels are recognized to limit the section under consideration. Afterwards, the hotspots are searched in the cropped temperature matrix.

4.1 Detection of the Solar Panels

While the platform moves along the rows of solar panels, the thermographic data is recorded at the same time. Thanks to the arm, the recording angle and image section remain largely constant during data recording, resulting in a simplified approach to panel detection. The already calculated temperature matrix of the images generated in this way is used for the hotspot detection.

Before the temperature matrix is displayed graphically as a grayscale image, it is necessary to restrict the temperature range. This is selected so that the edges of the solar panel are clearly distinguishable from the background in the grayscale image (Fig. 3a). This image serves as the starting point for edge detection, which is achieved by the targeted selection of parameters such as the threshold values for the upper and lower edge thickness (Fig. 3b). The detected edges are further used to identify the panel corner points (Fig. 3c). To detect the corner points, in addition to adjusting the parameters, it is necessary to divide the image into sectors in which the corner points of the solar panel are assumed to be located. However, this only works by keeping the angle and area of the solar panels in the pictures similar. The result is a cut-out solar panel that is analysed for hotspots in the next processing step (Fig. 3d).

4.2 Hotspot Detection

To detect the hotspots, the image section of the solar panel calculated in Fig. 3d is transferred to the temperature matrix to take only the temperatures on the panel surface into account. A histogram is created from this matrix section, which compares temperature ranges with the number of matrix values. The maxima of the histogram represent the largest areas within a temperature range, which corresponds to the base temperature of the panel. Larger clusters of points that are warmer than the panel base temperature are hotspots and get displayed graphically for the user (Fig. 4). In this way, each panel of the solar park is checked and provided with the position data of the GNSS system.

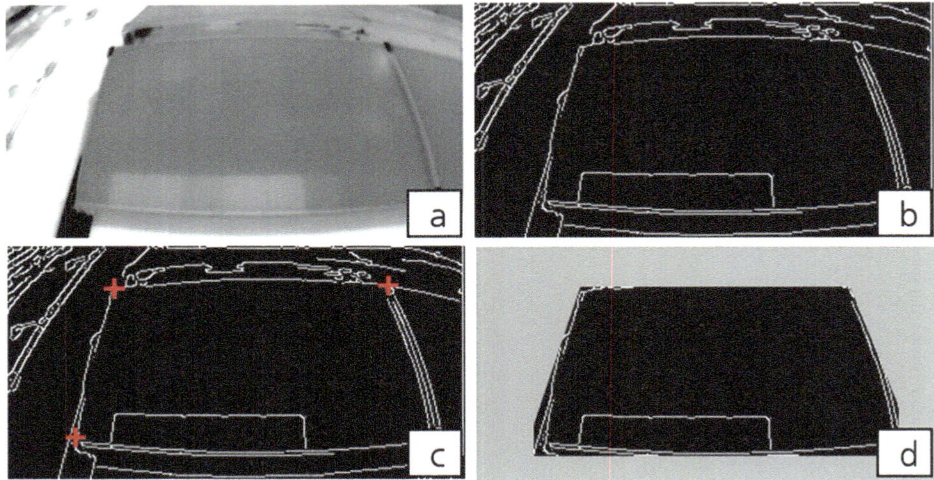

Fig. 3 Panel detection process (original image (**a**), edge detection (**b**), corner detection and filter (**c**) and cropped image section (**d**))

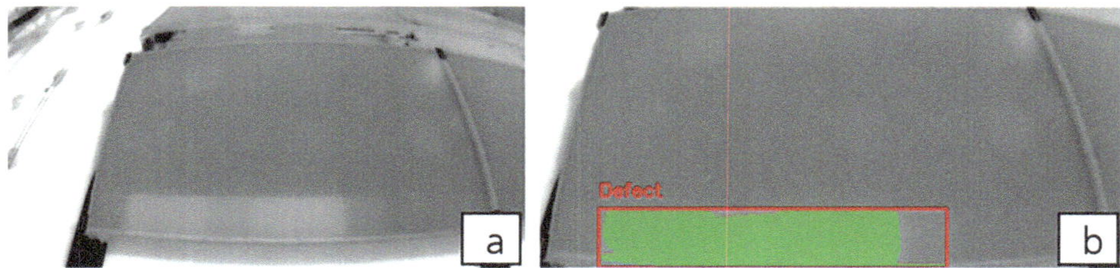

Fig. 4 Image of solar panel with hotspot (original (**a**) and after image processing (**b**))

4.3 Discussion of the Evaluation Process of Thermographic Images

The hotspot detection in the temperature data offers significant advantages over detection using conventional image processing. For example, more precise limits can be set for critical hotspots and further information can be obtained to gain conclusions about the panel condition. The use of the temperature matrix appears to be promising for automated defect detection, as it can be used to implement both image processing approaches and alternative evaluation methods.

The only drawback of the current software version is the sometimes-unreliable panel recognition using classic image processing methods. In addition to the temperature range of the thermographic image, the parameters of the individual image processing procedures offer many possibilities for adjustments and optimisation. However, with an AI-based approach to panel detection, the software should become more robust. The overall reliability of the evaluation software still needs to be evaluated in larger test series.

As shown in Kirubakaran et al. [19], other concepts from classical image processing may be more robust. Also successful AI-based approach to detect warmer regions on solar panels with neural networks with an accuracy of 99.02% and a precision of 91.67% has already been developed [20].

The approach for hotspot detection in the separated panels is significantly more independent of parameters and appears to achieve reliable results in the initial investigations. Nevertheless, a systematic error is made, as distortion effects caused by the lens of the thermographic camera and the angle of view are not taken into account. As a result, the pixels have a different projection area, but this is negligible for this application if the recording angle is between 10° and 20° and the solar panel is centred. This can be maintained continuously with the current platform and arm design when travelling slowly in various solar parks. For larger unevenness or inspection with a lever arm over approximately 1.5 m, an active control against unevenness must still be set. This requires further motorisation of the arm.

5 Conclusions and Future Works

The system presented is an extension of the conventional, mostly manual, or drone-based inspection of solar panels. It enables automated inspection with simultaneous gras mowing in many solar parks built. Furthermore, a single system can be used to carry out inspection and cleaning tasks thanks to its modular design.

The system is under constant development and further tasks include the implementation of event-based thermographic camera control as well as an active control of the arm via the data from the stereographic camera and the further development of the existing inspection and cleaning system. A large-scale evaluation in various solar parks is also still pending.

For the detection of hotspots in solar panels a non-AI-based approach is presented, which is implemented in form of a program with user interface. Algorithmically, the programme can be divided into panel detection and hotspot detection. While the panel detection seems to be complicated to use automatically due to the large number of parameters, the hotspots are reliable detected with the presented approach.

The primary goal in the further development of the evaluation software will therefore aim at improving the reliability and robustness of panel detection, as this is the base for further analyses. In addition, the inclusion of weather data in the inspection process and the assessment of the condition of the solar modules is conceivable. There is still a lot of research to be done to estimate the solar panel status from non-destructive and non-tactile testing.

Furthermore, an adaptation of the evaluation to solar parks additionally recorded with drones appears to make sense, as these enable a significantly higher speed in data recording and also make solar plants that are difficult to access, such as those on the roofs of houses, accessible.

References

1. Energy Institute: Statistical Review of World Energy 2023. Energy Institute (2023)
2. Constantin, Alexandru-Ionel, et al.: Importance of Preventive Maintenance in Solar Energy Systems and Fault Detection for Solar Panels based on Thermal Images. Electrotehnica, Electronica, Automatica (EEA) (71),01–12 (2023).
3. Meribout, Mahmoud, et al.: Solar panel inspection techniques and prospects. Measurement (209),112466 (2023).
4. de Oliveira, A. K. V., Aghaei, M., Rüther, R.: Automatic inspection of photovoltaic power plants using aerial infrared thermography: a review. Energies 2022, 15, 2055 (2022).
5. Lee, D.H., Park, J.H.: Developing inspection methodology of solar energy plants by thermal infrared sensor on board unmanned aerial vehicles. Energies **12**(15), 2928 (2019)
6. Pedersen, S.M., Fountas, S., Have, H., Blackmore, B.S.: Agricultural robots—system analysis and economic feasibility. Precision Agric. **7**, 295–308 (2006)
7. Vector, Autonomous & radio-controlled mowers, https://www.vectormachines.nl/en/. Last access 09 Jul 2024
8. Patil, P. A.; Bagi, J. S.; Wagh, M. M.: A review on cleaning mechanism of solar photovoltaic panel. In: 2017 International Conference on Energy, Communication, Data Analytics and Soft Computing (ICECDS), p. 250–256. IEEE (2017).
9. Derakhshandeh, Javad Farrokhi, et al.: A comprehensive review of automatic cleaning systems of solar panels. Sustainable Energy Technologies and Assessments 47 101518 (2021).
10. Antonelli, M.G., Zobel, P.B., De Marcellis, A., Palange, E.: Autonomous robot for cleaning photovoltaic panels in desert zones. Mechatronics **68**, 102372 (2020)
11. FLIR: Thermal imaging guidebook for building and renewable energy applications. FLIR (2011)
12. Glavaš, Hrvoje, et al.: Infrared thermography in inspection of photovoltaic panels. In: Žagar, D. et al. (eds.) International Conference on Smart Systems and Technologies 2017 (SST 2017), pp. 63–68. IEEE, Red Hook (2017)
13. PV Modul-Größen im Überblick. https://www.energie-experten.org/erneuerbare-energien/photovoltaik/solarmodule/groesse. Last accessed 14 Mar 2024
14. Roboter Plattform—HAWE Hydraulik. https://www.hawe.com/de-de/produkte/roboter-plattform/. Last accessed 15 Mar 2024
15. Robot Makers GmbH—Intelligent Mobile Machines. https://robotmakers.de/. Last accessed 15 Mar 2024
16. PIR uc 605 | IR-Kameramodul von InfraTec GmbH. https://www.infratec.de/thermografie/waermebildkameras/pir-uc-605/. Last accessed 18 Mar 2024
17. ZED 2—AI Stereo Camera | Stereolabs, https://www.stereolabs.com/products/zed-2. Last accessed 18 Mar 2024

18. Universal Robots A/S: Benutzerhandbuch; UR10/CB3; Version 3.3.0. Universal Robots A/S (2016)
19. Kirubakaran, Victor, et al.: Infrared thermal images of solar PV panels for fault identification using image processing technique. Int. J. Photoenergy (2022)
20. Herraiz, Á.H., Marugán, A.P., Márquez, F.P.G.: Photovoltaic plant condition monitoring using thermal images analysis by convolutional neural network-based structure. Renew. Energy **153**, 334–348 (2020)

Challenges of Mobile Robots in Motion: A Comparative Trajectory Accuracy Evaluation of AMCL Using Motion Capture Systems

Hauke Heeren, Lukas Lachmayer, and Annika Raatz

Abstract

This paper explores the precision of trajectory estimations provided by *AMCL*. By using a motion capture system for detailed error measurements and investigating the effects of motion parameters and occlusions on localization accuracy, this work fills a significant gap in previous research. It provides a nuanced understanding of the factors that influence *AMCL* accuracy and lays the groundwork for the development of more robust localization approaches to effectively navigate robotic systems on trajectories in complex environments such as construction sites. The results highlight the importance of path velocity and environmental occlusion for localization and point to directions for future research to improve the reliability and precision of mobile robot localization.

Keywords

Mobile robot • AMCL • Mocap • Trajectory • Accuracy • Localization • Monte Carlo • Particle filter • Repeatability • Movement • Occlusion

1 Introduction

Their flexible usage and the achievable degree of automation are currently leading to an increase in the number of mobile robot platforms observable in the construction industry. Applications include mobile printing platforms, handling and assembly tasks. In the field of additive manufacturing and cooperative handling with multiple robots, achieving high

H. Heeren (✉) · L. Lachmayer · A. Raatz
Institute of Assembly Technology and Robotics, Leibniz University Hannover, Garbsen, Germany
e-mail: heeren@match.uni-hannover.de
URL: https://www.match.uni-hannover.de/en/

M.-C. Wanner et al. (eds.), *Annals of Scientific Society for Assembly, Handling and Industrial Robotics 2024*, https://doi.org/10.1007/978-3-031-91463-8_33

accuracy in robot localization is crucial. In additive manufacturing, highly accurate layer-stacking is relevant to ensure proper layer adhesion to enable the production of components. In more comprehensive manufacturing processes, precise localization ensures compliance with the exact tolerances required for efficient machining and inspection of workpieces. Furthermore, in scenarios involving the collaborative transport of objects by multiple robots, improved localization accuracy is essential to minimize the forces acting on workpieces due to misalignment, thereby maintaining material integrity.

Most accurate localization methods rely on external tracking systems, including the use of fixed markers [1, 2] or extensive camera setups [3]. However, these methodologies encounter substantial obstacles in large-scale, dynamic environments such as construction sites, where the practicalities of marker deployment and the continuous calibration of camera systems due to environmental disturbances, like vibrations, present significant challenges. Therefore, another approach is to use the printed contour as a reference instead of markers [4]. However, even though referencing to the component is proposed, the described approach requires additional sensors. Both methods for increasing localization accuracy therefore require either additional measurement technology or an external reference. Focusing on onboard localization, the desire for higher accuracy in the localization of mobile robots has particularly motivated the improvement of the Monte Carlo localization algorithm by adapting the particle size using KLD sampling to Fox's *Adaptive Monte Carlo Localization* (*AMCL*) [5]. In the context of cooperative tasks, although collective localization techniques have been presented, they introduce disadvantages such as increased complexity and communication overhead [6].

This paper focuses on improving the understanding of the accuracy of existing on-board mobile robot localization using *AMCL*, especially in motion. By analyzing the error characteristics of *AMCL*, including rotational errors during motion and the influence of linear and angular acceleration and velocity on localization accuracy, this study provides a comprehensive investigation of localization accuracy beyond the conditions commonly considered in the literature. In addition, the effects of occluded fields of view on localization accuracy, which are particularly important in cooperative handling and manufacturing scenarios, are investigated.

2 Related Work

To evaluate the localization accuracy, various approaches have been derived. For example, the positioning accuracy of mobile robots at standstill was investigated by Hennes et al. in simulation [7] and by James et al. in the real world [8]. Both studies consider only the distance and do not divide the accuracy into Cartesian dimensions. Röwekämper et al. investigated the position accuracy of mobile robots in detail [9]. However, here too the accuracy is only measured at taught-in reference positions while the robot is stationary. In contrast, Dudzik measured the accuracy in motion by calculating the Hausdorff distance to the path recorded by a motion capture system [10]. By using the Hausdorff distance, no temporal aspects are considered. The path is examined purely spatially, thus overlooking potential tracking errors and kinematic effects on the accuracy measurement. Considering temporal aspects, Ivanjko et al. compared their improved localization algorithm to the Monte Carlo Algorithm, demonstrating how important a thorough evaluation of *AMCL* is, to be the baseline for further accuracy comparisons [11]. Instead, Yu et al. measured the accuracy in motion by comparing the scanning contour of the LiDAR to a map contour for different positions at a singular, static speed. However, their investigation was confined to the accumulated positional errors in x and y. The rotational error was not taken into consideration [1]. A graphic evaluation of the *AMCL* localization accuracy compared to Motion Capture data was done by Sustarevas et al. With a holonomic platform, it is shown, how crucial accurate tracking of the robot for printing-in-motion is. When using *AMCL*, an error of up to 0.12 m compared to the planned trajectory of the printing nozzle was computed using a motion capture ground truth [3]. Although the use of a motion capture system is a more accurate approach than that of Yu et al., the position controllers used and the added complexity by tracking of the printing nozzle instead of the platform itself allow only a very limited assessment of the localization accuracy. However, Vanhie-Van Gerwen et al. compared the localization estimate of UAVs directly to a motion capture ground truth, but their focus was on various sensor configurations [12]. No other insights into the parameters influencing the localization accuracy were given. The current state-of-the-art reveals a common oversight as it focuses on the absolute positional error (x/y coordinates) without sufficient attention to orientation or the dynamic errors introduced by mobile robot motion. This critical gap in current research highlights the need for a more comprehensive evaluation of robot localization accuracy, such as provided by our investigations.

3 Approach

To derive our evaluation methodology and for our investigations we use our mobile robot manipulator, shown in Fig. 1b. Like tracked vehicles, the robot uses a differential-drive and the movement is limited to two degrees of freedom. For the localization algorithm, the LiDAR data, along with the motor encoder data fused with an inertial measurement unit, serve as input. To measure the error of the robot localization, a motion capture system is used as a reference. The motion capture system used consists of nine *Qualisys Arqus A12* cameras and is therefore sufficient to cover the whole test area of approx. 215 m^2. To get the best possible ground truth, the motion capture system is calibrated right before use. The average residual is stated as 0.5mm. To not influence the accuracy of the *AMCL* by false mapping, the map seen in Fig. 1a is recorded right before the start of the experiments.

For determination of the pose error, the robot's pose estimate via *AMCL* as well as the ground truth via the motion capture system is measured on the whole trajectory. To evaluate different influences on the trajectory accuracy, several measurements will be taken. Not

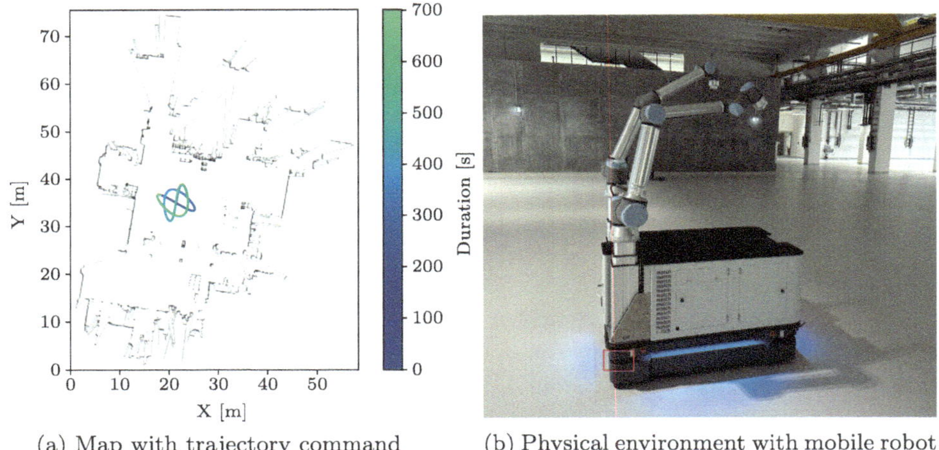

(a) Map with trajectory command (b) Physical environment with mobile robot

Fig. 1 Test environment featuring a robot equipped with two diagonally placed SICK microScan3 LiDAR (marked by box) which cover the entire surroundings

only the overall accuracy while in motion, but also the influence of different translational as well as rotational velocities and accelerations on the trajectory accuracy shall be evaluated. Therefore we propose using the following four different measurement setups:

Lissajous: The mobile robot follows a trajectory as seen in Fig. 1a.

This trajectory is derived from Eq. 1 for Lissajous figures, with the parameters $A_x = A_y = 4m$, $\omega_x = 3S^{-1}$ and $\omega_y = 2S^{-1}$. Factor K is varied and influences the speed at which the Lissajous figure is traversed.

$$t \mapsto \begin{pmatrix} A_x \sin\left(K\omega_x t\right) \\ A_y \sin\left(K\omega_y t\right) \end{pmatrix} \tag{1}$$

By using a Lissajous figure, a combination of different linear and rotational speeds and accelerations is given that covers all typical parts of the trajectory of a mobile robot.

Lissajous with second robot: To measure the influence of a mobile robot with an occluded field of view, another mobile robot follows a trajectory parallel to the robot used for measurements. This simulates a cooperative handling/manufacturing task. A formation planner is used for this purpose. The robot used for the measurements is in the centre of the formation and follows the Lissajous path from Fig. 1a. The second robot's target position is 1.8 m left of the formation centre in the direction of travel.

Translation: To detect dependencies of translational velocities and accelerations on the accuracy, the robot follows a linear path. For a comparable study, only the minimum distance of all acceleration setpoints until the robot reaches its maximum velocity is taken into account for the accelerations. To measure the influence of velocity, the robot travels on a 7.5m long linear path with static velocities.

Rotation: For the rotation, measurements will be taken with the same principle as for translation. Instead of linear setpoints, rotational setpoints for acceleration and velocity will be specified. For static velocities, a complete turn will be executed.

4 Evaluation

Within this chapter, we compare the pose estimate of *AMCL* with the measurement of the Motion Capture System during the mobile robot movement. Additionally, the influences of an occluded field of view for the 2D LiDAR scanners, the acceleration and the speed of the mobile platforms are elaborated. Therefore, six experiments, alined with the setups derived in Sect. 3 are evaluated. The difference between the Motion Capture pose and the *AMCL* pose estimate is calculated for the translational and rotational error. This analysis not only encompasses the spatial, but also the temporal aspect of the trajectory, by comparing the aligned data for each timestamp. Considering the *AMCL* pose estimation frequency of 10 Hz, a dense dataset across the whole trajectory for each measurement setup is provided, which is therefore evaluated of its statistical properties. The analysis includes boxplots where the whiskers indicate the minimal and maximal values, and the box represents the interquartile range.

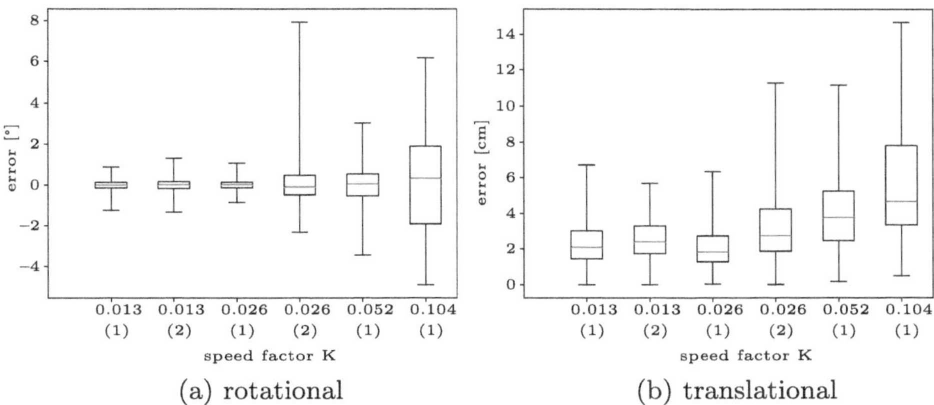

Fig. 2 Lissajous trajectory error for single (1) and dual (2) robot formation: Errors increase with higher speed settings or occluded field of view

Lissajous: Fig. 2 displays the trajectory error depending on the speed setting for the Lissajous trajectory. For the translational error, the euclidean distance is displayed. Therefore the minimum error in Fig. 2b is 0 m. For rotational and translational error, no influence of the speed is recognizable, comparing $K = 0.013$ to $K = 0.026$. For higher speed settings, the standard-deviation as well as the maximum errors are increasing. The translational mean error is increasing due to the nature of the euclidean distance. This leads to the assumption that either the robot's onboard *AMCL* motion update or the *AMCL* sensor update are influenced by the speed setting. We further assume potential slippage due to higher accelerations leading to errors in the motion update. At higher velocities, the resolution of the environment measured by the LiDAR decreases and would lead to errors in the sensor update. These two influences are evaluated using the measurement setups *Translation* and *Rotation*.

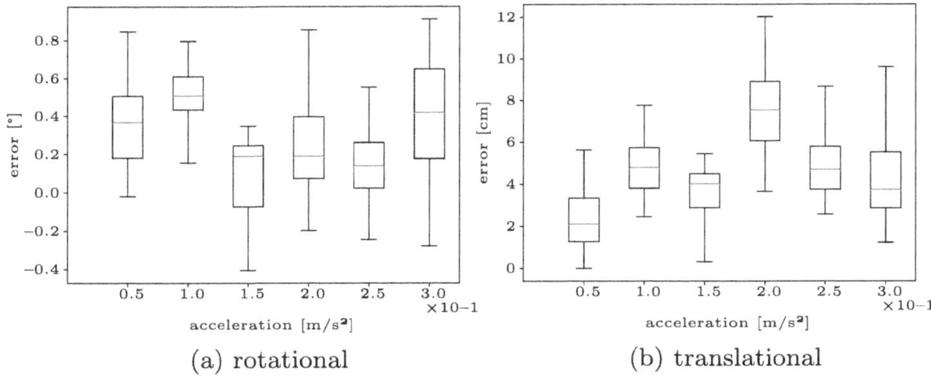

Fig. 3 Trajectory error for linear acceleration: No influence is evident

Lissajous with second robot: As can be seen in Fig. 2, both the translational and rotational errors increase when two robots (2) are used relative to just one robot (1) at the same speed. This is expected because the measured robot has fewer sensor readings to compare to the recorded map and therefore to localize itself in the environment. The increase in errors is even more pronounced at higher speeds. For $K = 0.013$, the translational mean error increases by a factor of 1.13, and the rotational error range increases by a factor of 1.25. At the higher speed setting of $K = 0.026$, the translational mean error increases by a factor of 1.53 and the rotational error range by a factor of 5.36.

Translation: To evaluate various translation parameters, we moved our robot straight forward in the Y direction. As can be obtained in Fig. 3, the trajectory error is not influenced by the acceleration of the mobile platform. Neither the measurements for the translation nor the rotational error show any significant difference between the numerous acceleration settings. However, When comparing the trajectory error in Fig. 4a to the velocity in Fig. 4b, for some accelerations a slight correlation can be recognized. Therefore only the error in Y-axis is shown in Fig. 5b, but no clear correlation can be seen. We conclude that the trajectory error is greater in the direction of the movement, as the absolute error for all measurement series in the Y direction is greater than in the X direction. For the rotational error in Fig. 5a, the standard deviation increases slightly with higher velocity settings. This leads to the assumption that the velocity is not high enough to see any effects on the localization accuracy.

The sensor readings still have sufficient resolution to detect the environment and localize the robot.

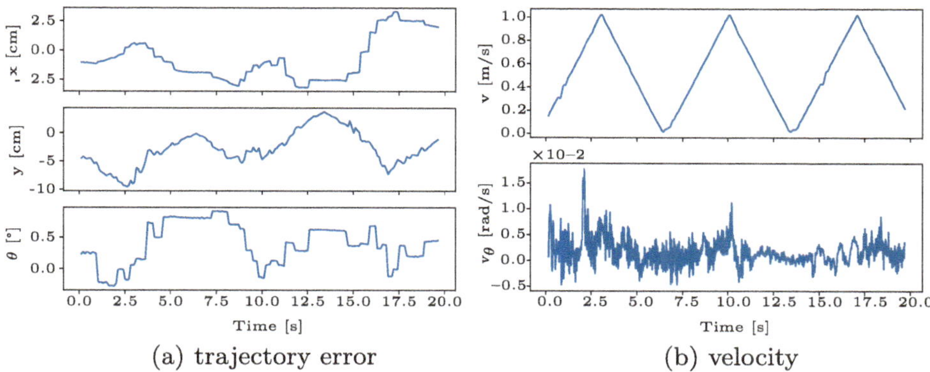

(a) trajectory error (b) velocity

Fig. 4 Linear acceleration of $0.3\,\mathrm{ms}^{-1}$: Slight correlations between velocity and error is observed

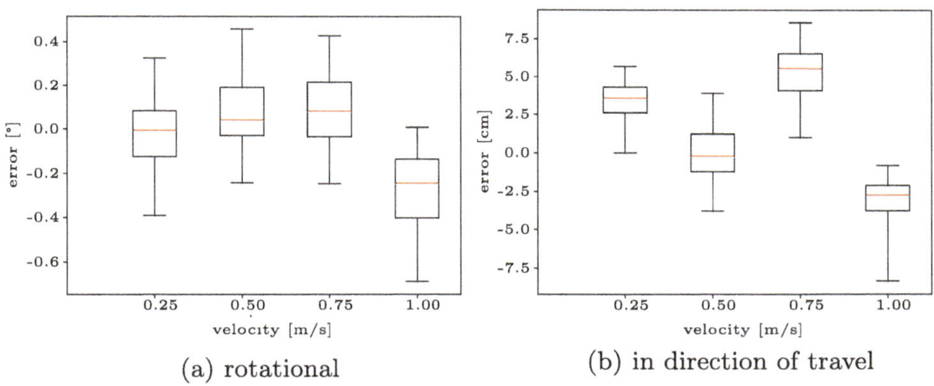

(a) rotational (b) in direction of travel

Fig. 5 Trajectory error for linear velocities: No correlation for translational error in direction of travel. For the rotational error, the standard deviation increases

Rotation: As for the linear acceleration, the localization accuracy is not influenced by the rotational acceleration of the mobile platform. As can be seen in Fig. 6. However, a correlation between the velocity and the accuracy can be recognized much better in both Figs. 7 and 8. Here too, the error is particularly noticeable in the direction of movement. The absolute error in the rotation direction increases with the velocity. This can be explained by the higher relative velocity of the detected objects through the rotation compared to the translation and should therefore increase with the distance.

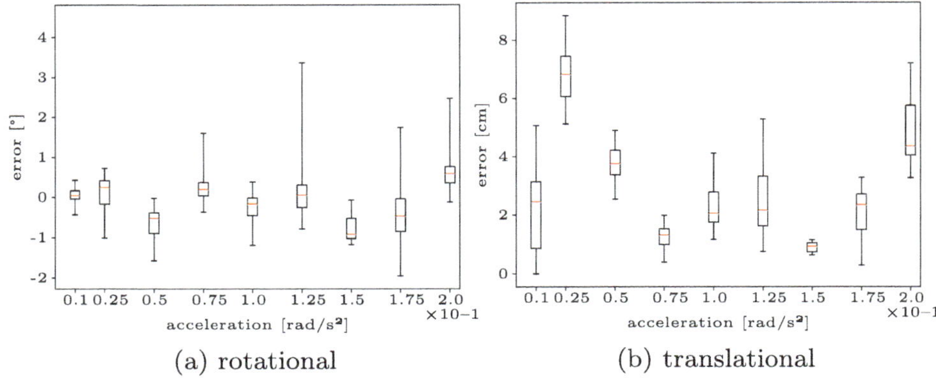

Fig. 6 Trajectory error for angular accelerations: No influence is evident

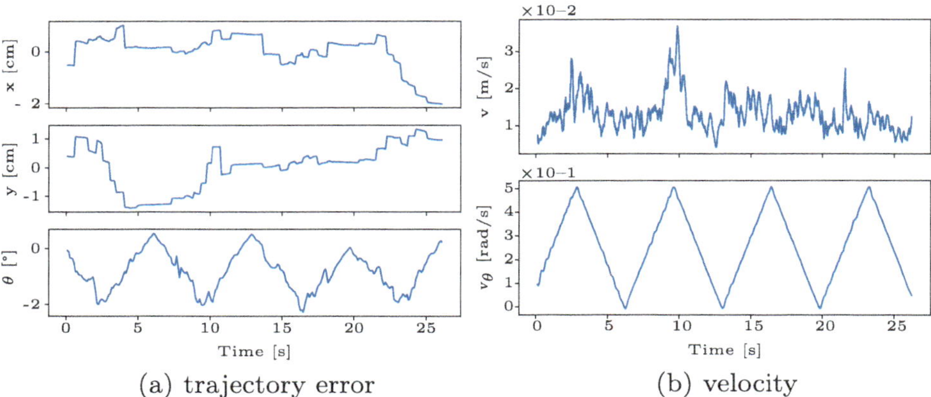

Fig. 7 Angular acceleration of 0.15 rad s^{-1}: Correlation between velocity and error is observed

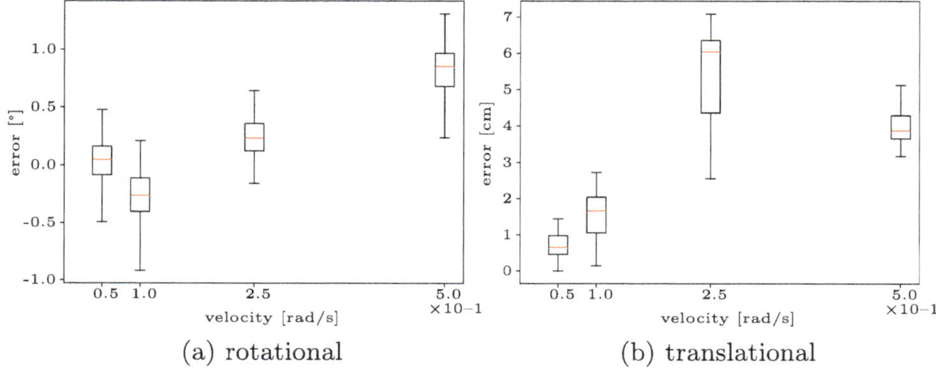

Fig. 8 Trajectory error for angular velocities: The absolute error increases especially in direction of movement

5 Conclusion and Outlook

In this work, a comparison of the localization accuracy of *AMCL* based on Motion Capture System data was presented. Through a series of experiments, the impact of an occluded field of view for the lidar scanners and different accelerations and velocities on the accuracy of the *AMCL* was analyzed. This analysis was not limited to the translational error, where a maximum error of 141.4 mm was observed, but the rotational error, reaching up to 7.4°, was also evaluated. One of the notable findings of this investigation is the clear relationship between trajectory speed and localization accuracy. The data shows that an increase in trajectory speed correlates with an increase in localization error.

The analysis also accounted for the effects of translational and rotational velocities on localization accuracy. While translational velocity showed some correlation with localization accuracy in the recorded error over time, this relationship was not confirmed by the statistical analysis. Conversely, the effect of rotational velocity on localization accuracy was both visually and statistically significant, highlighting its crucial role in the accuracy of robot localization. Interestingly, within the experimental limits, acceleration did not affect localization accuracy. Nevertheless, its role cannot be entirely ruled out. The potential for acceleration-induced slippage as a function of ground conditions, which could lead to errors in the motion update, is a topic for future research. In addition, the experimental scenarios in which the lidar's field of view was obscured by another robot driving in formation showed a noticeable impact on localization accuracy. This result emphasizes how important it is to consider objects that are not in the recorded map, especially in scenarios in which a close formation is driven or cooperative tasks are performed.

Looking to the future offers a wealth of research opportunities to further unravel the complexity of robot localization. One key area of interest is the in-depth investigation of the influence of acceleration on localization accuracy. For use in different environments, like construction sides and factory halls, it will be critical to explore the nuances of how different acceleration profiles coupled with variable ground conditions affect slip and consequently localization accuracy. This investigation should cover a wide range of ground types, ranging from firm to loose, as found on construction sites, and examine various vehicle types, including tracked vehicles.

Furthermore, the results regarding the influence of objects on the localization accuracy open up another promising starting point for investigations. The influence of static objects that are not considered in the map and objects appearing in the field of view, for example through additive manufacturing processes, should also be investigated. Understanding how unforeseen environmental features interact with the localization process can significantly improve the robustness and adaptability of localization algorithms.

In conclusion, while this work has provided valuable insights into the facets of mobile robot localization accuracy, it also lays the foundation for a wide range of future investigations.

Acknowledgements The authors gratefully acknowledge the partial funding by the Deutsche Forschungsgemeinschaft (DFG – German Research Foundation) – Project no. 414265976. The authors would like to thank the DFG for the support within the SFB/Transregio 277 – Additive manufacturing in construction. (Subproject B04)
The research building SCALE—Scalable Production Systems of the Future and the testing equipment "mobile assembly platforms" were funded by the Federal Ministry of Education and Research (BMBF) and zukunft.niedersachsen, a funding program of the Ministry for Science and Culture of Lower Saxony (MWK) and the Volkswagen Foundation.

References

1. Yu L., Li M., Pan G.: Indoor localization based on fusion of apriltag and adaptive Monte Carlo. In: 2021 IEEE 5th Information Technology,Networking,Electronic and Automation Control Conference (ITNEC) (2021). https://doi.org/10.1109/ITNEC52019.2021.9587205
2. Tiryaki M.E., Zhang X., Pham Q.-C.: Printing-while-moving: a new paradigm for large-scale robotic 3D Printing. In: 2019 IEEE/RSJ International Conference on Intelligent Robots and Systems (IROS) (2019). https://doi.org/10.1109/IROS40897.2019.8967524
3. Sustarevas, J., Kanoulas, D., Julier, S.: Autonomous mobile 3D printing of large-scale trajectories. In: 2022 IEEE/RSJ International Conference on Intelligent Robots and Systems (IROS) (2022). https://doi.org/10.1109/IROS47612.2022.9982274
4. Lachmayer, L., Recker, T., Raatz, A.: Contour tracking control for mobile robots applicable to large-scale assembly and additive manufacturing in construction. Procedia CIRP (2022). https://doi.org/10.1016/j.procir.2022.02.163
5. Fox, D.: Adapting the sample size in particle filters through KLD-sampling. Int. J. Robot. Res. (2003). https://doi.org/10.1177/0278364903022012001
6. Rone, W., Ben-Tzvi, P.: Mapping, localization and motion planning in mobile multi-robotic systems. Robotica (2013). https://doi.org/10.1017/S0263574712000021
7. Hennes, D., Claes, D., Meeussen, W., Tuyls, K.: Multi-robot collision avoidance with localization uncertainty. In: International Conference on Autonomous Agents and Multiagent Systems (2012). https://doi.org/10.5555/2343576.2343597
8. James, J., Clarke, G., Mathew, R., Mulkeen, B., Papakostas, N.: On Reducing the Localisation Error of Modern Mobile Robotic Platforms. Procedia CIRP (2022). https://doi.org/10.1016/j.procir.2022.09.067
9. Röwekämper, J., Sprunk, C., Tipaldi, G.D., Stachniss, C., Pfaff, P., Burgard, W.: On the position accuracy of mobile robot localization based on particle filters combined with scan matching. In: IEEE/RSJ International Conference on Intelligent Robots and Systems (2012). https://doi.org/10.1109/IROS.2012.6385988
10. Dudzik, S.: Application of the motion capture system to estimate the accuracy of a wheeled mobile robot localization. Energies (2020). https://doi.org/10.3390/en13236437
11. Ivanjko, E., Kitanov, A., Petrovic, I.: Model based Kalman filter mobile robot self-localization. Robot Localization and Map Building (2010). https://doi.org/10.5772/9256
12. Vanhie-Van, G.J., Geebelen, K., Wan, J., Joseph, W., Hoebeke, J., De Poorter, E.: Indoor drone positioning: Accuracy and cost trade-off for sensor fusion. IEEE Trans. Veh. Technol. (2021). https://doi.org/10.1109/TVT.2021.3129917

Implementation and Performance Evaluation of a Curve Fitting Algorithm for Path Planning of a Gluing Process Using k-Means Clustering

Till Braun, Wilko Flügge, and Jan Sender

Abstract

As part of the path planning of a mobile gluing robot, the task is to reconstruct the geometry of the joining point of the component from a disordered point cloud in order to apply the adhesive in the desired position. Curve fitting using the least squares method is a simple and widely used approach, but it is computationally intensive with a large number of data points or parameters. For the reconstruction of open and closed curves from unorganised point clouds, an improved approach for the classical least-squares fitting using B-splines is proposed, implemented in Python and tested on several data sets. The results of the Gauss–Newton algorithm used to minimise the objective function were compared with the Sequential Least Squares Programming (SLSQP) algorithm implemented in the SciPy library in terms of convergence speed and achievable Root Mean Square Error (RSME). By determining the initial values of the control points of the B-spline curve with the k-Means algorithm, the speed and stability of the least squares fitting could be improved.

T. Braun (✉)
Fraunhofer Institute for Large Structures in Production Engineering IGP, Rostock, Germany
e-mail: till.braun@igp.fraunhofer.de

W. Flügge
Chair of Manufacturing Technology, University of Rostock, Rostock, Germany
e-mail: wilko.fluegge@uni-rostock.de

J. Sender
Chair of Production Organziation and Logistics, University of Rostock, Rostock, Germany
e-mail: jan.sender@uni-rostock.de

M.-C. Wanner et al. (eds.), *Annals of Scientific Society for Assembly, Handling and Industrial Robotics 2024*, https://doi.org/10.1007/978-3-031-91463-8_34

Keywords

B • Spline curve fitting • Gauss • Newton • Method • k • Means clustering

1 Introduction

Welding, riveting and screwing are currently the most established joining methods in the shipbuilding industry. They are used for joining larger assemblies such as blocks, but also for the assembly of elements in ship equipment [1]. Until now, gluing has mainly been used to produce structural connections for larger assemblies in boatbuilding, where different materials are often used than in shipbuilding. In shipbuilding, however, gluing has so far been limited to the joining of smaller equipment elements such as brackets, rails for seats or window panes [2]. Undesirable effects, such as the risk of crevice corrosion during riveting, can be avoided by the material-locking connection of the components achieved by gluing. From a design perspective, the key advantage is the ability to combine non-metallic materials such as fibre-reinforced composites with conventional materials such as steel. In the future, this will enable the increased use of lightweight materials, which can increase the range of ships and reduce emissions [2].

The aim of the joint project "smartBOND", funded by the Federal Ministry of Economics and Climate Protection (BMWK), is therefore to provide the necessary means to increase the use of gluing in shipbuilding. This is to be achieved, among other things, by partially automating the gluing process with the involvement of qualified personnel. In addition to the goal of a repeatable process, the main focus is on the quality management of the process required by DIN ISO 9001. The gluing process is to be implemented using a mobile robot in a shipbuilding environment. A user-guided procedure was developed in [3] for the geometric referencing of the robot to the joining geometry. After the robot reaches the joining point, it is first recorded in a measuring run with a pattern projection system. The operator then uses the three-dimensional measurement data to define the component coordinate system (CCS) from a voxel representation of the point cloud captured by the optical 3D sensor.

Subsequently, the actual geometry of the joint must be extracted from the unorganised point cloud in order to realise the path planning of the robot for pre-treatment (grinding and wiping) and adhesive application. Therefore, a curve fitting has to be performed. Figure 1 shows an example application for an open contour to be recognised. The components welded to the girder, the so-called hot eyes, which serve as load attachment points in the ship for temporary suspension of equipment, are to be bonded in future. By defining the CCS in the plane machining surface, the problem of curve fitting in the given application scenarios can be reduced to the two-dimensional case.

In this paper, we present an approach for curve fitting from unorganized point clouds using k-Means clustering and a Gauss–Newton (GN) algorithm to identify the control points of a B-spline. Furthermore, we compare the GN approach with Sequential Least

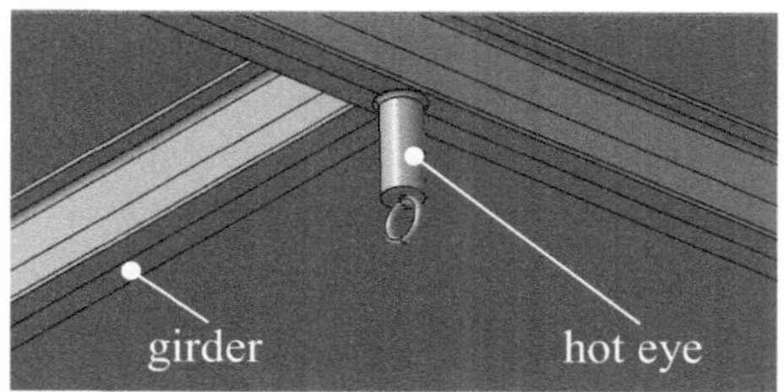

Fig. 1 Girder with hot eye

Squares Programming (SLSQP). The paper is organized as follows: First, previous work on the topic of curve fitting is presented, followed by a more detailed description of the proposed method. In the following chapter, the developed method is tested on several noisy and non-noisy point clouds. The results of our method are then compared with the SLSQP. Finally, the results obtained and the limitations of the methods are discussed.

2 Related Work

Curve and surface reconstruction is an important topic in the field of reverse engineering and other industrial applications. The problem of data fitting is particularly important in the area of generating CAD models from point clouds. B-splines are often used in CAD programs and curve fitting methods for the geometric description of curves and surfaces [4]. The curve reconstruction from point clouds can be divided into open or closed contours. Furthermore, the point clouds can be noisy and the curves can branch or self-intersect [5].

A widely used approach for curve fitting is the least squares fitting, which is well described by Wang et al. They distinguish between point distance minimisation (PDM) and tangent distance minimisation (TDM) [6]. Since the PDM method has a slow but stable convergence and the TDM converges fast but unstable, they introduce a new objective function, the squared distance minimization (SDM). The iterative determination of the footpoints leads to the biggest time consumption of the method [6].

Wan and Yin follow the curve fitting approach by interpolating intermediate points using cubic B-splines [7]. The interpolation points are determined by a time-efficient k-d tree and the search for the k-nearest neighbors of the vertices of the topological structure. Although this method is suitable for three-dimensional curve fitting, it can lead to problems if the point cloud is not dense enough and too few neighboring points are found [7].

To handle the varying density of point clouds along the curve, Ötztürk and Hasirci present an algorithm for dynamic bandwidth selection and subsequent regression of the local point sets of noisy point clouds [8].

Quan and Chen present a generalised shape fitting function to improve the fitting of geometric primitives. By choosing five parameters, polygons and ellipses can be generated and iteratively approximated even to incomplete measurement data. The method described provides better results than comparable methods, but is limited to the primitive shapes described in [9] and is therefore not suitable for arbitrary geometric shapes.

To handle large data sets with unordered points, Peng et al. present a curve reconstruction method based on k-Means clustering and principal component analysis of the point cloud [10]. To pre-process the point cloud, a K-means++ algorithm is used to segment the point cloud into k clusters. Similar to [8], a principal component analysis is performed for the identified clusters and a local regression vector is determined by the eigenvector. The regression vector can then be used to identify and delete those points as outliers whose distance to their projection points on the regression line is above a specified threshold. By retaining the projection points, the curve is thinned. Finally, the center points of the clusters are used to recursively approximate the curve using a four-point subdivision scheme [10]. Recent works such as [11] and [12] aim to improve the quality of the curve fitting by parameterizing a B-spline entirely by neural networks.

Our goal is to adapt the method described in [10] for determining the centre points of individual clusters for the classical least-squares curve fitting, in order to obtain a faster but robust curve fit from unorganised point clouds for the application of path planning of a mobile gluing robot. Instead of approximating the center points of the clusters and the thinned curve by the curve subdivision, we want to use the center points as starting values for the parameter identification. By improving the choice of starting parameters, we hope to improve the slow but stable convergence of the PDM method observed by [6] in order to maintain the comparatively simple calculation of the error term. Furthermore, the goal is to reduce the time required to calculate the footpoints by using a k-d tree instead of the Newton-like procedure described by [6].

3 Preliminaries

3.1 B-Spline Curves

In this paper we use rational B-spline curves for the fitting. A B-spline basis function can be described by the Cox-de Boor recurrence formula [4]. Let $\underline{U} = \{u_0, \ldots, u_m\}$ be a node vector consisting of m non-decreasing nodes of real numbers with $u_i \leq u_{i+1}$ and $i = 0, \ldots, m-1$, which describe the basis function of the B-spline of degree p. The basis functions, starting with $N_{i,0}(u)$, are calculated according to Eq. 1 and 2 [4].

$$N_{i,0}(u) = \begin{cases} 1 & if \ u_i \leq u < u_{i+1} \\ 0 & otherwise \end{cases} \tag{1}$$

$$N_{i,\mathrm{p}}(u) = \frac{u - u_i}{u_{i+p} - u_i} N_{i,\mathrm{p}-1}(u) + \frac{u_{i+p+1} - u}{u_{i+p+1} - u_{i+1}} N_{i+1,\mathrm{p}-1}(u) \tag{2}$$

Using the i basis functions and the matrix of control points $\underline{\boldsymbol{P}}$ a B-spline curve $\underline{C}(u)$ of degree p can now be calculated according to Eq. 3.

$$\underline{C}(u) = \sum_{i=0}^{n} N_{i,p}(u)\underline{P}_i \tag{3}$$

In our work we use curves with cubic B-splines ($p = 3$). Changing the control points changes the curve locally. The aim is therefore to identify suitable values for $\underline{\boldsymbol{P}}$, to ensure that the resulting B-spline curve optimally approximates the measured data.

3.2 Least Squares Fitting

In our case, the control points are determined by iteratively minimizing the function of the error squares. For this purpose, the next point of the curve $\underline{C}$ is first determined as the foot point $\underline{P}_f$ for each point of the data set $\underline{\boldsymbol{X}} = \left[\underline{X}_0, \underline{X}_1, \ldots, \underline{X}_n\right]^{\boldsymbol{T}}$ using a k-d tree, as shown in Fig. 2.

According to [6] in Eq. 4, this results in the quadratic error term $e_{PD,k}$ of the PDM method, which is iteratively minimized. Furthermore, the vector of residuals $\underline{f}$ or the objective function $\hat{f}$ is calculated by neglecting the regularization term f_s according to Eqs. 5 and 6.

$$e_{PD,k} = \left|\left|\underline{P}_f(u_k) - \underline{X}_k\right|\right|^2 \tag{4}$$

$$\underline{f} = \left[\ldots, e_{PD,k-1}, e_{PD,k}, e_{PD,k+1}, \ldots\right]^T \tag{5}$$

$$\hat{f} = \frac{1}{2}\sum_k e_{PD,k} \tag{6}$$

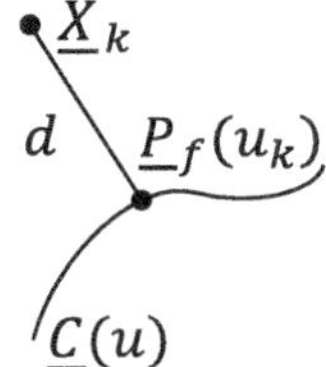

Fig. 2 Foot point $\underline{P}_f(u_k)$ of the data point $\underline{X}_k$ at the location u_k of the curve $\underline{C}(u)$

3.3 Gauss–Newton Method

The Gauss–Newton (GN) method provides a simple approach for the minimization of nonlinear objective functions, which is approximated by a second-degree Taylor polynomial. One advantage of the GN method is the less computationally intensive fitting of the Hessian matrix while neglecting the calculation of the partial second-degree derivatives [13]. The partial derivatives of the Jacobian matrix $\underline{\boldsymbol{J}}$ from Eq. 7 are calculated by varying the B-spline curve or the matrix of its initial control points $\underline{\boldsymbol{P}}_0$. These are varied by the increments $\underline{D} = [D_0, D_1, \ldots, D_m]^T$ with $\underline{\boldsymbol{P}}_p = \underline{\boldsymbol{P}}_0 + \underline{D}$ according to Eqs. 3 and 8.

$$\underline{\boldsymbol{J}} = \left[\frac{\partial \underline{f}}{\partial p_0}, \frac{\partial \underline{f}}{\partial p_1}, \ldots, \frac{\partial \underline{f}}{\partial p_m}\right] \tag{7}$$

$$\frac{\partial \underline{f}}{\partial p_i} \approx \frac{\sum_{i=1}^{m} N_i(u)\underline{P}_{0,i} - \sum_{i=1}^{m} N_i(u)\underline{P}_{p,i}}{\underline{D}_i} \tag{8}$$

The updated parameters $\underline{\boldsymbol{P}}^{k+1}$ of the k+1th iteration are calculated with the GN iteration step from Eq. 9. The parameter α describes the factor of the step size.

$$\underline{\boldsymbol{P}}^{k+1} = \underline{\boldsymbol{P}}^k - \alpha^k \left(\underline{\boldsymbol{J}}^T \underline{\boldsymbol{J}}\right)^{-1} \underline{\boldsymbol{J}}^T \underline{f} \tag{9}$$

A threshold value for $\hat{f}$ is defined as the termination criterion. Alternatively, the procedure ends when the difference in the residuals between two iteration steps is below a specified threshold value.

3.4 K-Means Clustering

The k-Means clustering is an unsupervised learning method used in data analysis and pattern recognition [14]. Similar to least squares fitting, the sum of squared distances is used to minimize the distance d between the data points $\underline{\boldsymbol{X}}$ and the center point $\underline{\mu}$ of each cluster. Equation 10 describes the objective function to be minimized [14].

$$f = \sum_{k=1}^{K} \sum_{\underline{X}_i \in C_k} d\left(\underline{X}_i, \underline{\mu}_k\right)^2 \tag{10}$$

Figure 3 shows an example of the result of k-Means clustering for a test data set of an application with a closed contour, the bonding of window panes on a cruise ship. The number of clusters k was determined as $k = 15$ as part of the parameter study in Chap. 4. The starting values $\underline{\boldsymbol{P}}_0$ correspond to the centres of the identified clusters.

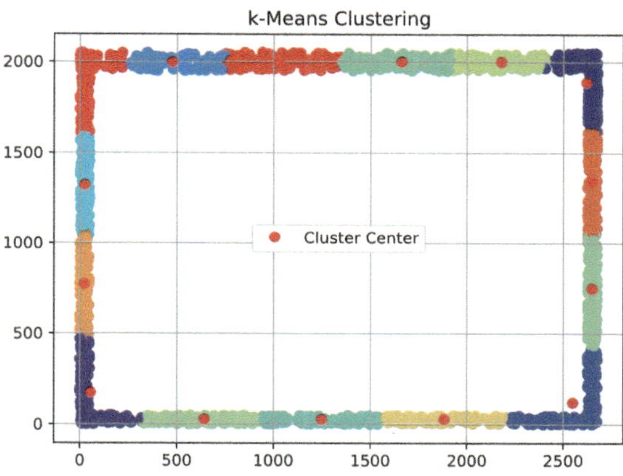

Fig. 3 k-Means clustering of the frame structure of a window pane with 15 clusters

4 Implemented Procedure

The curve fitting algorithm is implemented in Python using the theoretical principles explained above. Algorithm 1 represents the pseudo code of the procedure.

Algorithm 1 Pseudo code of the implemented procedure.

```
1. Calculate k Cluster with k-Means Algorithm
2. Set P0 to center points of clusters
3. IF curve shape is closed: sort P0
4. Start Gauss-Newton Method
   4.1. For i in range of maximal Iterations:
      4.1.1. Calculate previous Residuum f0
      4.1.2. Calculate partial derivatives and Jacobian J
      4.1.3. Set alpha = 1
      4.1.4. WHILE True:
         4.1.4.1.Update parameter P0 to Pn
         4.1.4.2.IF Curve shape is closed set last p con-
                 trol points to first p control points
         4.1.4.3.Calculate new Residuum fn
         4.1.4.4.IF break criteria are met: break
         4.1.4.5.ELSE: Set alpha /= 2
      4.1.5. Set f0 = fn
   4.2. Return optimized Control Points Popt and fn
5. Calculate B-Spline Curve with Popt
```

The SLSQP implementation of SciPy is used as a reference for the GN method [15]. To test the method, test data sets of different open and closed geometries of non-noisy and noisy point clouds are generated. These are generated by adding Gaussian noise with $\mu = 0$ and $\sigma^2 = 0,01$. The function $y = \sin(x)$ serves as an open contour. In addition

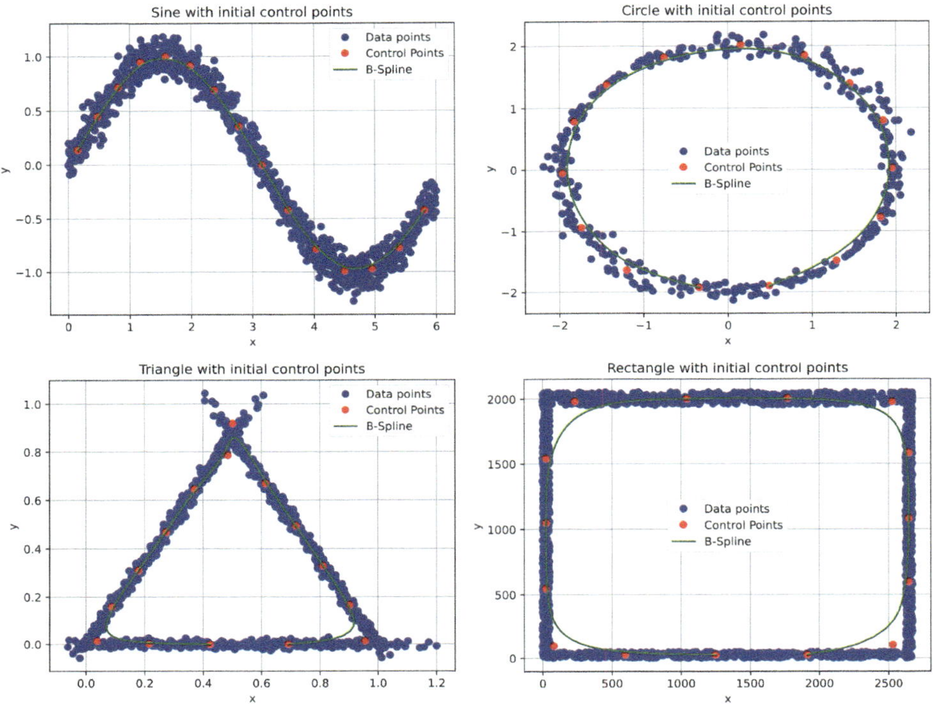

Fig. 4 Noisy test data sets with starting values selected by k-Means and the initial B-spline

to the of window frames use case (noisy rectangle), we test a triangular geometry, as contours with sharp corners pose a particular challenge to the fitting [5]. Figure 4 shows a selection of the noisy test data sets.

First, a parameter study is carried out to determine a suitable step size $\underline{D}$ and number of clusters k. For this purpose, the step sizes are varied between $\underline{D} = [1e-06; 1e+02]$ and $k = [10; 50]$. In order to assess the influence of the starting values $\underline{\boldsymbol{P}}_0$ selected by the k-Means algorithm on the convergence of the method, the starting values are selected in a further series of tests using an uninformed selection. To do this, a straight line is drawn through the mean Y value for the open contour. A circle is used accordingly for the closed contours. To assess the quality of the curve fitting, we record the Root Mean Squared Error (RMSE) as well as the number of interation steps n_{iter} and the time t of the procedure.

5 Results

Initially, the parameter study was used to determine $\underline{D} = 1e - 03$ as a suitable step size and $k = 15$ for most data sets. Table 1 and Table 2 show the results of the methods when applied to the test data sets. Here μ represents the mean value of the test series per dataset. The non-noisy curves are approximated almost exactly (see Fig. 5).

The uninformed start values led to unstable solutions with self-intersections or other errors in the contour in 15.6% of the runs. This could not be observed with the starting values selected by k-Means.

Compared to the SQLSP, slightly higher RMSE values were achieved with significantly shorter running times. Increasing the number of control points/ clusters does not have a significant effect on the runtime up to $k = 50$, as fewer iterations with smaller changes are required.

Table 1 Approximated datasets with 1000 points, $\underline{D} = 1e - 03$ and 15 control points

Gauss–Newton								
Improvements	RMSE		21,23%		Iter	16,39%	Time [s]	4,49%
	RMSE k-Means		RMSE Uninf		Iter k-Means	Iter Uninf	t k-Means	t uninf
	μ	σ	μ	σ	μ	μ	μ	μ
Sine	1.73E + 01	3.73E − 01	2.61E + 01	3.37E + 00	7.6	22.6	2.282	6.352
Circle	3.91E + 00	2.86E − 01	3.16E + 00	4.34E − 16	7	8	0.787	0.895
Circle (Noise)	2.96E + 01	1.44E + 00	3.81E + 01	2.07E + 00	7.4	23.8	2.223	6.591
Triangle	1.53E + 00	2.34E − 01	1.00E + 00	0.00E + 00	24	10	6.602	2.897
Triangle (Noise)	5.46E + 00	1.07E − 01	5.90E + 00	5.88E − 01	7	12	2.096	3.434
Rectangle	3.60E + 00	5.01E − 01	3.74E + 00	4.34E − 16	18.4	9	10.367	5.334
μ	1.02E + 01	4.90E − 01	1.30E + 01	1.01E + 00	11.9	14.2	4.1	4.3

Table 2 SQLSP with 15 Control Points

Sequential Quadratic Programming								
Improvements	RMSE		11,86%		Iter	4,94%	Time [s]	6,15%
	RMSE k-Means		RMSE Uninf		Iter k-Means	Iter Uninf	t k-Means	t uninf
	μ	σ	μ	σ	μ	μ	μ	μ
Sine	1.59E + 01	5.69E − 01	1.66E + 01	1.15E + 00	51.8	65.6	14	17
Circle	3.13E + 00	1.89E − 01	3.16E + 00	4.34E − 16	41.2	23	4	2
Circle (Noise)	2.54E + 01	8.46E − 01	3.13E + 01	2.67E + 00	53.2	64.8	14	17
Triangle	1.00E + 00	0.00E + 00	1.00E + 00	0.00E + 00	43.2	47	12	13
Triangle (Noise)	5.04E + 00	1.88E − 01	5.37E + 00	5.72E − 01	48	56.2	13	15
Rectangle	2.56E + 00	2.02E − 01	2.65E + 00	0.00E + 00	47.4	43	26	24
μ	8.82E + 00	3.32E − 01	1.00E + 01	7.33E − 01	47.5	49.9	13.8	14.7

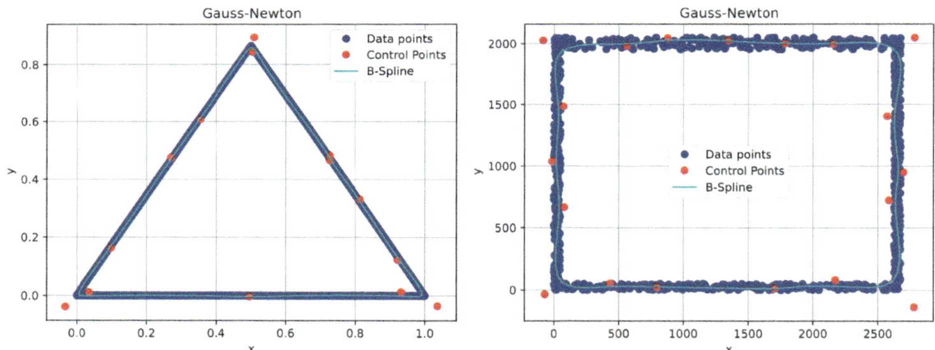

Fig. 5 Triangle and noised rectangle fitting with 15 Control Points and d = 1e−03

6 Discussion

The presented approach improved the stability and speed of the conventional least squares curve fitting and will therefore be used for the path planning of the gluing process. Although the method approximates the part contour, the achievable accuracy of the method depends on the choice of the parameters of the step size and the number of control or start points. Therefore, approaches for automatic parameterisation such as [11, 12] are of further interest.

Acknowledgements This Project is supported by the Federal Ministry for Economic Affairs and Climate Action (BMWK) on the basis of a decision by the German Bundestag.

References

1. Deutscher Verband für Schweißen und verwandte Verfahren e.V.: Betriebsfestigkeitsgerechte Gestaltung von Ausrüstungselementen für den Schiffbau. Merkblatt DVS 3501 (2016)
2. Center of maritime Technologies gGmbH (CMT): Untersuchungen zum aktuellen internationalen Stand der Technik bezüglich der Entwicklung und Anwendung von Klebverbindungen in der Schiffsfertigung (2020)
3. Braun, T., Schack, C., Flügge, W.: Vergleich von optischen 3D-Sensoren zur Feinreferenzierung eines Klebprozesses für schiffbauliche Komponenten. In: Rothstock, S., Hohnhäuser, B., Krueger, D., Pochanke, M. (eds.) 3D-NordOst 2023: 25. Anwendungsbezogener Workshop zur Erfassung, Modellierung, Verarbeitung und Auswertung von 3D-Daten, vol. 1, ISBN: 978-3-942709-33-0, Gesellschaft zur Förderung angewandter Informatik e.V. (GFaI), Berlin (2024)
4. Piegl, L., Tiller, W.: The NURBS Book. 2nd edn. ISBN: 978-3-540-61545-3. Springer, Berlin, Heidelberg (1997)
5. Hasirci, Z., Öztürk, M.: The comparison of region growing algorithms with using EMST for point clouds. In: 2015 38th International Conference on Telecommunications and Signal Processing (TSP), pp. 1–5. IEEE. ISBN: 978-1-4799-8498-5 (2015). https://doi.org/10.1109/TSP.2015.7296396
6. Wang, W., Pottmann, H., Liu, Y.: Fitting B-spline curves to point clouds by curvature-based squared distance minimization. ACM Transactions on Graphics, vol. 25, pp. 214–238, ISSN: 0730-0301 (2006). https://doi.org/10.1145/1138450.1138453
7. Wan, Y., Yin, S.: Three-dimensional curve fitting based on cubic B-spline interpolation curve. In: 2014 7th International Congress on Image and Signal Processing., pp. 765–770. IEEE (2014). https://doi.org/10.1109/CISP.2014.7003880
8. Ötztürk, M., Hasirci, Z.: A novel dynamic bandwidth selection method for thinning noisy point clouds. Turk. J. Electr. Eng. Comput. Sci. **21**, 2239–2258 (2013). https://doi.org/10.3906/elk-1203-114
9. Quan, Y., Chen, S.: Optimised Least Squares Approach for Accurate Polygon and Ellipse Fitting. arXiv preprint (2023). arXiv:2307.06528
10. Peng, K., Tan, J., Zhang, G.: A method of curve reconstruction based on point cloud clustering and PCA. Symmetry **14**. ISSN: 2073–8994 (2022). https://doi.org/10.3390/sym14040726
11. Laube, P., Franz, M.O., Umlauf, G.: Deep learning parametrization for B-spline curve fitting. In: 2018 International Conference on 3D Vision (3DV), pp. 691–699. IEEE, (2018). https://doi.org/10.1109/3DV.2018.00084
12. Rodas, D.E.R.: Using Neural Networks to Generate and Optimize Parameterization of Point Data for Fitting Spline Curves and Surfaces. Doctoral thesis, Johannes Kepler University Linz, Linz (2023)

13. Nocedal, J., Wright, S.J.: Numerical Optimization. 2nd edn. Springer New York. ISBN: 978-0-387-40065-5 (2006). https://doi.org/10.1007/978-0-387-40065-5
14. Botsch, B.: Maschinelles Lernen—Grundlagen und Anwendungen. 1st edn. Springer Spektrum Berlin, Heidelberg, ISBN: 978-3-662-67277-8 (2023). https://doi.org/10.1007/978-3-662-67277-8
15. Virtanen, P., Gommers, R., Oliphant, T.E. et al.: SciPy 1.0: fundamental algorithms for scientific computing in Python. Nat. Methods **17**, 261–272 (2020). https://doi.org/10.1038/s41592-019-0686-2

ExoPowerCheck: A Performance Test Bench to Measure Support Behavior of Exoskeletons with Single-DoF Support

Luis Dangel, Benjamin Reimeir, and Robert Weidner

Abstract

Exoskeletons for industrial applications offer a promising approach in reducing physical strain, a risk factor for work-related musculoskeletal diseases. However, comparability of different devices is complicated as various physiological, kinematic, and kinetic measurement methods are common practice for assessment. The objective of this study was to assess the validity of a self-developed performance test bench to measure static and dynamic support behavior of industrial exoskeletons with single degree-of-freedom support. Two upper-limb exoskeletons (Comau Mate—CM, Levitate Airframe—LA) were used for reliability assessment. Root mean squared error between observed data and valid reference was below 5% for static and between 10 and 20% ($\dot{\theta} = 100$—$400°/s2$) for dynamic measurements. Test—retest reliability was high for both tested exoskeletons based on $ICC_{CM} = 0.99$ and $ICC_{LA} = 0.98$ with a $TEM_{CM} = 0.09$ Nm (2.4%) and $TEM_{LA} = 0.2$ Nm (5.6%). Likewise, inter-operator reliability was good with ICC = 0.95 and a TEM = 0.25 Nm (5.8%). Overall, validity and static reliability criteria for industrial exoskeleton assessment were well satisfied. Dynamic

L. Dangel · B. Reimeir (✉) · R. Weidner
Chair of Production Technology, Institute of Mechatronics, University of Innsbruck, Innsbruck, Austria
e-mail: benjamin.reimeir@uibk.ac.at

B. Reimeir
Department of Sport Science, University of Innsbruck, Innsbruck, Austria

R. Weidner
Laboratory of Manufacturing Technology, Helmut-Schmidt-University/University of the Federal Armed Forces Hamburg, Hamburg, Germany

M.-C. Wanner et al. (eds.), *Annals of Scientific Society for Assembly, Handling and Industrial Robotics 2024*, https://doi.org/10.1007/978-3-031-91463-8_35

reliability for movements with high angular accelerations ($\dot{\theta}$ > 100°/s2) needs further optimization. Continuous measurements of support behavior form the fundamental basics for reliable biomechanical models of human-exoskeleton interactions.

Keywords

Exoskeleton • Support behavior • Performance • Industrial ergonomics

1 Introduction

1.1 Background

Exoskeleton technology has emerged as a promising approach to reduce physical strain and enhance productivity in various industries, particularly in manual labor-intensive sectors [1]. Especially working scenarios in production and logistics pose high physical strain on workers during tasks like lifting and carrying loads or overhead assembly work [2]. Hence, body areas showing the highest prevalence of work-related musculoskeletal diseases (MSD) are the shoulder and lower back region [2]. Exoskeletons offer physical support to the human body, augmenting movements through the application of torques and forces along different joints [3]. However, the diversity of exoskeleton designs and functionalities poses challenges in evaluating their performance and effectiveness especially regarding the frequently proclaimed goals of reducing excessive joint loads and muscular strain, both risk factors for work-related MSD. Exoskeleton designs vary from passive devices using elastic elements or springs for mechanical support to active systems employing external powered actuators for enhanced strength augmentation, targeting different body regions such as lower limb or upper body exoskeletons [4]. The lack of standardized evaluation protocols combined with factors such as variations in user-anthropometry, physical condition, and individual preferences further complicates the comparative assessment of exoskeletons for industrial applications. Manufacturing enterprises continuously seek potential solutions for mitigating work-related musculoskeletal disorders [5]. Consequently, within the industrial context, the decision to procure and deploy exoskeletons relies on standardized metrics regarding their effects on MSD risk.

1.2 State of the Art

While some studies focus on objective measures such as biomechanical loading and generated supportive force [6, 7], subjective factors like user satisfaction [8] and perceived exertion and fatigue [9, 10] also play a significant role in determining the overall effectiveness of exoskeletons. Tracing the chain of causality to its root, the torque generated by the device around the supported joint serves as an objective starting point in the mechanical assessment of exoskeletons. Various test benches have been developed on that basis to measure the generated torque as a function of the joint angle [11, 12]. Among the various testing setups, a notable approach entails the utilization of a mannequin-based framework and an isokinetic dynamometer for static and dynamic analysis of hip supportive Exoskeletons [11]. Another setup to test shoulder exoskeletons consists of two motor-controlled rotating arms connected through a force sensor facilitates static and dynamic measurements [12]. Additionally, there is a setup replicating the human body structure, equipped with actuators at hip and knee joints, suitable for testing hip and lumbar spine-supporting exoskeletons [13]. One further approach for electric motor-driven exoskeletons is to achieve torque measurement by monitoring the motor's current consumption, allowing measurements while wearing the exoskeleton [14]. However, regarding precision this method can't keep up with previous mentioned methods. Lastly, a simplified measurement method using a digital force sensor and angle sensor for torque measurement is only suitable for static measurements. This method compromises reliability and objectivity due to its reliance on the operator [15].

Most testing devices mentioned are designed for specific exoskeleton models, lacking adaptable interfaces, geometric dimensions, and a standardized protocol for comparison. Additionally, few devices enable dynamic testing. As a result, accommodating different exoskeleton types on a single test bench is unfeasible. Testing various exoskeleton models on the same test bench would significantly improve the comparability of test results. Furthermore, there is interest in analyzing the supportive torque characteristics of exoskeletons during human movement.

1.3 Aim of the Study

The developed test bench aims to provide a valid method to compare different exoskeleton models supporting the same body region. It records torque curves across one supported Degree of Freedom (DoF) joint, covering various commercially available exoskeletons, which mostly have a single supported DoF and additional passive unsupported DoFs to allow more complex movements of the joint. The validity test aims for an error below 5% compared to reference and measurement repeatability should be comparable or better than proposed solutions in literature.

2 Testbench and Methods

2.1 Concept and Structure of the Test Bench

Functioning: A fundamental component of the test bench constitutes the basic structure, which establishes a connection to the actuator, drive shaft, and measuring unit (see Fig. 1). The torso interfaces of the exoskeleton are mounted on the basic structure of the test

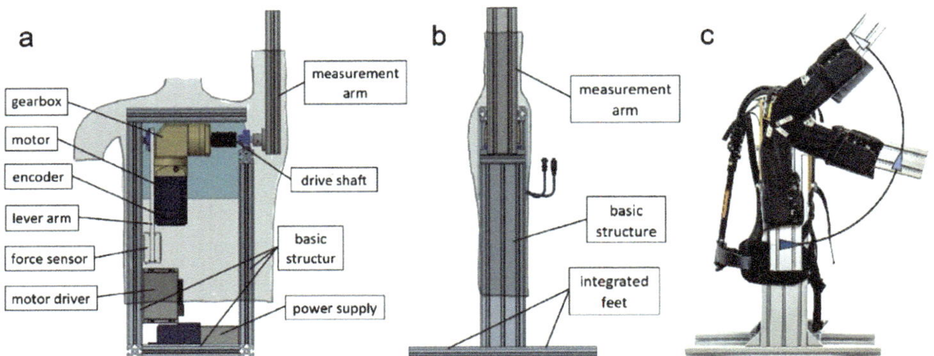

Fig. 1 Test bench from **a** front and **b** side view with the structural main parts labelled. The grey shaded overlay indicates a human upper body which served as a design basis. The dashed-dotted line and the cross represent the axis of rotation **c** for an exemplary mounted exoskeleton

bench, and the limb interfaces are fixed on the measurement arm which is driven by the actuator. The basic structure accommodates the motor-gearbox unit and the force sensor. The shaft, driven by the motor-gearbox unit, transmits torque and movement to the measurement arm. The motor-gearbox unit, along with the shaft and measurement arm, is rotationally mounted within the basic structure to enable torque measurement using a force sensor. Its rotational movement is restrained by a lever arm connected to the basic structure, which incorporates a force sensor at the junction. Consequently, the total torque exerted around the axis of the shaft is absorbed by the force sensor. The resultant torque $\boldsymbol{\tau}(\boldsymbol{\theta}, \dot{\theta}) = \mathbf{F}(\boldsymbol{\theta}, \dot{\theta}) \times \mathbf{d}$, where $\mathbf{F}(\boldsymbol{\theta}, \dot{\theta})$ represents the force captured by the force sensor and **d** the length of the lever arm. This approach allows the implementation of a uniaxial force sensor which stays static during measurements.

Sensors: The sensor system integrated into the test bench comprises a force sensor for torque measurement and a position encoder for precise positioning of the measuring arm. The force sensor utilized is a six-axis force-torque sensor (*Schunk FTN-Mini-45 SI-580–20*). This sensor employs silicon strain gauges to measure applied loads across all six degrees of freedom (F_x, F_y, F_z, M_x, M_y, M_z). Amplification and analog–digital conversion are realized by an external unit (*Netbox*). For this test bench, force measurement is only needed in one direction, thus utilizing a single sensing direction of the sensor would be sufficient. Utilizing the six-axis force-torque sensor minimizes excessive errors in force derivation. The position encoder is seamlessly integrated within the stepper motor assembly, forming a closed-loop configuration with the motor driver, ensuring precise positioning accuracy (Fig. 2).

Software and Control: To conduct measurements, the measuring arm follows prescribed movements while simultaneously recording measurement data. A system is implemented

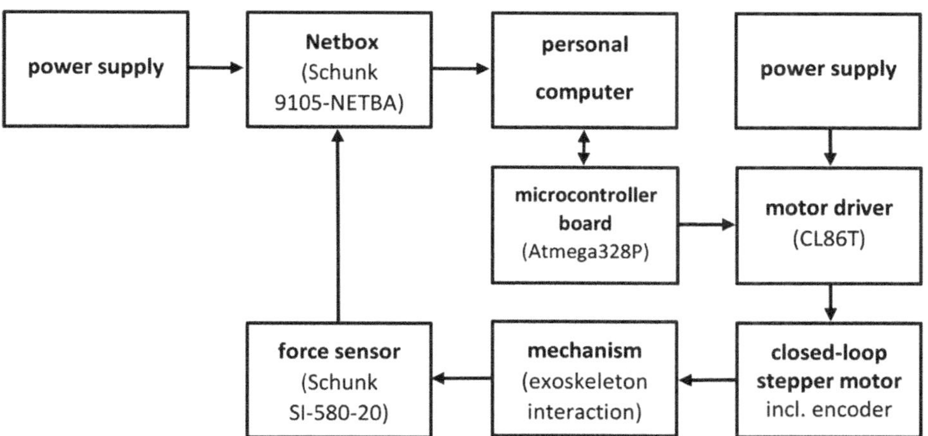

Fig. 2 A schematic overview of the integration and interactions of the hardware components

using a development board (featuring the *ATmega328P* microcontroller) in conjunction with a stepper motor driver to control the motor. This actuator is configured as a closed-loop stepper motor, exhibiting a holding torque of 9 Nm, augmented by a subsequent 10:1 planetary gearbox to amplify the output torque. User inputs are intended to be facilitated via a personal computer (PC), necessitating communication between the development board and the PC.

For the data exchange between the PC and the microcontroller board, the serial interface of the development board is utilized. A *Python3*-program provides the user interface and handles the reading, processing, and visualization of the measurement data. Depending on user inputs, commands are relayed to the *C/C* + + -program implemented on the development board to realize the control of the stepper motor. During a measurement the current motor position and the measurement value from the force sensor is received by the *Python3*-program at a predetermined frequency.

Operation: At initial stage of a measurement, the movable measuring arm is manually set to its zero configuration, followed by the selection of the operating mode. Basic option allows for individual configuration of the movements executed by the measuring arm. This operating mode is intended for conducting measurements with simple movements between two points and is parameterized by a user-defined target position, target velocity, and acceleration. To create more complex motion sequences as encountered in real-word tasks involving an exoskeleton, motion data can be utilized as motion directives. A specific mode facilitates this function, enabling the reading and execution of pre-existing motion data.

The test bench's mobile components, possessing inherent weight and moment of inertia, generate torque moments during accelerated movements, affecting measurement accuracy. To address this, the control system conducts a pre-measurement calibration without the exoskeleton to compensate for inertia-induced errors. The resulting data are filtered, graphically visualized, and stored for further analysis.

2.2 Study Design for the Verification of Quality Criteria

Validity: To assess the validity of the measured outcome of the test bench $\boldsymbol{\tau}$ $(\boldsymbol{\theta}, \dot{\theta})$, two validity tests were performed in an experimental setup. A quasi-static test to measure the validity of $\boldsymbol{\tau}$ $(\boldsymbol{\theta})$ and a dynamic test investigating the influence of angular acceleration $\ddot{\theta}$. The setup consisted of full range 0° to 180° motions of the measurement arm to assess validity over the whole measurement range. Calibrated weights were used at a fixed lever arm of 300 mm to apply a reference torque. The validity test compared the measured outcome of the experiment with the calculated analytical values.

Static validity test. Control configurations of the test bench were set to an acceleration of $\ddot{\theta}$ = 50°/s2 until measurement arm velocity of $\dot{\theta}$ = 5°/s was reached. Seven different weights (0.25–20 kg) were tested each with ten repetitions. The weights apply a peak torque between 0.7 and 58.9 Nm, covering the generated support torque of most commercial shoulder- and back-supporting exoskeletons.

Dynamic validity test. Three different control configurations were tested ($\ddot{\theta}$ = 100°/s2, 250°/s2, 400°/s2). A constant acceleration was applied until the measurement arm reached $\boldsymbol{\theta}$ = 90° followed by a constant deceleration until the turning point ($\boldsymbol{\theta}$ = 180°). A 5 kg-weight was applied and ten repetitions were conducted and averaged for each configuration.

Intra-operator reliability: The static support characteristics $\boldsymbol{\tau}$ $(\boldsymbol{\theta}, \dot{\theta} = 5°/s)$, of two commercial shoulder supporting exoskeletons—*Levitate Airframe* (LA; settings: range 90° & assistance level 3) and *Comau Mate* (CM; settings: assistance level 8)—were measured. Depending on the exoskeleton design the measurement arm performed a 0°–120° motion (LA) or a 0°–160° motion (CM). Each exoskeleton was tested 20 times, whereby the device was fully unmounted from the test bench after each repetition.

Inter-operator reliability: Six operators participated and performed a test procedure according to ExoPowerCheck user instructions. Each operator conducted ten measurement repetitions with the CM exoskeleton and predetermined control configurations ($\boldsymbol{\theta}$ = 0° to 160°, $\dot{\theta}$ = 5°/s) to investigate the influence of different non-trained operators on the outcomes of the test bench. A test procedure took about 45 min per operator.

Data processing and statistical analysis: The raw data was filtered using a 5th order low pass Butterworth filter with a dynamic cut off frequency depending on the control configurations. Measurement with $\dot{\theta} = 5°/s$ arm velocity were filtered with a 0.2 Hz cut off frequency. For higher velocities the cut off frequency was adjusted accordingly.

Each measured condition was calculated by averaging ten measurement repetitions. Static and dynamic validity results are reported as root mean squared (RMSE) and maximum errors (MaxE) between a measured condition and the analytical values. Intra-operator reliability was analyzed by a test–retest design and is reported using Pearson's r and absolute and relative typical error of measurement (TEM). Intraclass Correlation Coefficients (ICC) were fit using single-rating, absolute-agreement, two-way mixed-effects models for the test–retest design ($ICC_{A,1}$ under McGraw and Wong (1996) conventions). Inter-operator reliability was calculated based on the data collected from six selected test bench operators and is also reported as $ICC_{A,1}$ and a comparison of within- vs between-operator variance.

Results are presented as averaged time series data ± standard deviation or means over the averaged time series. Data processing and statistical analysis was performed using Matlab R2021b.

3 Results

3.1 Validity Analysis

Relative errors between the measured values and the analytical reference for the static validity test are shown in Fig. 3. Relative RMSE was between 1.0 and 4.5% across the tested loads (0.25–20 kg). MaxE that occurred at any time within a measurement was between 1.8 and 11.5% of the respective load.

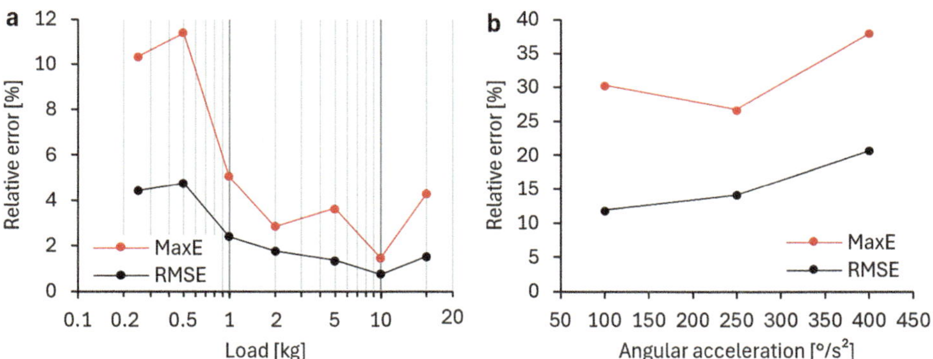

Fig. 3 Averaged relative error from 10-repetition-measurements of **a** seven different load conditions under quasi-static control configurations ($\dot{\theta} = 5°/s$) and of **b** three different control configurations ($\ddot{\theta} = 100–400°/s2$). The load was constant for this test with 5 kg

The highest relative error with 4.5 and 11.5% for RMSE and MaxE respectively, was observed in the 0.5 kg-condition, which led to a peak torque (at $\boldsymbol{\theta} = 90°$) of ~ 1.5 Nm, while the lowest relative error was measured at the 10 kg-condition (corresponds to ~ 30 Nm).

Relative RMSE for the three different configurations tested in the dynamic validity test were within 10–20% between the measured and the analytical outcomes (see Fig. 3b). With increasing accelerations, hence increasing mass inertia, the relative RMSE increased as well. Relative MaxE was between 26 and 38%, with its lowest value at a control configuration of $\ddot{\theta} = 250°/s2$. The 5 kg-load used for the dynamic validity test resulted in an analytical peak torque of 16.5 Nm, which is within the range of commercial exoskeletons.

3.2 Intra-Operator Reliability

The number of measurement runs that need to be performed to deduce a stable measurement was concluded to be at least ten. This is based on an analysis of intra-operator variance between single measurement runs. Figure 4 shows the relationship between the RMSE (calculated to a 20-run-average) and the number of runs included in a measurement. With an increasing number of measurement runs, fluctuations and noise are cancelled out and the measurement error to the 20-run-average decreases. Beyond eight measurement runs the error decreases linearly, indicating a decline in information gain from additional observations.

Test–Retest analysis of the CM exoskeleton showed a Pearson's r = 0.9974 and an ICC = 0.9968 (95% CI: 0.9944—0.9979). The TEM averaged over a whole measurement was 0.09 Nm (2.4%). Reliability results for the LA exoskeleton revealed r = 0.9935 and ICC = 0.9896 (95% CI: 0.9890—0.9902). The average TEM was higher than for CM with 0.20 Nm (5.6%). Figure 5 shows the results of the measurement for CM and LA over the full measurement range of the respective exoskeletons as mean over ten repetitions ± standard deviation.

3.3 Inter-Operator Reliability

High inter-operator reliability was observed with an ICC of 0.9538 (95% CI: 0.9509—0.9565) and a between-operator averaged TEM of 0.25 Nm (relative TEM = 5.8%). Figure 6 shows the individual measurements and the group mean over the whole measurement range ($\boldsymbol{\theta} = 0$—$160°$) with the corresponding standard deviation of the mean. Between-operator deviations were elevated at the start-/endpoint ($\boldsymbol{\theta} = 0°$) and the turning point ($\boldsymbol{\theta} = 180°$) of the measurement. Lower deviations appeared around an elevation angle of 70°–80°.

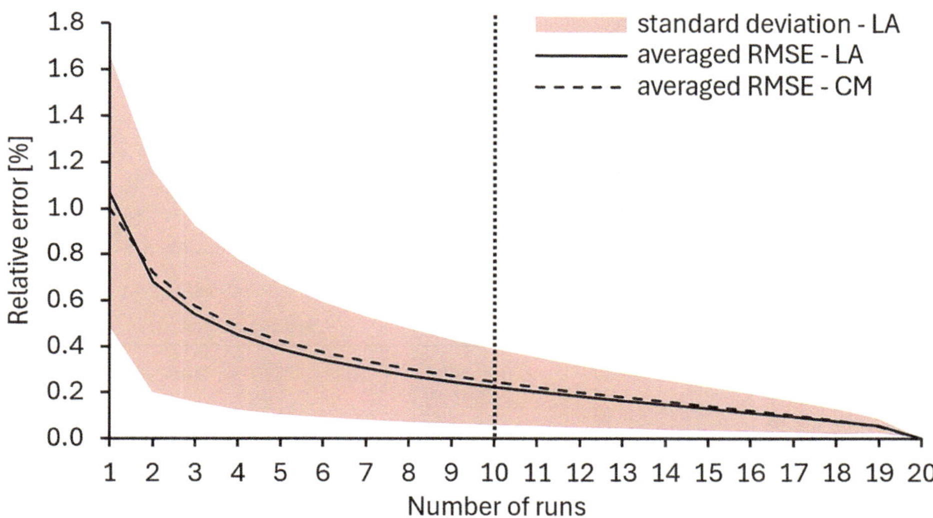

Fig. 4 Simulated relation between RMSE of averaged measurements and the number of repetitions included to calculate that average. Δ *Relative error* levels off at around eight to ten runs. The dotted line shows the chosen number of runs for this study

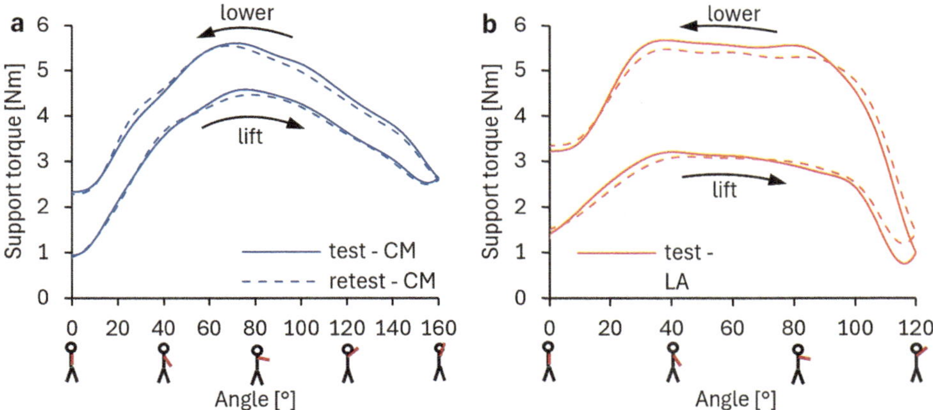

Fig. 5 The CM (**a**) and the LA (**b**) static support behavior measured in a test–retest design. Black arrows indicate the movement direction of the measurement arm ($\dot{\theta} = 5°/s$)

An analysis of variances showed higher between-operator than within-operator variance by 12% (between sum of squares = 3.6094, within sum of squares = 3.2238).

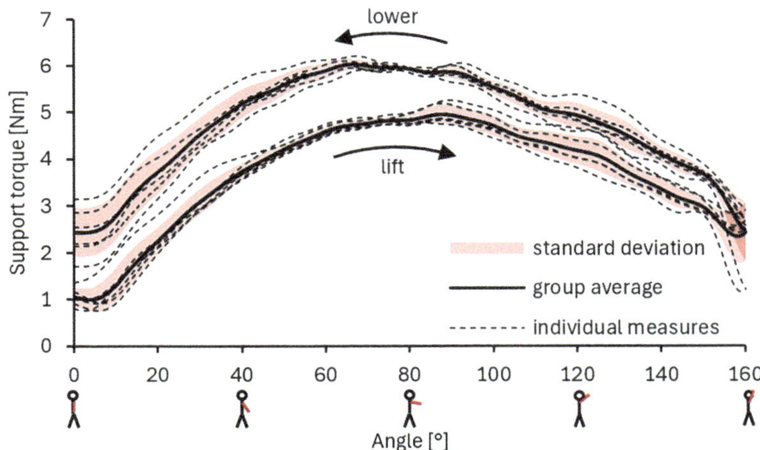

Fig. 6 The averaged CM static support behavior over all six operators in the inter-operator reliability test. Black arrows indicate the movement direction of the measurement arm ($\dot{\theta} = 5°/s$)

4 Discussion

In summary, the investigation of measurement performance in the static validity test reveals an average relative error below 5%. Thus, a high level of static validity can be observed across the entire measurement range. A higher average relative error between 10 and 20%, occurs during dynamic measurement with high accelerations between 100 and 400°/s^2. A probable cause might lie in the bearing of the force sensor, which was coupled to the mechanism with a small slack to allow proper reorientation during push-/ pull scenarios to minimize force-loss due to torque-carryover. In this case, a redesign of the sensor bearing could improve the dynamic reliability of the test bench. The test–retest reliability (intra-operator reliability) demonstrates a very stable test result, as evidenced by the low typical error averaged over the whole measurement (CM: 2.4%; LA: 5.6%).

The testbench described in this study demonstrates higher intra-operator reliability compared to a previously developed model based on the LA assessment [16]. Their setup utilized a dynamometer for torque assessment of LA (settings: range 90°, assistance level: 3), resulting in an ICC of 0.97 (95% CI: 0.94–0.99). In contrast, this study revealed an ICC_{LA} of 0.9986 (95% CI: 0.9985–0.9987).

Furthermore, the quasi-static torques $\boldsymbol{\tau}(\boldsymbol{\theta})$ of LA and CM determined in this study are comparable to measurements from other testing setups. However, only flexion results of the exoskeleton joint are considered, as extension data are mainly lacking in literature so far. Comparison of LA torques (settings: range 90°, assistance level 3) shows alignment with literature findings. The maximal torques $\boldsymbol{\tau}_{\mathbf{max}}(\boldsymbol{\theta})$ align closely at approximately 5.5 Nm, while a decline of torque is observed towards the peripheries. Admittedly, there is a noticeable shift in the torque profile, suggesting a potential difference of the defined zero angle reference [16]. For CM, a similar measurement result for the quasi-static torque $\boldsymbol{\tau}(\boldsymbol{\theta})$ can also be found, displaying a very close agreement with the results depicted in this study. Here, the maximal torques $\boldsymbol{\tau}_{\mathbf{max}}(\boldsymbol{\theta})$ for both slightly above 5.5 Nm, and the decline of torque at the peripheries of 0° and 160° to approximately 2.5 Nm exhibits substantial resemblance [12].

The involvement of various test stand operators introduces additional variabilities, leading to a lower inter-operator reliability, indicated by an ICC_{CM} of 0.9538 (intra-operator reliability: $ICC_{CM} = 0.9996$). The additional variability is primarily attributed to the attachment of the exoskeleton to the testbench, as further parameters are not subject to operator influence due to the computer-controlled operation of the testbench. Consequently, the importance of detailed training of the test stand operator should be emphasized. Despite various operators, the characteristic torque curves of different exoskeletons can be distinctly differentiated from one another. Nevertheless, it is important to ensure that the interfaces are attached correctly and that the axis of rotation of the supported joint is precisely aligned with the rotation axis of the testbench.

A limitation to the presented testbench is that it enables investigation of the exoskeleton-joint only about one DoF, as rotation of the measurement arm and torque detection is restricted exclusively to one axis. Consequently, the testbench is confined to measuring exoskeletons wherein the respective joint is supported to rotate about a single axis or, in cases of models with multiple supported DoF, only one can be examined. During measurements, the measuring arm achieves a maximum velocity of 420°/s, with exoskeletons supporting a maximum torque of 80 Nm being permissible for the testbench. Measurements involving acceleration above $100°/s^2$ should be approached with caution, as they entail a relatively substantial margin of error.

5 Conclusion

In general, the proposed test bench satisfied the validity and reliability criteria for the assessment of industrial exoskeletons. Nevertheless, dynamic reliability should be optimized to enable accurate assessments of dynamic support behaviors. A thorough introduction in the operations of the test bench is an important element to enable stable and reliable measurements. The testbench enables generalizable and comparable measurements between different exoskeleton models supporting the same body region. The insights obtained through the investigation of continuous dynamic measurements in concentric and eccentric movements, especially from human motion capture data, constitute a valuable foundation for further development and research in the field. This data can be integrated into biomechanical simulations, thereby enhancing and expanding their capabilities. Furthermore, the acquired insights can propel the advancement of more demand-oriented designs of exoskeletons.

References

1. Vallée, A.: Exoskeleton technology in nursing practice: assessing effectiveness, usability, and impact on nurses' quality of work life, a narrative review. BMC Nurs. (2024)
2. Roquelaure, Y.: Musculoskeletal disorders and psychosocial factors at work. SSRN Electron. J. (2018). https://doi.org/10.2139/ssrn.3316143
3. Auer, S., Tröster, M., Schiebl, J., Iversen, K., Chander, D.S., Damsgaard, M., Dendorfer, S.: Biomechanical assessment of the design and efficiency of occupational exoskeletons with the anybody modeling system. Zeitschrift für Arbeitswissenschaft **76**(4), 440–449 (2022)
4. Hoffmann, N., Prokop, G., Weidner, R.: Methodologies for evaluating exoskeletons with industrial applications. Ergonomics **65**(2), 276–295 (2022)
5. Smets, M.: A field evaluation of arm-support exoskeletons for overhead work applications in automotive assembly. IISE Trans. Occup. Ergon. Hum. Factors **7**(3–4), 192–198 (2019)
6. Koopman, A.S., Kingma, I., Faber, G.S., de Looze, M.P., van Dieën, J.H.: Effects of a passive exoskeleton on the mechanical loading of the low back in static holding tasks. J. Biomech. **83**, 97–103 (2019)
7. Rathore, A., Wilcox, M., Ramirez, D.Z.M., Loureiro, R., Carlson, T.: Quantifying the human-robot interaction forces between a lower limb exoskeleton and healthy users. In: 2016 38th Annual International Conference of the IEEE Engineering in Medicine and Biology Society (EMBC), pp. 586–589 (2016)
8. Ingraham, K.A., Tucker, M., Ames, A.D., Rouse, E.J., Shepherd, M.K.: Leveraging user preference in the design and evaluation of lower-limb exoskeletons and prostheses. Curr. Opin. Biomed. Eng. (2023)
9. De Bock, S., Ampe, T., Rossini, M., Tassignon, B., Lefeber, D., Rodriguez-Guerrero, C.: Passive shoulder exoskeleton support partially mitigates fatigue-induced effects in overhead work. Appl. Ergon. (2023)
10. Argubi-Wollesen, A., Weidner, R.: Biomechanical analysis: adapting to users' physiological preconditions and demands. In: Developing Support Technologies: Integrating Multiple Perspectives to Create Assistance that People Really Want, pp. 47–61 (2018)
11. Madinei, S., Kim, S., Park, J.H., Srinivasan, D., Nussbaum, M.A.: A novel approach to quantify the assistive torque profiles generated by passive back-support exoskeletons. J. Biomech. **145** (2022)
12. Hartmann, V.N., Rinaldi, D.D.M., Taira, C., Forner-Cordero, A.: Industrial upper-limb exoskeleton characterization: paving the way to new standards for benchmarking. Machines **9**(12), 362 (2021)
13. Nabeshima, C., Ayusawa, K., Hochberg, C., Yoshida, E.: Standard performance test of wearable robots for lumbar support. IEEE Robot. Autom. Lett. **3**(3) (2018)
14. Ito, T., Ayusawa, K., Yoshida, E., Kobayashi, H.: Evaluation of active wearable assistive devices with human posture reproduction using a humanoid robot. Adv. Robot. **32**(12), 635–645 (2018)
15. Tang, L., Liu, G., Yang, M., Li, F., Ye, F., Li, C.: Joint design and torque feedback experiment of rehabilitation robot. Adv. Mech. Eng. **12**(5) (2020)
16. Watterworth, M.W., Dharmaputra, R., Porto, R., Cort, J.A., La Delfa, N.J.: Equations for estimating the static supportive torque provided by upper-limb exoskeletons. Appl. Ergon. **113** (2023)

BY

Author Index

M.-C. Wanner et al. (eds.), *Annals of Scientific Society for Assembly, Handling and Industrial Robotics 2024*, https://doi.org/10.1007/978-3-031-91463-8

J

K

L

M

N

O

P

R

S

T

V

GPSR Compliance

The European Union's (EU) General Product Safety Regulation (GPSR) is a set of rules that requires consumer products to be safe and our obligations to ensure this.

If you have any concerns about our products, you can contact us on ProductSafety@springernature.com

In case Publisher is established outside the EU, the EU authorized representative is:

Springer Nature Customer Service Center GmbH
Europaplatz 3
69115 Heidelberg, Germany

Batch number: 10391208

Printed by Printforce, the Netherlands